multiply	by	to obtain
in/ft	1/0.012	mm/m
in^2	6.4516	cm^2
in^3	1/1728	ft^3
J	9.4778×10^{-4}	Btu
J	6.2415×10^{18}	eV
J	0.73756	ft-lbf
J	1.0	N·m
J/s	1.0	W
kg	2.2046	lbm
kg/m^3	0.06243	lbm/ft^3
kip	4.4480	kN
kip	1000.0	lbf
kip	4448.0	N
kip/ft	14.594	kN/m
kip/ft^2	47.880	kPa
kJ	0.94778	Btu
kJ	737.56	ft-lbf
kJ/kg	0.42992	Btu/lbm
kJ/kg·K	0.23885	Btu/lbm-°R
km	3280.8	ft
km	0.62138	mi
km/h	0.62138	mi/hr
kN	0.2248	kips
kN·m	0.73757	ft-kips
kN/m	0.06852	kips/ft
kPa	9.8692×10^{-3}	atm
kPa	0.14504	lbf/in^2
kPa	1000.0	Pa
kPa	0.02089	kips/ft^2
kPa	0.01	bar
ksi	6.8948×10^{6}	Pa
ksi	6894.8	kPa
kW	737.56	ft-lbf/sec
kW	44,250	ft-lbf/min
kW	1.3410	hp
kW	3413.0	Btu/hr
kW	0.9483	Btu/s
kW-hr	3413.0	Btu
kW-hr	3.60×10^{6}	J
L	1/102,790	ac-in
L	1000.0	cm^3
L	0.03531	ft^3
L	0.26417	gal
L	61.024	in^3
L	0.0010	m^3
L/s	2.1189	ft^3/min
L/s	15.850	gal/min
lbf	0.001	kips
lbf	4.4482	N
lbf/ft^2	0.01414	in Hg
lbf/ft^2	0.19234	in water
lbf/ft^2	0.00694	lbf/in^2
lbf/ft^2	47.880	Pa
lbf/ft^2	5.0×10^{-4}	tons/ft^2
lbf/in^2	0.06805	atm
lbf/in^2	144.0	lbf/ft^2
lbf/in^2	2.308	ft water
lbf/in^2	27.70	in water
lbf/in^2	2.0370	in Hg
lbf/in^2	6894.8	Pa
lbf/in^2	0.00050	tons/in^2
lbf/in^2	0.0720	tons/ft^2
lbm	7000.0	grains
lbm	453.59	g
lbm	0.45359	kg
lbm	4.5359×10^{5}	mg
lbm	5.0×10^{-4}	tons (mass)
lbm (of water)	0.12	gal (of water)
lbm/ac-ft-day	0.02296	lbm/1000 ft^3-day
lbm/ft^3	0.016018	g/cm^3
lbm/ft^3	16.018	kg/m^3
lbm/1000 ft^3-day	43.560	lbm/ac-ft-day
lbm/1000 ft^3-day	133.68	lbm/MG-day
lbm/MG	0.0070	grains/gal
lbm/MG	0.11983	mg/L
lbm/MG-day	0.00748	lbm/1000 ft^3-day
leagues	4428.0	m
m	1.0×10^{10}	angstroms
m	[illegible]	[illegible]
m	[illegible]	[illegible]
m	[illegible]	[illegible]
m	[illegible]	[illegible]
m/s	[illegible]	[illegible]
m^2	2.4711×10^{-4}	ac
m^2	10.764	ft^2
m^2	1/10,000	hectare
mi^2-in	53.3	ac-ft
mi^2-in/day	26.89	ft^3/s
m^3	8.1071×10^{-4}	ac-ft
m^3/m·d	80.5196	gal/day-ft
m^3/m^2·d	24.542	gal/day-ft^2
Meinzer unit	1.0	gal/day-ft^2
mg	2.2046×10^{-6}	lbm
mg/L	1.0	ppm
mg/L	0.05842	grains/gal
mg/L	8.3454	lbm/MG
MG	1.0×10^{6}	gal
MG/ac-day	22.968	gal/ft^2-day
MGD	1.5472	ft^3/sec
MGD	1×10^{6}	gal/day
MGD/ac (mgad)	22.957	gal/day-ft^2
mi	5280.0	ft
mi	80.0	chains
mi	1.6093	km
mi (statute)	0.86839	miles (nautical)
mi	320.0	rods
mi/hr	1.4667	ft/sec
mi^2	640.0	acres
micron	1.0×10^{-6}	m
micron	0.001	mm
mil (angular)	0.05625	degrees
mil (angular)	3.375	min
min (angular)	0.29630	mils
min (angular)	2.90888×10^{-4}	radians
min (time, mean solar)	60	s
mm	1/25.4	in
mm	1000.0	microns
mm/m	0.012	in/ft
MPa	1.0×10^{6}	Pa
N	0.22481	lbf
N	1.0×10^{5}	dynes
N·m	0.73756	ft-lbf
N·m	8.8511	in-lbf
N·m	1.0	J
N/m^2	1.0	Pa
oz	28.353	g
Pa	0.001	kPa
Pa	1.4504×10^{-7}	ksi
Pa	1.4504×10^{-4}	lbf/in^2
Pa	0.02089	lbf/ft^2
Pa	1.0×10^{-6}	MPa
Pa	1.0	N/m^2
ppm	0.05842	grains/gal
radian	$180/\pi$	degrees (angular)
radian	3437.7	min (angular)
rod	0.250	chain
rod	16.50	ft
rod	1/320	mi
s (time)	1/86,400	day (mean solar)
s (time)	1.1605×10^{-5}	day (sidereal)
s (time)	1/60	min
therm	1.0×10^{5}	Btu
ton (force)	2000.0	lbf
ton (mass)	2000.0	lbm
ton/ft^2	2000.0	lbf/ft^2
ton/ft^2	13.889	lbf/in^2
W	3.413	Btu/hr
W	0.73756	ft-lbf/sec
W	1.3410×10^{-3}	hp
W	1.0	J/s
yd	1/22	chain
yd	0.91440	m
yd^3	27	ft^3
yd^3	201.97	gal

PE Civil
Companion
for the Sixteenth Edition

Michael R. Lindeburg, PE

PPI2PASS.COM
A **KAPLAN** COMPANY

Professional Publications, Inc. • Belmont, California

PE CIVIL COMPANION FOR THE SIXTEENTH EDITION

Current release of this edition: 1

Release History

date	edition number	revision number	update
Aug 2018	1	1	New book.

Printed in the United States of America.

PPI
1250 Fifth Avenue, Belmont, CA 94002
(650) 593-9119
ppi2pass.com

ISBN: 978-1-59126-628-0

F E D C B A

Table of Contents

Introduction

The *PE Civil Companion* is an ideal accompaniment to both the *PE Civil Reference Manual* and *PE Civil Practice Problems*. This convenient volume of valuable supporting resources can be used side by side with both titles to pinpoint information quickly during study and on exam day.

Start from any chapter in *PE Civil Reference Manual* and use the *Companion's* comprehensive index to navigate easily to related coverage in other chapters. Thousands of entries cover all topics and key terms in the *PE Civil Reference Manual*.

The *Companion* offers over 100 appendices of essential support material. These civil engineering charts, diagrams, codes, and tables can also be used to solve practice problems in *PE Civil Practice Problems*.

The *Companion* also contains a glossary of over 550 common civil engineering terms to help you grasp key concepts.

Rely on the *Companion* to support you in your exam preparation goals.

Appendices

APPENDIX 1.A
Conversion Factors

multiply	by	to obtain
acres	0.40468	hectares
	43,560.0	square feet
	1.5625×10^{-3}	square miles
ampere-hours	3600.0	coulombs
angstrom units	3.937×10^{-9}	inches
	1×10^{-4}	microns
astronomical units	1.496×10^{8}	kilometers
atmospheres	76.0	centimeters of mercury
atomic mass unit	9.3149×10^{8}	electron-volts
	1.4924×10^{-10}	joules
	1.6605×10^{-27}	kilograms
BeV (also GeV)	1×10^{9}	electron-volts
Btu	3.93×10^{-4}	horsepower-hours
	778.2	foot-pounds
	1055.1	joules
	2.931×10^{-4}	kilowatt hours
	1×10^{-5}	therms
Btu/hr	0.2161	foot-pounds/sec
	3.929×10^{-4}	horsepower
	0.2931	watts
bushels	2150.4	cubic inches
calories, gram (mean)	3.9683×10^{-3}	Btu (mean)
centares	1.0	square meters
centimeters	1×10^{-5}	kilometers
	1×10^{-2}	meters
	10.0	millimeters
	3.281×10^{-2}	feet
	0.3937	inches
chains	792.0	inches
coulombs	1.036×10^{-5}	faradays
cubic centimeters	0.06102	cubic inches
	2.113×10^{-3}	pints (U.S. liquid)
cubic feet	0.02832	cubic meters
	7.4805	gallons
cubic feet/min	62.43	pounds H_2O/min
cubic feet/sec	448.831	gallons/min
	0.64632	millions of gallons per day
cubits	18.0	inches
days	86,400.0	seconds
degrees (angle)	1.745×10^{-2}	radians
degrees/sec	0.1667	revolutions/min
dynes	1×10^{-5}	newtons
electron-volts	1.0735×10^{-9}	atomic mass units

multiply	by	to obtain
feet	30.48	centimeters
	0.3048	meters
	1.645×10^{-4}	miles (nautical)
	1.894×10^{-4}	miles (statute)
feet/min	0.5080	centimeters/sec
feet/sec	0.592	knots
	0.6818	miles/hr
foot-pounds	1.285×10^{-3}	Btu
	5.051×10^{-7}	horsepower-hours
	3.766×10^{-7}	kilowatt-hours
foot-pound/sec	4.6272	Btu/hr
	1.818×10^{-8}	horsepower
	1.356×10^{-3}	kilowatts
furlongs	660.0	feet
	0.125	miles (statute)
gallons	0.1337	cubic feet
	3.785	liters
gallons H_2O	8.3453	pounds H_2O
gallons/min	8.0208	cubic feet/hr
	0.002228	cubic feet/sec
GeV (also BeV)	1×10^{9}	electron-volts
grams	1×10^{-3}	kilograms
	3.527×10^{-2}	ounces (avoirdupois)
	3.215×10^{-2}	ounces (troy)
	2.205×10^{-3}	pounds
hectares	2.471	acres
	1.076×10^{5}	square feet
horsepower	2545.0	Btu/hr
	42.44	Btu/min
	550	foot-pounds/sec
	0.7457	kilowatts
	745.7	watts
horsepower-hours	2545.0	Btu
	1.976×10^{-6}	foot-pounds
	0.7457	kilowatt-hours
hours	4.167×10^{-2}	days
	5.952×10^{-3}	weeks
inches	2.540	centimeters
	1.578×10^{-5}	miles
inches, H_2O	5.199	pounds force/ft^2
	0.0361	psi
	0.0735	inches, mercury
inches, mercury	70.7	pounds force/ft^2
	0.491	pounds force/in^2

APPENDIX 1.A *(continued)*
Conversion Factors

multiply	by	to obtain	multiply	by	to obtain
	1×10^{-9}	BeV (also GeV)		13.60	inches, H_2O
	1.60218×10^{-19}	joules	joules	6.705×10^{9}	atomic mass units
	1.78266×10^{-36}	kilograms		9.478×10^{-4}	Btu
	1×10^{-6}	MeV		1×10^{7}	ergs
faradays/sec	96,485	amperes		6.2415×10^{18}	electron-volts
fathoms	6.0	feet		1.1127×10^{-17}	kilograms
kilograms	6.0221×10^{26}	atomic mass units	pascal-sec	1000	centipoise
	5.6096×10^{35}	electron-volts		10	poise
	8.9875×10^{16}	joules		0.02089	pound force-sec/ft^2
	2.205	pounds		0.6720	pound mass/ft-sec
kilometers	3281.0	feet		0.02089	slug/ft-sec
	1000.0	meters	pints (liquid)	473.2	cubic centimeters
	0.6214	miles		28.87	cubic inches
kilometers/hr	0.5396	knots		0.125	gallons
kilowatts	3412.9	Btu/hr		0.5	quarts (liquid)
	737.6	foot-pounds/sec	poise	0.002089	pound-sec/ft^2
	1.341	horsepower	pounds	0.4536	kilograms
kilowatt-hours	3413.0	Btu		16.0	ounces
knots	6076.0	feet/hr		14.5833	ounces (troy)
	1.0	nautical miles/hr		1.21528	pounds (troy)
	1.151	statute miles/hr	pounds/ft^2	0.006944	pounds/in^2
light years	5.9×10^{12}	miles	pounds/in^2	2.308	feet, H_2O
links (surveyor)	7.92	inches		27.7	inches, H_2O
liters	1000.0	cubic centimeters		2.037	inches, mercury
	61.02	cubic inches		144	pounds/ft^2
	0.2642	gallons (U.S. liquid)	quarts (dry)	67.20	cubic inches
	1000.0	milliliters	quarts (liquid)	57.75	cubic inches
	2.113	pints		0.25	gallons
MeV	1×10^{6}	electron-volts		0.9463	liters
meters	100.0	centimeters	radians	57.30	degrees
	3.281	feet		3438.0	minutes
	1×10^{-3}	kilometers	revolutions	360.0	degrees
	5.396×10^{-4}	miles (nautical)	revolutions/min	6.0	degrees/sec
	6.214×10^{-4}	miles (statute)	rods	16.5	feet
	1000.0	millimeters		5.029	meters
microns	1×10^{-6}	meters	rods (surveyor)	5.5	yards
miles (nautical)	6076	feet	seconds	1.667×10^{-2}	minutes
	1.853	kilometers	square meters/sec	1×10^{6}	centistokes
	1.1516	miles (statute)		10.76	square feet/sec
miles (statute)	5280.0	feet		1×10^{4}	stokes
	1.609	kilometers	slugs	32.174	pounds mass
	0.8684	miles (nautical)	stokes	0.0010764	square feet/sec
miles/hr	88.0	feet/min	tons (long)	1016.0	kilograms
milligrams/liter	1.0	parts/million		2240.0	pounds

APPENDIX 1.A *(continued)*
Conversion Factors

multiply	by	to obtain	multiply	by	to obtain
milliliters	1×10^{-3}	liters		1.120	tons (short)
millimeters	3.937×10^{-2}	inches	tons (short)	907.1848	kilograms
newtons	1×10^{5}	dynes		2000.0	pounds
ounces	28.349527	grams		0.89287	tons (long)
	6.25×10^{-2}	pounds	watts	3.4129	Btu/hr
ounces (troy)	1.09714	ounces (avoirdupois)		1.341×10^{-3}	horsepower
parsecs	3.086×10^{13}	kilometers	yards	0.9144	meters
	1.9×10^{13}	miles		4.934×10^{-4}	miles (nautical)
				5.682×10^{-4}	miles (statute)

APPENDIX 1.B
Common SI Unit Conversion Factors

multiply	by	to obtain
AREA		
circular mil	506.7	square micrometer
square foot	0.0929	square meter
square kilometer	0.3861	square mile
square meter	10.764	square foot
	1.196	square yard
square micrometer	0.001974	circular mil
square mile	2.590	square kilometer
square yard	0.8361	square meter
ENERGY		
Btu (international)	1.0551	kilojoule
erg	0.1	microjoule
foot-pound	1.3558	joule
horsepower-hour	2.6485	megajoule
joule	0.7376	foot-pound
	0.10197	meter·kilogram force
kilogram·calorie (international)	4.1868	kilojoule
kilojoule	0.9478	Btu
	0.2388	kilogram·calorie
kilowatt·hour	3.6	megajoule
megajoule	0.3725	horsepower-hour
	0.2778	kilowatt-hour
	0.009478	therm
meter·kilogram force	9.8067	joule
microjoule	10.0	erg
therm	105.506	megajoule
FORCE		
dyne	10.0	micronewton
kilogram force	9.8067	newton
kip	4448.2	newton
micronewton	0.1	dyne
newton	0.10197	kilogram force
	0.0002248	kip
	3.597	ounce force
	0.2248	pound force
ounce force	0.2780	newton
pound force	4.4482	newton
HEAT		
Btu/ft^2-hr	3.1546	watt/m^2
Btu/hr-ft^2-°F	5.6783	watt/m^2·°C
Btu/ft^3	0.0373	megajoule/m^3
Btu/ft^3-°F	0.06707	megajoule/m^3·°C
Btu/hr	0.2931	watt
Btu/lbm	2326	joule/kg
Btu/lbm-°F	4186.8	joule/kg·°C
Btu-in/hr-ft^2-°F	0.1442	watt/meter·°C
joule/kg	0.000430	Btu/lbm
joule/kg·°C	0.0002388	Btu/lbm-°F
megajoule/m^3	26.839	Btu/ft^3
megajoule/m^3·°C	14.911	Btu/ft^3-°F
watt	3.4121	Btu/hr
watt/m·°C	6.933	Btu-in/hr-ft^2-°F
watt/m^2	0.3170	Btu/ft^2-hr
watt/m^2·°C	0.1761	Btu/hr-ft^2-°F

APPENDIX 1.B *(continued)*
Common SI Unit Conversion Factors

multiply	by	to obtain
LENGTH		
angstrom	0.1	nanometer
foot	0.3048	meter
inch	25.4	millimeter
kilometer	0.6214	mile
	0.540	mile (nautical)
meter	3.2808	foot
	1.0936	yard
micrometer	1.0	micron
micron	1.0	micrometer
mil	0.0254	millimeter
mile	1.6093	kilometer
mile (nautical)	1.852	kilometer
millimeter	0.0394	inch
	39.370	mil
nanometer	10.0	angstrom
yard	0.9144	meter
MASS		
grain	64.799	milligram
gram	0.0353	ounce (avoirdupois)
	0.03215	ounce (troy)
kilogram	2.2046	pound mass
	0.068522	slug
	0.0009842	ton (long—2240 lbm)
	0.001102	ton (short—2000 lbm)
milligram	0.0154	grain
ounce (avoirdupois)	28.350	gram
ounce (troy)	31.1035	gram
pound mass	0.4536	kilogram
slug	14.5939	kilogram
ton (long—2240 lbm)	1016.047	kilogram
ton (short—2000 lbm)	907.185	kilogram
PRESSURE		
bar	100.0	kilopascal
inch, H_2O (20°C)	0.2486	kilopascal
inch, Hg (20°C)	3.3741	kilopascal
kilogram force/cm^2	98.067	kilopascal
kilopascal	0.01	bar
	4.0219	inch, H_2O (20°C)
	0.2964	inch, Hg (20°C)
	0.0102	kilogram force/cm^2
	7.528	millimeter, Hg (20°C)
	0.1450	pound force/cm^2
	0.009869	standard atmosphere (760 torr)
	7.5006	torr
millimeter, Hg (20°C)	0.13332	kilopascal
pound force/in^2	6.8948	kilopascal
standard atmosphere (760 torr)	101.325	kilopascal
torr	0.13332	kilopascal
POWER		
Btu (international)/hr	0.2931	watt
foot-pound/sec	1.3558	watt
horsepower	0.7457	kilowatt
kilowatt	1.341	horsepower
	0.2843	ton of refrigeration

APPENDIX 1.B (*continued*)
Common SI Unit Conversion Factors

multiply	by	to obtain
meter·kilogram force/sec	9.8067	watt
ton of refrigeration	3.517	kilowatt
watt	3.4122	Btu (international)/hr
	0.7376	foot-pound/sec
	0.10197	meter·kilogram force/sec
TEMPERATURE		
Celsius	$\frac{9}{5}°C+32°$	Fahrenheit
Fahrenheit	$\frac{5}{9}(°F-32°)$	Celsius
Kelvin	$\frac{9}{5}$	Rankine
Rankine	$\frac{5}{9}$	Kelvin
TORQUE		
gram force·centimeter	0.098067	millinewton·meter
kilogram force·meter	9.8067	newton·meter
millinewton	10.197	gram force·centimeter
newton·meter	0.10197	kilogram force·meter
	0.7376	foot-pound
	8.8495	inch-pound
foot-pound	1.3558	newton·meter
inch-pound	0.1130	newton·meter
VELOCITY		
feet/sec	0.3048	meters/sec
kilometers/hr	0.6214	miles/hr
meters/sec	3.2808	feet/sec
	2.2369	miles/hr
miles/hr	1.60934	kilometers/hr
	0.44704	meters/sec
VISCOSITY		
centipoise	0.001	pascal·sec
centistoke	1×10^{-6}	square meter/sec
pascal·sec	1000	centipoise
square meter/sec	1×10^{6}	centistoke
VOLUME (capacity)		
cubic centimeter	0.06102	cubic inch
cubic foot	28.3168	liter
cubic inch	16.3871	cubic centimeter
cubic meter	1.308	cubic yard
cubic yard	0.7646	cubic meter
gallon (U.S. fluid)	3.785	liter
liter	0.2642	gallon (U.S. fluid)
	2.113	pint (U.S. fluid)
	1.0567	quart (U.S. fluid)
	0.03531	cubic foot
milliliter	0.0338	ounce (U.S. fluid)
ounce (U.S. fluid)	29.574	milliliter
pint (U.S. fluid)	0.4732	liter
quart (U.S. fluid)	0.9464	liter

APPENDIX 1.B *(continued)*
Common SI Unit Conversion Factors

multiply	by	to obtain
VOLUME FLOW (gas-air)		
cubic meter/sec	2119	standard cubic foot/min
liter/sec	2.119	standard cubic foot/min
microliter/sec	0.000127	standard cubic foot/hr
milliliter/sec	0.002119	standard cubic foot/min
	0.127133	standard cubic foot/hr
standard cubic foot/min	0.0004719	cubic meter/sec
	0.4719	liter/sec
	471.947	milliliter/sec
standard cubic foot/hr	7866	microliter/sec
	7.8658	milliliter/sec
VOLUME FLOW (liquid)		
gallon/hr (U.S. fluid)	0.001052	liter/sec
gallon/min (U.S. fluid)	0.06309	liter/sec
liter/sec	951.02	gallon/hr (U.S. fluid)
	15.850	gallon/min (U.S. fluid)

Appendices

APPENDIX 7.A
Mensuration of Two-Dimensional Areas

Nomenclature
A total surface area
b base
c chord length
d distance
h height
L length
p perimeter
r radius
s side (edge) length, arc length
θ vertex angle, in radians
ϕ central angle, in radians

Circular Sector

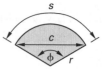

$$A = \tfrac{1}{2}\phi r^2 = \tfrac{1}{2}sr$$
$$\phi = \frac{s}{r}$$
$$s = r\phi$$
$$c = 2r\sin\left(\frac{\phi}{2}\right)$$

Triangle

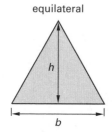 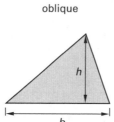

equilateral

$$A = \tfrac{1}{2}bh = \frac{\sqrt{3}}{4}b^2$$
$$h = \frac{\sqrt{3}}{2}b$$

right

$$A = \tfrac{1}{2}bh$$
$$H^2 = b^2 + h^2$$

oblique

$$A = \tfrac{1}{2}bh$$

Parabola

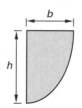

$$A = \tfrac{2}{3}bh$$

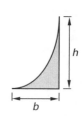

$$A = \tfrac{1}{3}bh$$

Circle

$$p = 2\pi r$$
$$A = \pi r^2 = \frac{p^2}{4\pi}$$

Circular Segment

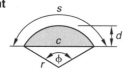

$$A = \tfrac{1}{2}r^2(\phi - \sin\phi)$$
$$\phi = \frac{s}{r} = 2\left(\arccos\frac{r-d}{r}\right)$$
$$c = 2r\sin\left(\frac{\phi}{2}\right)$$

Ellipse

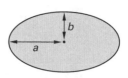

$$A = \pi ab$$
$$p \approx 2\pi\sqrt{\tfrac{1}{2}(a^2+b^2)} \quad \left[\begin{array}{c}\text{Euler's}\\\text{upper bound}\end{array}\right]$$

APPENDIX 7.A *(continued)*
Mensuration of Two-Dimensional Areas

Trapezoid

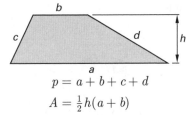

$$p = a + b + c + d$$
$$A = \tfrac{1}{2}h(a+b)$$

If $c = d$, the trapezoid is isosceles.

Parallelogram

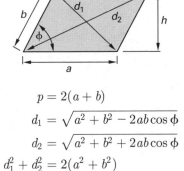

$$p = 2(a+b)$$
$$d_1 = \sqrt{a^2 + b^2 - 2ab\cos\phi}$$
$$d_2 = \sqrt{a^2 + b^2 + 2ab\cos\phi}$$
$$d_1^2 + d_2^2 = 2(a^2 + b^2)$$
$$A = ah = ab\sin\phi$$

If $a = b$, the parallelogram is a rhombus.

Regular Polygon (*n* equal sides)

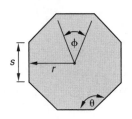

$$\phi = \frac{2\pi}{n}$$
$$\theta = \frac{\pi(n-2)}{n} = \pi - \phi$$
$$p = ns$$
$$s = 2r\tan\frac{\theta}{2}$$
$$A = \tfrac{1}{2}nsr$$

sides	name	area (A) when diameter of inscribed circle = 1	area (A) when side = 1	radius (r) of circumscribed circle when side = 1	length (L) of side when radius (r) of circumscribed circle = 1	length (L) of side when perpendicular to circle = 1	perpendicular (p) to center when side = 1
3	triangle	1.299	0.433	0.577	1.732	3.464	0.289
4	square	1.000	1.000	0.707	1.414	2.000	0.500
5	pentagon	0.908	1.720	0.851	1.176	1.453	0.688
6	hexagon	0.866	2.598	1.000	1.000	1.155	0.866
7	heptagon	0.843	3.634	1.152	0.868	0.963	1.038
8	octagon	0.828	4.828	1.307	0.765	0.828	1.207
9	nonagon	0.819	6.182	1.462	0.684	0.728	1.374
10	decagon	0.812	7.694	1.618	0.618	0.650	1.539
11	undecagon	0.807	9.366	1.775	0.563	0.587	1.703
12	dodecagon	0.804	11.196	1.932	0.518	0.536	1.866

Appendices

APPENDIX 7.B
Mensuration of Three-Dimensional Volumes

Nomenclature
A surface area
b base
h height
r radius
R radius
s side (edge) length
V internal volume

Sphere

$$V = \tfrac{4}{3}\pi r^3 = \tfrac{4}{3}\pi\left(\frac{d}{2}\right)^3 = \tfrac{1}{6}\pi d^3$$

$$A = 4\pi r^2$$

Right Circular Cone (excluding base area)

$$V = \tfrac{1}{3}\pi r^2 h = \tfrac{1}{3}\pi\left(\frac{d}{2}\right)^2 h = \tfrac{1}{12}\pi d^2 h$$

$$A = \pi r\sqrt{r^2 + h^2}$$

Right Circular Cylinder (excluding end areas)

$$V = \pi r^2 h$$

$$A = 2\pi rh$$

Spherical Segment (spherical cap)

Surface area of a spherical segment of radius r cut out by an angle θ_0 rotated from the center about a radius, r, is

$$A = 2\pi r^2(1 - \cos\theta_0)$$

$$\omega = \frac{A}{r^2} = 2\pi(1 - \cos\theta_0)$$

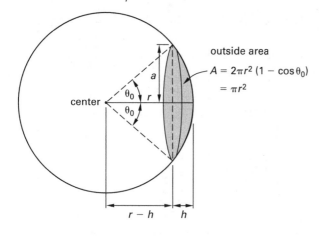

$$V_{\text{cap}} = \tfrac{1}{6}\pi h(3a^2 + h^2)$$
$$= \tfrac{1}{3}\pi h^2(3r - h)$$
$$a = \sqrt{h(2r - h)}$$

Paraboloid of Revolution

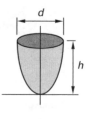

$$V = \tfrac{1}{8}\pi d^2 h$$

Torus

$$A = 4\pi^2 Rr$$
$$V = 2\pi^2 Rr^2$$

Regular Polyhedra (identical faces)

name	number of faces	form of faces	total surface area	volume
tetrahedron	4	equilateral triangle	$1.7321s^2$	$0.1179s^3$
cube	6	square	$6.0000s^2$	$1.0000s^3$
octahedron	8	equilateral triangle	$3.4641s^2$	$0.4714s^3$
dodecahedron	12	regular pentagon	$20.6457s^2$	$7.6631s^3$
icosahedron	20	equilateral triangle	$8.6603s^2$	$2.1817s^3$

The radius of a sphere inscribed within a regular polyhedron is

$$r = \frac{3V_{\text{polyhedron}}}{A_{\text{polyhedron}}}$$

APPENDIX 9.A
Abbreviated Table of Indefinite Integrals
(In each case, add a constant of integration. All angles are measured in radians.)

General Formulas

1. $\displaystyle\int dx = x$

2. $\displaystyle\int c\,dx = c\int dx$

3. $\displaystyle\int (dx + dy) = \int dx + \int dy$

4. $\displaystyle\int u\,dv = uv - \int v\,du$ $\quad\left[\begin{array}{l}\text{integration by parts; } u \text{ and } v \text{ are}\\ \text{functions of the same variable}\end{array}\right]$

Algebraic Forms

5. $\displaystyle\int x^n\,dx = \frac{x^{n+1}}{n+1} \quad [n \neq -1]$

6. $\displaystyle\int x^{-1}\,dx = \int \frac{dx}{x} = \ln\left|x\right|$

7. $\displaystyle\int (ax + b)^n\,dx = \frac{(ax+b)^{n+1}}{a(n+1)} \quad [n \neq -1]$

8. $\displaystyle\int \frac{dx}{ax+b} = \frac{1}{a}\ln(ax+b)$

9. $\displaystyle\int \frac{x\,dx}{ax+b} = \frac{1}{a^2}(ax+b-b\ln(ax+b))$

10. $\displaystyle\int \frac{x\,dx}{(ax+b)^2} = \frac{1}{a^2}\left[\frac{b}{ax+b}+\ln(ax+b)\right]$

11. $\displaystyle\int \frac{dx}{x(ax+b)} = \frac{1}{b}\ln\left(\frac{x}{ax+b}\right)$

12. $\displaystyle\int \frac{dx}{x(ax+b)^2} = \frac{1}{b(ax+b)}+\frac{1}{b^2}\ln\left(\frac{x}{ax+b}\right)$

13. $\displaystyle\int \frac{dx}{x^2+a^2} = \frac{1}{a}\arctan\left(\frac{x}{a}\right)$

14. $\displaystyle\int \frac{dx}{a^2-x^2} = \frac{1}{a}\operatorname{arctanh}\left(\frac{x}{a}\right)$

15. $\displaystyle\int \frac{x\,dx}{ax^2+b} = \frac{1}{2a}\ln(ax^2+b)$

16. $\displaystyle\int \frac{dx}{x(ax^n+b)} = \frac{1}{bn}\ln\left(\frac{x^n}{ax^n+b}\right)$

17. $\displaystyle\int \frac{dx}{ax^2+bx+c} = \frac{1}{\sqrt{b^2-4ac}}\ln\left(\frac{2ax+b-\sqrt{b^2-4ac}}{2ax+b+\sqrt{b^2-4ac}}\right) \quad [b^2 > 4ac]$

18. $\displaystyle\int \frac{dx}{ax^2+bx+c} = \frac{2}{\sqrt{4ac-b^2}}\arctan\left(\frac{2ax+b}{\sqrt{4ac-b^2}}\right) \quad [b^2 < 4ac]$

19. $\displaystyle\int \sqrt{a^2-x^2}\,dx = \frac{x}{2}\sqrt{a^2-x^2}+\frac{a^2}{2}\arcsin\left(\frac{x}{a}\right)$

20. $\displaystyle\int x\sqrt{a^2-x^2}\,dx = -\frac{1}{3}(a^2-x^2)^{3/2}$

APPENDIX 9.A *(continued)*
Abbreviated Table of Indefinite Integrals
(In each case, add a constant of integration. All angles are measured in radians.)

21. $\displaystyle \int \frac{dx}{\sqrt{a^2 - x^2}} = \arcsin\left(\frac{x}{a}\right)$

22. $\displaystyle \int \frac{x\,dx}{\sqrt{a^2 - x^2}} = -\sqrt{a^2 - x^2}$

APPENDIX 10.A
Laplace Transforms

$f(t)$	$\mathcal{L}\big(f(t)\big)$	$f(t)$	$\mathcal{L}\big(f(t)\big)$
$\delta(t)$ [unit impulse at $t=0$]	1	$1-\cos at$	$\dfrac{a^2}{s(s^2+a^2)}$
$\delta(t-c)$ [unit impulse at $t=c$]	e^{-cs}	$\cosh at$	$\dfrac{s}{s^2-a^2}$
1 or u_0 [unit step at $t=0$]	$\dfrac{1}{s}$	$t\cos at$	$\dfrac{s^2-a^2}{(s^2+a^2)^2}$
u_c [unit step at $t=c$]	$\dfrac{e^{-cs}}{s}$	t^n [n is a positive integer]	$\dfrac{n!}{s^{n+1}}$
t [unit ramp at $t=0$]	$\dfrac{1}{s^2}$	e^{at}	$\dfrac{1}{s-a}$
rectangular pulse, magnitude M, duration a	$\left(\dfrac{M}{s}\right)(1-e^{-as})$	$e^{at}\sin bt$	$\dfrac{b}{(s-a)^2+b^2}$
triangular pulse, magnitude M, duration $2a$	$\left(\dfrac{M}{as^2}\right)(1-e^{-as})^2$	$e^{at}\cos bt$	$\dfrac{s-a}{(s-a)^2+b^2}$
sawtooth pulse, magnitude M, duration a	$\left(\dfrac{M}{as^2}\right)(1-(as+1)e^{-as})$	$e^{at}t^n$ [n is a positive integer]	$\dfrac{n!}{(s-a)^{n+1}}$
sinusoidal pulse, magnitude M, duration π/a	$\left(\dfrac{Ma}{s^2+a^2}\right)(1+e^{-\pi s/a})$	$1-e^{-at}$	$\dfrac{a}{s(s+a)}$
$\dfrac{t^{n-1}}{(n-1)!}$	$\dfrac{1}{s^n}$	$e^{-at}+at-1$	$\dfrac{a^2}{s^2(s+a)}$
$\sin at$	$\dfrac{a}{s^2+a^2}$	$\dfrac{e^{-at}-e^{-bt}}{b-a}$	$\dfrac{1}{(s+a)(s+b)}$
$at-\sin at$	$\dfrac{a^3}{s^2(s^2+a^2)}$	$\dfrac{(c-a)e^{-at}-(c-b)e^{-bt}}{b-a}$	$\dfrac{s+c}{(s+a)(s+b)}$
$\sinh at$	$\dfrac{a}{s^2-a^2}$	$\dfrac{1}{ab}+\dfrac{be^{-at}-ae^{-bt}}{ab(a-b)}$	$\dfrac{1}{s(s+a)(s+b)}$
$t\sin at$	$\dfrac{2as}{(s^2+a^2)^2}$	$t\sinh at$	$\dfrac{2as}{(s^2-a^2)^2}$
$\cos at$	$\dfrac{s}{s^2+a^2}$	$t\cosh at$	$\dfrac{s^2+a^2}{(s^2-a^2)^2}$

Appendices

APPENDIX 11.A
Areas Under the Standard Normal Curve
(0 to z)

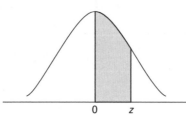

z	0	1	2	3	4	5	6	7	8	9
0.0	0.0000	0.0040	0.0080	0.0120	0.0160	0.0199	0.0239	0.0279	0.0319	0.0359
0.1	0.0398	0.0438	0.0478	0.0517	0.0557	0.0596	0.0636	0.0675	0.0714	0.0754
0.2	0.0793	0.0832	0.0871	0.0910	0.0948	0.0987	0.1026	0.1064	0.1103	0.1141
0.3	0.1179	0.1217	0.1255	0.1293	0.1331	0.1368	0.1406	0.1443	0.1480	0.1517
0.4	0.1554	0.1591	0.1628	0.1664	0.1700	0.1736	0.1772	0.1808	0.1844	0.1879
0.5	0.1915	0.1950	0.1985	0.2019	0.2054	0.2088	0.2123	0.2157	0.2190	0.2224
0.6	0.2258	0.2291	0.2324	0.2357	0.2389	0.2422	0.2454	0.2486	0.2518	0.2549
0.7	0.2580	0.2612	0.2642	0.2673	0.2704	0.2734	0.2764	0.2794	0.2823	0.2852
0.8	0.2881	0.2910	0.2939	0.2967	0.2996	0.3023	0.3051	0.3078	0.3106	0.3133
0.9	0.3159	0.3186	0.3212	0.3238	0.3264	0.3289	0.3315	0.3340	0.3365	0.3389
1.0	0.3413	0.3438	0.3461	0.3485	0.3508	0.3531	0.3554	0.3577	0.3599	0.3621
1.1	0.3643	0.3665	0.3686	0.3708	0.3729	0.3749	0.3770	0.3790	0.3810	0.3830
1.2	0.3849	0.3869	0.3888	0.3907	0.3925	0.3944	0.3962	0.3980	0.3997	0.4015
1.3	0.4032	0.4049	0.4066	0.4082	0.4099	0.4115	0.4131	0.4147	0.4162	0.4177
1.4	0.4192	0.4207	0.4222	0.4236	0.4251	0.4265	0.4279	0.4292	0.4306	0.4319
1.5	0.4332	0.4345	0.4357	0.4370	0.4382	0.4394	0.4406	0.4418	0.4429	0.4441
1.6	0.4452	0.4463	0.4474	0.4484	0.4495	0.4505	0.4515	0.4525	0.4535	0.4545
1.7	0.4554	0.4564	0.4573	0.4582	0.4591	0.4599	0.4608	0.4616	0.4625	0.4633
1.8	0.4641	0.4649	0.4656	0.4664	0.4671	0.4678	0.4686	0.4693	0.4699	0.4706
1.9	0.4713	0.4719	0.4726	0.4732	0.4738	0.4744	0.4750	0.4756	0.4761	0.4767
2.0	0.4772	0.4778	0.4783	0.4788	0.4793	0.4798	0.4803	0.4808	0.4812	0.4817
2.1	0.4821	0.4826	0.4830	0.4834	0.4838	0.4842	0.4846	0.4850	0.4854	0.4857
2.2	0.4861	0.4864	0.4868	0.4871	0.4875	0.4878	0.4881	0.4884	0.4887	0.4890
2.3	0.4893	0.4896	0.4898	0.4901	0.4904	0.4906	0.4909	0.4911	0.4913	0.4916
2.4	0.4918	0.4920	0.4922	0.4925	0.4927	0.4929	0.4931	0.4932	0.4934	0.4936
2.5	0.4938	0.4940	0.4941	0.4943	0.4945	0.4946	0.4948	0.4949	0.4951	0.4952
2.6	0.4953	0.4955	0.4956	0.4957	0.4959	0.4960	0.4961	0.4962	0.4963	0.4964
2.7	0.4965	0.4966	0.4967	0.4968	0.4969	0.4970	0.4971	0.4972	0.4973	0.4974
2.8	0.4974	0.4975	0.4976	0.4977	0.4977	0.4978	0.4979	0.4979	0.4980	0.4981
2.9	0.4981	0.4982	0.4982	0.4983	0.4984	0.4984	0.4985	0.4985	0.4986	0.4986
3.0	0.4987	0.4987	0.4987	0.4988	0.4988	0.4989	0.4989	0.4989	0.4990	0.4990
3.1	0.4990	0.4991	0.4991	0.4991	0.4992	0.4992	0.4992	0.4992	0.4993	0.4993
3.2	0.4993	0.4993	0.4994	0.4994	0.4994	0.4994	0.4994	0.4995	0.4995	0.4995
3.3	0.4995	0.4995	0.4996	0.4996	0.4996	0.4996	0.4996	0.4996	0.4996	0.4997
3.4	0.4997	0.4997	0.4997	0.4997	0.4997	0.4997	0.4997	0.4997	0.4997	0.4998
3.5	0.4998	0.4998	0.4998	0.4998	0.4998	0.4998	0.4998	0.4998	0.4998	0.4998
3.6	0.4998	0.4998	0.4999	0.4999	0.4999	0.4999	0.4999	0.4999	0.4999	0.4999
3.7	0.4999	0.4999	0.4999	0.4999	0.4999	0.4999	0.4999	0.4999	0.4999	0.4999
3.8	0.4999	0.4999	0.4999	0.4999	0.4999	0.4999	0.4999	0.4999	0.4999	0.4999
3.9	0.5000	0.5000	0.5000	0.5000	0.5000	0.5000	0.5000	0.5000	0.5000	0.5000

APPENDIX 11.B
Chi-Squared Distribution

degrees of freedom	probability of exceeding the critical value, α				
	0.10	0.05	0.025	0.01	0.001
1	2.706	3.841	5.024	6.635	10.828
2	4.605	5.991	7.378	9.210	13.816
3	6.251	7.815	9.348	11.345	16.266
4	7.779	9.488	11.143	13.277	18.467
5	9.236	11.070	12.833	15.086	20.515
6	10.645	12.592	14.449	16.812	22.458
7	12.017	14.067	16.013	18.475	24.322
8	13.362	15.507	17.535	20.090	26.125
9	14.684	16.919	19.023	21.666	27.877
10	15.987	18.307	20.483	23.209	29.588
11	17.275	19.675	21.920	24.725	31.264
12	18.549	21.026	23.337	26.217	32.910
13	19.812	22.362	24.736	27.688	34.528
14	21.064	23.685	26.119	29.141	36.123
15	22.307	24.996	27.488	30.578	37.697
16	23.542	26.296	28.845	32.000	39.252
17	24.769	27.587	30.191	33.409	40.790
18	25.989	28.869	31.526	34.805	42.312
19	27.204	30.144	32.852	36.191	43.820
20	28.412	31.410	34.170	37.566	45.315
21	29.615	32.671	35.479	38.932	46.797
22	30.813	33.924	36.781	40.289	48.268

APPENDIX 11.C
Values of t_C for Student's t-Distribution
(ν degrees of freedom; confidence level C; $\alpha = 1 - C$; shaded area $= p$)

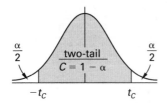
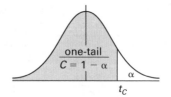

	two-tail, t_C									
	$t_{99\%}$	$t_{98\%}$	$t_{95\%}$	$t_{90\%}$	$t_{80\%}$	$t_{60\%}$	$t_{50\%}$	$t_{40\%}$	$t_{20\%}$	$t_{10\%}$
ν (df)	one-tail, t_C									
	$t_{99.5\%}$	$t_{99\%}$	$t_{97.5\%}$	$t_{95\%}$	$t_{90\%}$	$t_{80\%}$	$t_{75\%}$	$t_{70\%}$	$t_{60\%}$	$t_{55\%}$
1	63.657	31.821	12.706	6.314	3.078	1.376	1.000	0.727	0.325	0.158
2	9.925	6.965	4.303	2.920	1.886	1.061	0.816	0.617	0.289	0.142
3	5.841	4.541	3.182	2.353	1.638	0.978	0.765	0.584	0.277	0.137
4	4.604	3.747	2.776	2.132	1.533	0.941	0.741	0.569	0.271	0.134
5	4.032	3.365	2.571	2.015	1.476	0.920	0.727	0.559	0.267	0.132
6	3.707	3.143	2.447	1.943	1.440	0.906	0.718	0.553	0.265	0.131
7	3.499	2.998	2.365	1.895	1.415	0.896	0.711	0.549	0.263	0.130
8	3.355	2.896	2.306	1.860	1.397	0.889	0.706	0.546	0.262	0.130
9	3.250	2.821	2.262	1.833	1.383	0.883	0.703	0.543	0.261	0.129
10	3.169	2.764	2.228	1.812	1.372	0.879	0.700	0.542	0.260	0.129
11	3.106	2.718	2.201	1.796	1.363	0.876	0.697	0.540	0.260	0.129
12	3.055	2.681	2.179	1.782	1.356	0.873	0.695	0.539	0.259	0.128
13	3.012	2.650	2.160	1.771	1.350	0.870	0.694	0.538	0.259	0.128
14	2.977	2.624	2.145	1.761	1.345	0.868	0.692	0.537	0.258	0.128
15	2.947	2.602	2.131	1.753	1.341	0.866	0.691	0.536	0.258	0.128
16	2.921	2.583	2.120	1.746	1.337	0.865	0.690	0.535	0.258	0.128
17	2.898	2.567	2.110	1.740	1.333	0.863	0.689	0.534	0.257	0.128
18	2.878	2.552	2.101	1.734	1.330	0.862	0.688	0.534	0.257	0.127
19	2.861	2.539	2.093	1.729	1.328	0.861	0.688	0.533	0.257	0.127
20	2.845	2.528	2.086	1.725	1.325	0.860	0.687	0.533	0.257	0.127
21	2.831	2.518	2.080	1.721	1.323	0.859	0.686	0.532	0.257	0.127
22	2.819	2.508	2.074	1.717	1.321	0.858	0.686	0.532	0.256	0.127
23	2.807	2.500	2.069	1.714	1.319	0.858	0.685	0.532	0.256	0.127
24	2.797	2.492	2.064	1.711	1.318	0.857	0.685	0.531	0.256	0.127
25	2.787	2.485	2.060	1.708	1.316	0.856	0.684	0.531	0.256	0.127
26	2.779	2.479	2.056	1.706	1.315	0.856	0.684	0.531	0.256	0.127
27	2.771	2.473	2.052	1.703	1.314	0.855	0.684	0.531	0.256	0.127
28	2.763	2.467	2.048	1.701	1.313	0.855	0.683	0.530	0.256	0.127
29	2.756	2.462	2.045	1.699	1.311	0.854	0.683	0.530	0.256	0.127
30	2.750	2.457	2.042	1.697	1.310	0.854	0.683	0.530	0.256	0.127
40	2.705	2.423	2.021	1.684	1.303	0.851	0.681	0.529	0.255	0.126
60	2.660	2.390	2.000	1.671	1.296	0.848	0.679	0.527	0.254	0.126
120	2.617	2.358	1.980	1.658	1.289	0.845	0.677	0.526	0.254	0.126
∞	2.557	2.326	1.960	1.645	1.282	0.842	0.674	0.524	0.253	0.126

APPENDIX 11.D
Values of the Error Function and Complementary Error Function
(for positive values of x)

x	erf(x)	erfc(x)	x	erf(x)	erfc(x)	x	erf(x)	erfc(x)	x	erf(x)	erfc(x)	x	erf(x)	erfc(x)
0	0.0000	1.0000												
0.01	0.0113	0.9887	0.51	0.5292	0.4708	1.01	0.8468	0.1532	1.51	0.9673	0.0327	2.01	0.9955	0.0045
0.02	0.0226	0.9774	0.52	0.5379	0.4621	1.02	0.8508	0.1492	1.52	0.9684	0.0316	2.02	0.9957	0.0043
0.03	0.0338	0.9662	0.53	0.5465	0.4535	1.03	0.8548	0.1452	1.53	0.9695	0.0305	2.03	0.9959	0.0041
0.04	0.0451	0.9549	0.54	0.5549	0.4451	1.04	0.8586	0.1414	1.54	0.9706	0.0294	2.04	0.9961	0.0039
0.05	0.0564	0.9436	0.55	0.5633	0.4367	1.05	0.8624	0.1376	1.55	0.9716	0.0284	2.05	0.9963	0.0037
0.06	0.0676	0.9324	0.56	0.5716	0.4284	1.06	0.8661	0.1339	1.56	0.9726	0.0274	2.06	0.9964	0.0036
0.07	0.0789	0.9211	0.57	0.5798	0.4202	1.07	0.8698	0.1302	1.57	0.9736	0.0264	2.07	0.9966	0.0034
0.08	0.0901	0.9099	0.58	0.5879	0.4121	1.08	0.8733	0.1267	1.58	0.9745	0.0255	2.08	0.9967	0.0033
0.09	0.1013	0.8987	0.59	0.5959	0.4041	1.09	0.8768	0.1232	1.59	0.9755	0.0245	2.09	0.9969	0.0031
0.1	0.1125	0.8875	0.6	0.6039	0.3961	1.1	0.8802	0.1198	1.6	0.9763	0.0237	2.1	0.9970	0.0030
0.11	0.1236	0.8764	0.61	0.6117	0.3883	1.11	0.8835	0.1165	1.61	0.9772	0.0228	2.11	0.9972	0.0028
0.12	0.1348	0.8652	0.62	0.6194	0.3806	1.12	0.8868	0.1132	1.62	0.9780	0.0220	2.12	0.9973	0.0027
0.13	0.1459	0.8541	0.63	0.6270	0.3730	1.13	0.8900	0.1100	1.63	0.9788	0.0212	2.13	0.9974	0.0026
0.14	0.1569	0.8431	0.64	0.6346	0.3654	1.14	0.8931	0.1069	1.64	0.9796	0.0204	2.14	0.9975	0.0025
0.15	0.1680	0.8320	0.65	0.6420	0.3580	1.15	0.8961	0.1039	1.65	0.9804	0.0196	2.15	0.9976	0.0024
0.16	0.1790	0.8210	0.66	0.6494	0.3506	1.16	0.8991	0.1009	1.66	0.9811	0.0189	2.16	0.9977	0.0023
0.17	0.1900	0.8100	0.67	0.6566	0.3434	1.17	0.9020	0.0980	1.67	0.9818	0.0182	2.17	0.9979	0.0021
0.18	0.2009	0.7991	0.68	0.6638	0.3362	1.18	0.9048	0.0952	1.68	0.9825	0.0175	2.18	0.9980	0.0020
0.19	0.2118	0.7882	0.69	0.6708	0.3292	1.19	0.9076	0.0924	1.69	0.9832	0.0168	2.19	0.9980	0.0020
0.2	0.2227	0.7773	0.7	0.6778	0.3222	1.2	0.9103	0.0897	1.7	0.9838	0.0162	2.2	0.9981	0.0019
0.21	0.2335	0.7665	0.71	0.6847	0.3153	1.21	0.9130	0.0870	1.71	0.9844	0.0156	2.21	0.9982	0.0018
0.22	0.2443	0.7557	0.72	0.6914	0.3086	1.22	0.9155	0.0845	1.72	0.9850	0.0150	2.22	0.9983	0.0017
0.23	0.2550	0.7450	0.73	0.6981	0.3019	1.23	0.9181	0.0819	1.73	0.9856	0.0144	2.23	0.9984	0.0016
0.24	0.2657	0.7343	0.74	0.7047	0.2953	1.24	0.9205	0.0795	1.74	0.9861	0.0139	2.24	0.9985	0.0015
0.25	0.2763	0.7237	0.75	0.7112	0.2888	1.25	0.9229	0.0771	1.75	0.9867	0.0133	2.25	0.9985	0.0015
0.26	0.2869	0.7131	0.76	0.7175	0.2825	1.26	0.9252	0.0748	1.76	0.9872	0.0128	2.26	0.9986	0.0014
0.27	0.2974	0.7026	0.77	0.7238	0.2762	1.27	0.9275	0.0725	1.77	0.9877	0.0123	2.27	0.9987	0.0013
0.28	0.3079	0.6921	0.78	0.7300	0.2700	1.28	0.9297	0.0703	1.78	0.9882	0.0118	2.28	0.9987	0.0013
0.29	0.3183	0.6817	0.79	0.7361	0.2639	1.29	0.9319	0.0681	1.79	0.9886	0.0114	2.29	0.9988	0.0012
0.3	0.3286	0.6714	0.8	0.7421	0.2579	1.3	0.9340	0.0660	1.8	0.9891	0.0109	2.3	0.9989	0.0011
0.31	0.3389	0.6611	0.81	0.7480	0.2520	1.31	0.9361	0.0639	1.81	0.9895	0.0105	2.31	0.9989	0.0011
0.32	0.3491	0.6509	0.82	0.7538	0.2462	1.32	0.9381	0.0619	1.82	0.9899	0.0101	2.32	0.9990	0.0010
0.33	0.3593	0.6407	0.83	0.7595	0.2405	1.33	0.9400	0.0600	1.83	0.9903	0.0097	2.33	0.9990	0.0010
0.34	0.3694	0.6306	0.84	0.7651	0.2349	1.34	0.9419	0.0581	1.84	0.9907	0.0093	2.34	0.9991	0.0009
0.35	0.3794	0.6206	0.85	0.7707	0.2293	1.35	0.9438	0.0562	1.85	0.9911	0.0089	2.35	0.9991	0.0009
0.36	0.3893	0.6107	0.86	0.7761	0.2239	1.36	0.9456	0.0544	1.86	0.9915	0.0085	2.36	0.9992	0.0008
0.37	0.3992	0.6008	0.87	0.7814	0.2186	1.37	0.9473	0.0527	1.87	0.9918	0.0082	2.37	0.9992	0.0008
0.38	0.4090	0.5910	0.88	0.7867	0.2133	1.38	0.9490	0.0510	1.88	0.9922	0.0078	2.38	0.9992	0.0008
0.39	0.4187	0.5813	0.89	0.7918	0.2082	1.39	0.9507	0.0493	1.89	0.9925	0.0075	2.39	0.9993	0.0007
0.4	0.4284	0.5716	0.9	0.7969	0.2031	1.4	0.9523	0.0477	1.9	0.9928	0.0072	2.4	0.9993	0.0007
0.41	0.4380	0.5620	0.91	0.8019	0.1981	1.41	0.9539	0.0461	1.91	0.9931	0.0069	2.41	0.9993	0.0007
0.42	0.4475	0.5525	0.92	0.8068	0.1932	1.42	0.9554	0.0446	1.92	0.9934	0.0066	2.42	0.9994	0.0006
0.43	0.4569	0.5431	0.93	0.8116	0.1884	1.43	0.9569	0.0431	1.93	0.9937	0.0063	2.43	0.9994	0.0006
0.44	0.4662	0.5338	0.94	0.8163	0.1837	1.44	0.9583	0.0417	1.94	0.9939	0.0061	2.44	0.9994	0.0006
0.45	0.4755	0.5245	0.95	0.8209	0.1791	1.45	0.9597	0.0403	1.95	0.9942	0.0058	2.45	0.9995	0.0005
0.46	0.4847	0.5153	0.96	0.8254	0.1746	1.46	0.9611	0.0389	1.96	0.9944	0.0056	2.46	0.9995	0.0005
0.47	0.4937	0.5063	0.97	0.8299	0.1701	1.47	0.9624	0.0376	1.97	0.9947	0.0053	2.47	0.9995	0.0005
0.48	0.5027	0.4973	0.98	0.8342	0.1658	1.48	0.9637	0.0363	1.98	0.9949	0.0051	2.48	0.9995	0.0005
0.49	0.5117	0.4883	0.99	0.8385	0.1615	1.49	0.9649	0.0351	1.99	0.9951	0.0049	2.49	0.9996	0.0004
0.5	0.5205	0.4795	1	0.8427	0.1573	1.5	0.9661	0.0339	2	0.9953	0.0047	2.5	0.9996	0.0004

APPENDIX 14.A
Properties of Water at Atmospheric Pressure
(customary U.S. units)

temperature (°F)	density (lbm/ft³)	absolute viscosity (lbf-sec/ft²)	kinematic viscosity (ft²/sec)	surface tension (lbf/ft)	vapor pressure head[a,b,c] (ft)	bulk modulus (lbf/in²)
32	62.42	3.746×10^{-5}	1.931×10^{-5}	0.518×10^{-2}	0.20	293×10^3
40	62.43	3.229×10^{-5}	1.664×10^{-5}	0.514×10^{-2}	0.28	294×10^3
50	62.41	2.735×10^{-5}	1.410×10^{-5}	0.509×10^{-2}	0.41	305×10^3
60	62.37	2.359×10^{-5}	1.217×10^{-5}	0.504×10^{-2}	0.59	311×10^3
70	62.30	2.050×10^{-5}	1.059×10^{-5}	0.500×10^{-2}	0.84	320×10^3
80	62.22	1.799×10^{-5}	0.930×10^{-5}	0.492×10^{-2}	1.17	322×10^3
90	62.11	1.595×10^{-5}	0.826×10^{-5}	0.486×10^{-2}	1.62	323×10^3
100	62.00	1.424×10^{-5}	0.739×10^{-5}	0.480×10^{-2}	2.21	327×10^3
110	61.86	1.284×10^{-5}	0.667×10^{-5}	0.473×10^{-2}	2.97	331×10^3
120	61.71	1.168×10^{-5}	0.609×10^{-5}	0.465×10^{-2}	3.96	333×10^3
130	61.55	1.069×10^{-5}	0.558×10^{-5}	0.460×10^{-2}	5.21	334×10^3
140	61.38	0.981×10^{-5}	0.514×10^{-5}	0.454×10^{-2}	6.78	330×10^3
150	61.20	0.905×10^{-5}	0.476×10^{-5}	0.447×10^{-2}	8.76	328×10^3
160	61.00	0.838×10^{-5}	0.442×10^{-5}	0.441×10^{-2}	11.21	326×10^3
170	60.80	0.780×10^{-5}	0.413×10^{-5}	0.433×10^{-2}	14.20	322×10^3
180	60.58	0.726×10^{-5}	0.385×10^{-5}	0.426×10^{-2}	17.87	313×10^3
190	60.36	0.678×10^{-5}	0.362×10^{-5}	0.419×10^{-2}	22.29	313×10^3
200	60.12	0.637×10^{-5}	0.341×10^{-5}	0.412×10^{-2}	27.61	308×10^3
212	59.83	0.593×10^{-5}	0.319×10^{-5}	0.404×10^{-2}	35.38	300×10^3

[a] based on actual densities, not on standard "cold, clear water"

[b] can also be calculated from steam tables as $(p_{saturation})(12 \text{ in/ft})^2 (v_f)(g_c/g)$

[c] Multiply the vapor pressure head by the specific weight and divide by $(12 \text{ in/ft})^2$ to obtain lbf/in².

APPENDIX 14.B
Properties of Ordinary Liquid Water[a]
(SI units)

temperature (°C)	density (kg/m³)	absolute viscosity (Pa·s)	kinematic viscosity (m²/s)	vapor pressure[b] (kPa)	bulk modulus (isothermal) (kN/m²)
0					
0.01	999.79	1.791×10^{-3}	1.792×10^{-6}	0.6117	1.9640×10^6
1	999.90	1.7309×10^{-3}	1.731×10^{-6}	0.6571	1.9806×10^6
2	999.94	1.6734×10^{-3}	1.674×10^{-6}	0.7060	1.9942×10^6
3	999.97	1.6189×10^{-3}	1.619×10^{-6}	0.7581	2.0078×10^6
4	999.97	1.5672×10^{-3}	1.567×10^{-6}	0.8135	2.0209×10^6
5	999.97	1.5181×10^{-3}	1.518×10^{-6}	0.8726	2.0340×10^6
6	999.94	1.4714×10^{-3}	1.471×10^{-6}	0.9354	2.0465×10^6
7	999.90	1.4270×10^{-3}	1.427×10^{-6}	1.0021	2.0587×10^6
8	999.85	1.3847×10^{-3}	1.385×10^{-6}	1.0730	2.0707×10^6
9	999.78	1.3444×10^{-3}	1.345×10^{-6}	1.1483	2.0827×10^6
10	999.70	1.3059×10^{-3}	1.306×10^{-6}	1.2282	2.0940×10^6
11	999.61	1.2691×10^{-3}	1.270×10^{-6}	1.3130	2.1052×10^6
12	999.50	1.2340×10^{-3}	1.235×10^{-6}	1.4028	2.1160×10^6
13	999.38	1.2004×10^{-3}	1.201×10^{-6}	1.4981	2.1265×10^6
14	999.25	1.1683×10^{-3}	1.169×10^{-6}	1.5990	2.1370×10^6
15	999.10	1.1375×10^{-3}	1.139×10^{-6}	1.7058	2.1469×10^6
16	998.95	1.1081×10^{-3}	1.109×10^{-6}	1.8188	2.1569×10^6
17	998.78	1.0798×10^{-3}	1.081×10^{-6}	1.9384	2.1665×10^6
18	998.60	1.0527×10^{-3}	1.054×10^{-6}	2.0647	2.1755×10^6
19	998.41	1.0266×10^{-3}	1.028×10^{-6}	2.1983	2.1846×10^6
20	998.21	1.0016×10^{-3}	1.003×10^{-6}	2.3393	2.1933×10^6
21	998.00	0.9776×10^{-3}	0.980×10^{-6}	2.4882	2.2020×10^6
22	997.77	0.9544×10^{-3}	0.957×10^{-6}	2.6453	2.2101×10^6
23	997.54	0.9322×10^{-3}	0.934×10^{-6}	2.8111	2.2182×10^6
24	997.30	0.9107×10^{-3}	0.913×10^{-6}	2.9858	2.2260×10^6
25	997.05	0.8901×10^{-3}	0.893×10^{-6}	3.1699	2.2335×10^6
26	996.79	0.8702×10^{-3}	0.873×10^{-6}	3.3639	2.2407×10^6
27	996.52	0.8510×10^{-3}	0.854×10^{-6}	3.5681	2.2479×10^6
28	996.24	0.8325×10^{-3}	0.836×10^{-6}	3.7831	2.2547×10^6
29	995.95	0.8146×10^{-3}	0.818×10^{-6}	4.0092	2.2613×10^6
30	995.65	0.7974×10^{-3}	0.801×10^{-6}	4.2470	2.2678×10^6
31	995.34	0.7807×10^{-3}	0.784×10^{-6}	4.4969	2.2737×10^6
32	995.03	0.7646×10^{-3}	0.768×10^{-6}	4.7596	2.2796×10^6
33	994.70	0.7490×10^{-3}	0.753×10^{-6}	5.0354	2.2855×10^6
34	994.37	0.7339×10^{-3}	0.738×10^{-6}	5.3251	2.2907×10^6
35	994.03	0.7193×10^{-3}	0.724×10^{-6}	5.6290	2.2960×10^6
36	993.69	0.7052×10^{-3}	0.710×10^{-6}	5.9479	2.3013×10^6
37	993.33	0.6915×10^{-3}	0.696×10^{-6}	6.2823	2.3062×10^6
38	992.97	0.6783×10^{-3}	0.683×10^{-6}	6.6328	2.3108×10^6
39	992.60	0.6654×10^{-3}	0.670×10^{-6}	7.0002	2.3151×10^6
40	992.22	0.6530×10^{-3}	0.658×10^{-6}	7.3849	2.3193×10^6
41	991.83	0.6409×10^{-3}	0.646×10^{-6}	7.7878	2.3233×10^6
42	991.44	0.6292×10^{-3}	0.635×10^{-6}	8.2096	2.3272×10^6
43	991.04	0.6178×10^{-3}	0.623×10^{-6}	8.6508	2.3309×10^6
44	990.63	0.6068×10^{-3}	0.613×10^{-6}	9.1124	2.3345×10^6
45	990.21	0.5961×10^{-3}	0.602×10^{-6}	9.5950	2.3374×10^6
46	989.79	0.5857×10^{-3}	0.592×10^{-6}	10.099	2.3407×10^6
47	989.36	0.5755×10^{-3}	0.582×10^{-6}	10.627	2.3436×10^6
48	988.93	0.5657×10^{-3}	0.572×10^{-6}	11.177	2.3463×10^6
49	988.48	0.5562×10^{-3}	0.563×10^{-6}	11.752	2.3488×10^6
50	988.04	0.5469×10^{-3}	0.553×10^{-6}	12.352	2.3512×10^6
51	987.58	0.5378×10^{-3}	0.545×10^{-6}	12.978	2.3531×10^6
52	987.12	0.5290×10^{-3}	0.536×10^{-6}	13.631	2.3554×10^6
53	986.65	0.5204×10^{-3}	0.527×10^{-6}	14.312	2.3570×10^6
54	986.17	0.5121×10^{-3}	0.519×10^{-6}	15.022	2.3586×10^6
55	985.69	0.5040×10^{-3}	0.511×10^{-6}	15.762	2.3602×10^6
56	985.21	0.4961×10^{-3}	0.504×10^{-6}	16.533	2.3615×10^6
57	984.71	0.4884×10^{-3}	0.496×10^{-6}	17.336	2.3627×10^6
58	984.21	0.4809×10^{-3}	0.489×10^{-6}	18.171	2.3636×10^6
59	983.71	0.4735×10^{-3}	0.481×10^{-6}	19.041	2.3646×10^6
60	983.20	0.4664×10^{-3}	0.474×10^{-6}	19.946	2.3652×10^6
61	982.68	0.4594×10^{-3}	0.468×10^{-6}	20.888	2.3655×10^6

APPENDIX 14.B *(continued)*
Properties of Ordinary Liquid Water[a]
(SI units)

temperature (°C)	density (kg/m³)	absolute viscosity (Pa·s)	kinematic viscosity (m²/s)	vapor pressure[b] (kPa)	bulk modulus (isothermal) (kN/m²)
62	982.16	0.4527×10^{-3}	0.461×10^{-6}	21.867	2.3660×10^6
63	981.63	0.4460×10^{-3}	0.454×10^{-6}	22.885	2.3663×10^6
64	981.09	0.4396×10^{-3}	0.448×10^{-6}	23.943	2.3662×10^6
65	980.55	0.4333×10^{-3}	0.442×10^{-6}	25.042	2.3661×10^6
66	980.00	04271×10^{-3}	0.436×10^{-6}	26.183	2.3660×10^6
67	979.45	0.4211×10^{-3}	0.430×10^{-6}	27.368	2.3656×10^6
68	978.90	0.4152×10^{-3}	0.424×10^{-6}	28.599	2.3649×10^6
69	978.33	0.4095×10^{-3}	0.419×10^{-6}	29.876	2.3644×10^6
70	977.76	0.4039×10^{-3}	0.413×10^{-6}	31.201	2.3633×10^6
71	977.19	0.3984×10^{-3}	0.408×10^{-6}	32.575	2.3626×10^6
72	976.61	0.3931×10^{-3}	0.402×10^{-6}	34.000	2.3615×10^6
73	976.03	0.3879×10^{-3}	0.397×10^{-6}	35.478	2.3604×10^6
74	975.44	0.3827×10^{-3}	0.392×10^{-6}	37.009	2.3589×10^6
75	974.84	0.3777×10^{-3}	0.387×10^{-6}	38.595	2.3575×10^6
76	974.24	0.3729×10^{-3}	0.383×10^{-6}	40.239	2.3557×10^6
77	973.64	0.3681×10^{-3}	0.378×10^{-6}	41.941	2.3540×10^6
78	973.03	0.3634×10^{-3}	0.373×10^{-6}	43.703	2.3522×10^6
79	972.41	0.3588×10^{-3}	0.369×10^{-6}	45.527	2.3501×10^6
80	971.79	0.3544×10^{-3}	0.365×10^{-6}	47.414	2.3480×10^6
81	971.16	0.3500×10^{-3}	0.360×10^{-6}	49.367	2.3459×10^6
82	970.53	0.3457×10^{-3}	0.356×10^{-6}	51.387	2.3434×10^6
83	969.90	0.3415×10^{-3}	0.352×10^{-6}	53.476	2.3410×10^6
84	969.26	0.3374×10^{-3}	0.348×10^{-6}	55.635	2.3386×10^6
85	968.61	0.3333×10^{-3}	0.344×10^{-6}	57.867	2.3358×10^6
86	967.96	0.3294×10^{-3}	0.340×10^{-6}	60.173	2.3330×10^6
87	967.31	0.3255×10^{-3}	0.337×10^{-6}	62.556	2.3300×10^6
88	966.64	0.3218×10^{-3}	0.333×10^{-6}	65.017	2.3268×10^6
89	965.98	0.3180×10^{-3}	0.329×10^{-6}	67.558	2.3238×10^6
90	965.31	0.3144×10^{-3}	0.326×10^{-6}	70.182	2.3207×10^6
91	964.63	0.3109×10^{-3}	0.322×10^{-6}	72.890	2.3172×10^6
92	963.96	0.3074×10^{-3}	0.319×10^{-6}	75.684	2.3135×10^6
93	963.27	0.3039×10^{-3}	0.316×10^{-6}	78.568	2.3101×10^6
94	962.58	0.3006×10^{-3}	0.312×10^{-6}	81.541	2.3063×10^6
95	961.89	0.2973×10^{-3}	0.309×10^{-6}	84.608	2.3026×10^6
96	961.19	0.2941×10^{-3}	0.306×10^{-6}	87.771	2.2988×10^6
97	960.49	0.2909×10^{-3}	0.303×10^{-6}	91.030	2.2948×10^6
98	959.78	0.2878×10^{-3}	0.300×10^{-6}	94.390	2.2907×10^6
99	959.07	0.2847×10^{-3}	0.297×10^{-6}	97.852	2.2864×10^6
100	958.37	0.2818×10^{-3}	0.294×10^{-6}	101.42	2.2823×10^6

[a]Values are calculated using the National Institute of Standards and Technology (NIST) Standard Reference Database 23 (SRD 23): Reference Fluid Thermodynamic and Transport Properties Database (REFPROP) version 7.0 (IAPWS Formulation 1995).

[b]Saturation steam tables may also be used for vapor pressure.

APPENDIX 14.C
Viscosity of Water in Other Units
(customary U.S. units)

temperature	absolute viscosity	kinematic viscosity	
(°F)	(cP)	(cSt)	(SSU)
32	1.79	1.79	33.0
50	1.31	1.31	31.6
60	1.12	1.12	31.2
70	0.98	0.98	30.9
80	0.86	0.86	30.6
85	0.81	0.81	30.4
100	0.68	0.69	30.2
120	0.56	0.57	30.0
140	0.47	0.48	29.7
160	0.40	0.41	29.6
180	0.35	0.36	29.5
212	0.28	0.29	29.3

From *Hydraulic Handbook*, copyright © 1988, by Fairbanks Morse Pump.
Reproduced with permission.

APPENDIX 14.D
Properties of Air at Atmospheric Pressure
(customary U.S. units)

absolute temperature,[a] T_{abs} (°R)	temperature,[a] T (°F)	density,[b] ρ (lbm/ft³)	absolute (dynamic) viscosity,[c] μ (lbf-sec/ft²)	kinematic viscosity,[d] ν (ft²/sec)
300	−160	0.1322	2.378×10^{-7}	5.786×10^{-5}
350	−110	0.1133	2.722×10^{-7}	7.728×10^{-5}
400	−60	0.0992	3.047×10^{-7}	9.886×10^{-5}
450	−10	0.0881	3.355×10^{-7}	12.24×10^{-5}
460	0	0.0863	3.412×10^{-7}	12.72×10^{-5}
470	10	0.0845	3.471×10^{-7}	13.22×10^{-5}
480	20	0.0827	3.530×10^{-7}	13.73×10^{-5}
490	30	0.0810	3.588×10^{-7}	14.25×10^{-5}
492	32	0.0807	3.599×10^{-7}	14.35×10^{-5}
495	35	0.0802	3.616×10^{-7}	14.51×10^{-5}
500	40	0.0794	3.645×10^{-7}	14.77×10^{-5}
505	45	0.0786	3.674×10^{-7}	15.04×10^{-5}
510	50	0.0778	3.702×10^{-7}	15.30×10^{-5}
515	55	0.0771	3.730×10^{-7}	15.57×10^{-5}
520	60	0.0763	3.758×10^{-7}	15.84×10^{-5}
525	65	0.0756	3.786×10^{-7}	16.11×10^{-5}
528	68	0.0752	3.803×10^{-7}	16.28×10^{-5}
530	70	0.0749	3.814×10^{-7}	16.39×10^{-5}
535	75	0.0742	3.842×10^{-7}	16.66×10^{-5}
540	80	0.0735	3.869×10^{-7}	16.94×10^{-5}
545	85	0.0728	3.897×10^{-7}	17.21×10^{-5}
550	90	0.0722	3.924×10^{-7}	17.49×10^{-5}
555	95	0.0715	3.951×10^{-7}	17.78×10^{-5}
560	100	0.0709	3.978×10^{-7}	18.06×10^{-5}
570	110	0.0696	4.032×10^{-7}	18.63×10^{-5}
580	120	0.0684	4.085×10^{-7}	19.21×10^{-5}
590	130	0.0673	4.138×10^{-7}	19.79×10^{-5}
600	140	0.0661	4.190×10^{-7}	20.38×10^{-5}
610	150	0.0651	4.242×10^{-7}	20.98×10^{-5}
650	190	0.0610	4.448×10^{-7}	23.45×10^{-5}
660	200	0.0601	4.496×10^{-7}	24.06×10^{-5}
700	240	0.0567	4.693×10^{-7}	26.65×10^{-5}
710	250	0.0559	4.740×10^{-7}	27.28×10^{-5}
750	290	0.0529	4.930×10^{-7}	29.99×10^{-5}
760	300	0.0522	4.975×10^{-7}	30.65×10^{-5}
800	340	0.0496	5.159×10^{-7}	33.47×10^{-5}
850	390	0.0467	5.380×10^{-7}	37.09×10^{-5}
900	440	0.0441	5.594×10^{-7}	40.84×10^{-5}
950	490	0.0418	5.802×10^{-7}	44.71×10^{-5}
1000	540	0.0397	6.004×10^{-7}	48.70×10^{-5}
1050	590	0.0378	6.201×10^{-7}	52.81×10^{-5}
1100	640	0.0361	6.393×10^{-7}	57.04×10^{-5}
1150	690	0.0345	6.580×10^{-7}	61.37×10^{-5}
1200	740	0.0331	6.762×10^{-7}	65.82×10^{-5}
1250	790	0.0317	6.941×10^{-7}	70.37×10^{-5}
1300	840	0.0305	7.115×10^{-7}	75.03×10^{-5}
1350	890	0.0294	7.286×10^{-7}	79.78×10^{-5}
1400	940	0.0283	7.454×10^{-7}	84.64×10^{-5}
1450	990	0.0274	7.618×10^{-7}	89.60×10^{-5}
1500	1040	0.0264	7.779×10^{-7}	94.65×10^{-5}
1550	1090	0.0256	7.937×10^{-7}	99.79×10^{-5}
1600	1140	0.0248	8.093×10^{-7}	105.0×10^{-5}
1650	1190	0.0240	8.246×10^{-7}	110.4×10^{-5}
1700	1240	0.0233	8.396×10^{-7}	115.8×10^{-5}
1800	1340	0.0220	8.689×10^{-7}	126.9×10^{-5}
1900	1440	0.0209	8.974×10^{-7}	138.3×10^{-5}
2000	1540	0.0198	9.250×10^{-7}	150.1×10^{-5}
2100	1340	0.0189	9.519×10^{-7}	162.1×10^{-5}
2200	1740	0.0180	9.781×10^{-7}	174.5×10^{-5}
2300	1840	0.0172	10.04×10^{-7}	187.2×10^{-5}
2400	1940	0.0165	10.29×10^{-7}	200.2×10^{-5}
2500	2040	0.0159	10.53×10^{-7}	213.5×10^{-5}

APPENDIX 14.D *(continued)*
Properties of Air at Atmospheric Pressure
(customary U.S. units)

absolute temperature,[a] T_{abs} (°R)	temperature,[a] T (°F)	density,[b] ρ (lbm/ft³)	absolute (dynamic) viscosity,[c] μ (lbf-sec/ft²)	kinematic viscosity,[d] ν (ft²/sec)
2600	2140	0.0153	10.77×10^{-7}	227.1×10^{-5}
2800	2340	0.0142	11.23×10^{-7}	255.1×10^{-5}
3000	2540	0.0132	11.68×10^{-7}	284.1×10^{-5}
3200	2740	0.0124	12.11×10^{-7}	314.2×10^{-5}
3400	2940	0.0117	12.52×10^{-7}	345.3×10^{-5}

[a]Temperatures are rounded to the nearest whole degree, but all significant digits were used in calculations.
[b]Density is calculated from the ideal gas law using $p = 14.696$ lbf/in² and $R_{air} = 53.35$ ft-lbf/lbm-°R.
[c]Absolute viscosity is calculated from Sutherland's formula using $C_{Sutherland} = 109.1$°C and $\mu_0 = 0.14592$ kg/m·s, and is subsequently converted to customary U.S. units. Error is ~0 at 530°R. Error is expected to be less than 0.7% at 300°R and 0.2% at 3400°R.
[d]Kinematic viscosity is calculated as $\nu = \mu g_c/\rho$, with $g_c = 32.1742$ lbm-ft/lbf-sec².

APPENDIX 14.E
Properties of Air at Atmospheric Pressure
(SI units)

absolute temperature,[a] T_{abs} (K)	temperature,[a] T (°C)	density,[b] ρ (kg/m³)	absolute (dynamic) viscosity,[c] μ (kg/m·s or Pa·s)	kinematic viscosity,[d] ν (m²/s)
175	−98	2.015	1.190×10^{-5}	0.5905×10^{-5}
200	−73	1.764	1.336×10^{-5}	0.7576×10^{-5}
225	−48	1.568	1.475×10^{-5}	0.9408×10^{-5}
250	−23	1.411	1.607×10^{-5}	1.139×10^{-5}
273	0	1.292	1.723×10^{-5}	1.334×10^{-5}
275	2	1.283	1.733×10^{-5}	1.351×10^{-5}
283	10	1.247	1.772×10^{-5}	1.422×10^{-5}
293	20	1.204	1.821×10^{-5}	1.512×10^{-5}
300	27	1.176	1.854×10^{-5}	1.577×10^{-5}
303	30	1.164	1.868×10^{-5}	1.605×10^{-5}
313	40	1.127	1.915×10^{-5}	1.699×10^{-5}
323	50	1.092	1.961×10^{-5}	1.795×10^{-5}
325	52	1.086	1.970×10^{-5}	1.815×10^{-5}
333	60	1.060	2.006×10^{-5}	1.894×10^{-5}
343	70	1.029	2.051×10^{-5}	1.994×10^{-5}
350	77	1.008	2.082×10^{-5}	2.065×10^{-5}
353	80	0.9995	2.095×10^{-5}	2.096×10^{-5}
363	90	0.9720	2.138×10^{-5}	2.200×10^{-5}
373	100	0.9459	2.181×10^{-5}	2.306×10^{-5}
375	102	0.9409	2.190×10^{-5}	2.327×10^{-5}
400	127	0.8821	2.294×10^{-5}	2.600×10^{-5}
423	150	0.8342	2.386×10^{-5}	2.861×10^{-5}
450	177	0.7841	2.492×10^{-5}	3.178×10^{-5}
473	200	0.7460	2.579×10^{-5}	3.457×10^{-5}
500	227	0.7057	2.679×10^{-5}	3.796×10^{-5}
523	250	0.6747	2.762×10^{-5}	4.093×10^{-5}
550	277	0.6416	2.856×10^{-5}	4.452×10^{-5}
573	300	0.6159	2.935×10^{-5}	4.765×10^{-5}
600	327	0.5881	3.025×10^{-5}	5.143×10^{-5}
623	350	0.5664	3.100×10^{-5}	5.473×10^{-5}
650	377	0.5429	3.186×10^{-5}	5.868×10^{-5}
673	400	0.5244	3.258×10^{-5}	6.213×10^{-5}
700	427	0.5041	3.341×10^{-5}	6.626×10^{-5}
723	450	0.4881	3.410×10^{-5}	6.985×10^{-5}
750	477	0.4705	3.489×10^{-5}	7.415×10^{-5}
773	500	0.4565	3.556×10^{-5}	7.788×10^{-5}
800	527	0.4411	3.632×10^{-5}	8.234×10^{-5}
850	577	0.4152	3.771×10^{-5}	9.082×10^{-5}
900	627	0.3921	3.905×10^{-5}	9.958×10^{-5}
950	677	0.3715	4.035×10^{-5}	10.86×10^{-5}
1000	727	0.3529	4.161×10^{-5}	11.79×10^{-5}
1050	777	0.3361	4.284×10^{-5}	12.74×10^{-5}
1100	827	0.3208	4.403×10^{-5}	13.72×10^{-5}
1150	877	0.3069	4.520×10^{-5}	14.73×10^{-5}
1200	927	0.2941	4.634×10^{-5}	15.76×10^{-5}
1250	977	0.2823	4.745×10^{-5}	16.81×10^{-5}
1300	1027	0.2715	4.854×10^{-5}	17.88×10^{-5}
1350	1077	0.2614	4.961×10^{-5}	18.98×10^{-5}
1400	1127	0.2521	5.065×10^{-5}	20.09×10^{-5}
1500	1227	0.2353	5.269×10^{-5}	22.39×10^{-5}
1600	1327	0.2206	5.464×10^{-5}	24.77×10^{-5}
1700	1427	0.2076	5.654×10^{-5}	27.23×10^{-5}
1800	1527	0.1961	5.837×10^{-5}	29.77×10^{-5}
1900	1627	0.1858	6.015×10^{-5}	32.38×10^{-5}

[a]Temperatures are rounded to the nearest whole degree, but all significant digits were used in calculations.

[b]Density is calculated from the ideal gas law using $p = 101.325$ kPa; and $R_{air} = 287.058$ J/kg·K.

[c]Absolute viscosity is calculated from Sutherland's formula using $C_{Sutherland} = 109.1$°C and $\mu_0 = 0.14592$ kg/m·s. Error is ~0 at 293K. Error is expected to be less than 0.7% at 175K and 0.2% at 1900K.

[d]Kinematic viscosity is calculated as $\nu = \mu/\rho$.

APPENDIX 14.F
Properties of Common Liquids

liquid	temp (°F)	specific gravity*	absolute viscosity (cP)	kinematic viscosity		
				centistokes	SSU	ft²/sec
acetone	68	0.792		0.41		
alcohol, ethyl (ethanol)	68	0.789		1.52	31.7	1.65×10^{-5}
(C$_2$H$_5$OH)	104	0.772		1.2	31.5	
alcohol, methyl (methanol)	68	0.79				
(CH$_3$OH)	59			0.74		
	0	0.810				
ammonia	0	0.662		0.30		
benzene	60	0.885				
	32	0.899				
butane	−50			0.52		
	30			0.35		
	60	0.584				
carbon tetrachloride	68	1.595				
castor oil	68	0.96				1110×10^{-5}
	104	0.95		259–325	1200–1500	
	130			98–130	450–600	
ethylene glycol	0 (−18°C)	1.16	310	267	1210	2.88×10^{-3}
(mono, MEG)	40 (4.4°C)	1.145	48	42	195	4.51×10^{-4}
	68 (20°C)	1.115	17	15	77	1.64×10^{-5}
	140 (60°C)	1.107	5.2	4.7	41	5.06×10^{-5}
	200 (93°C)	1.084	1.4	1.3	32	1.39×10^{-5}
Freon-11	70	1.49		21.1	0.21	
Freon-12	70	1.33		21.1	0.27	
fuel oils, no. 1 to no. 6	60	0.82–0.95				
fuel oil no. 1	70			2.39–4.28	34–40	
	100			−2.69	32–35	
fuel oil no. 2	70			3.0–7.4	36–50	
	100			2.11–4.28	33–40	
fuel oil no. 3	70			2.69–5.84	35–45	
	100			2.06–3.97	32.8–39	
fuel oil no. 5A	70			7.4–26.4	50–125	
	100			4.91–13.7	42–72	
fuel oil no. 5B	70			26.4–	125–	
	100			13.6–67.1	72–310	
fuel oil no. 6	122			97.4–660	450–3000	
	160			37.5–172	175–780	
gasoline	60	0.728				0.73×10^{-5}
	80	0.719				0.66×10^{-5}
	100	0.710				0.60×10^{-5}
glycerine	68	1.261				
kerosene	60	0.78–0.82				
	68			2.17	35	
jet fuel (JP1, 3, 4, 5, 6)	−30			7.9	52	
	60	0.78–0.82				
mercury	70	13.55		21.1	0.118	
	100	13.55		37.8	0.11	
oil, SAE 5 to 150	60	0.88–0.94				
SAE-5W	0			1295 max	6000 max	
SAE-10W	0			1295–2590	6000–12,000	
SAE-20W	0			2590–10,350	12,000–48,000	
SAE-20	210			5.7–9.6	45–58	
SAE-30	210			9.6–12.9	58–70	
SAE-40	210			12.9–16.8	70–85	
SAE-50	210			16.8–22.7	85–110	

APPENDIX 14.F *(continued)*
Properties of Common Liquids

liquid	temp (°F)	specific gravity*	absolute viscosity (cP)	kinematic viscosity		
				centistokes	SSU	ft²/sec
propylene glycol	0 (−18°C)	1.063	~1600	~1500	~6900	~1.63×10^{-2}
	25 (−4°C)	1.054	~380	~360	~1600	~3.89×10^{-3}
	50 (10°C)	1.044	~130	~125	~570	~1.35×10^{-3}
	77 (25°C)	1.032	49	47	220	5.11×10^{-4}
	140 (60°C)	1.006	8.4	8.3	53	8.99×10^{-5}
	200 (93°C)	0.985	2.7	2.7	35	2.95×10^{-5}
saltwater (5%)	39	1.037				
	68			1.097	31.1	
saltwater (25%)	39	1.196				
	60	1.19		2.4	34	
seawater	59	1.025				

*Measured with respect to 60°F water.

APPENDIX 14.G
Viscosities Versus Temperature of Hydrocarbons, Lubricants, Fuels, and Heat Transfer Fluids

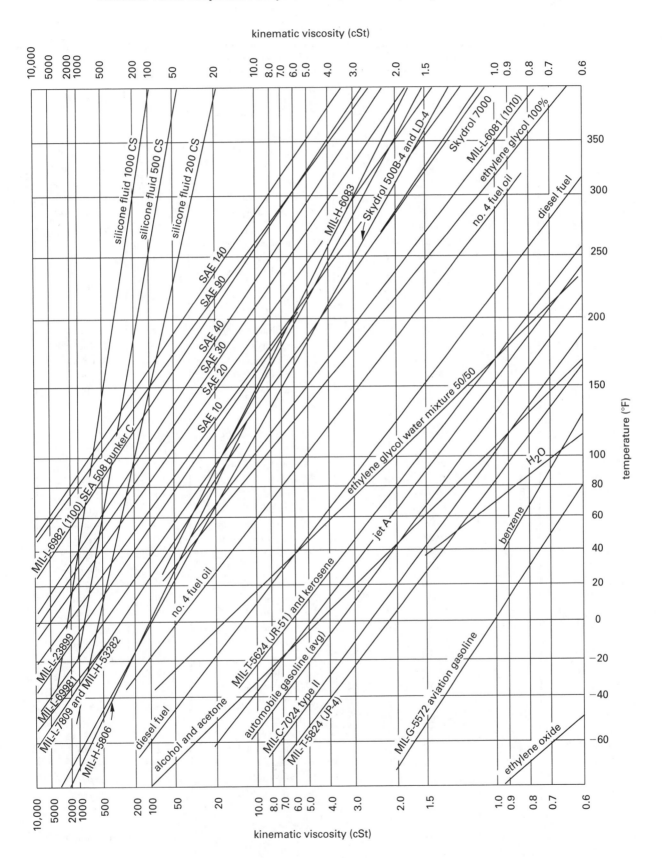

Appendices

APPENDIX 14.H
Properties of Uncommon Fluids

Typical fluid viscosities are listed in the following table. The values given for thixotropic fluids are effective viscosities at normal pumping shear rates. Effective viscosity can vary greatly with changes in solids content, concentration, and so on.

Viscous Behavior Type: N—Newtonian T—Thixotropic

fluid	specific gravity	viscosity CPS	viscous behavior type	fluid	specific gravity	viscosity CPS	viscous behavior type
reference—water	1.0	1.0	N	meat products			
adhesives				animal fat, melted	0.9	43 at 100°F	N
"box" adhesives	1±	3000	T	ground beef fat	0.9	11,000 at 60°F	T
PVA	1.3	100	T	meat emulsion	1.0	22,000 at 40°F	T
rubber and solvents	1.0	15,000	N	pet food	1.0	11,000 at 40°F	T
bakery				pork fat slurry	1.0	650 at 40°F	T
batter	1	2200	T	paint			
butter, melted	0.98	18 at 140°F	N	auto paint, metallic		200	T
egg, whole	0.5	60 at 50°F	N	solvents	0.8–0.9	0.5–10	N
emulsifier		20	T	titanium dioxide slurry		10,000	T
frosting	1	10,000	T	varnish	1.06	140 at 100°F	
lecithin		3250 at 125°F	T	turpentine	0.86	2 at 60°F	
77% sweetened condensed milk	1.3	10,000 at 77°F	N	paper and textile			
yeast slurry, 15%	1	180	T	black liquor	1.3	1100 at 122°F	
beer, wine				black liquor soap		7000 at 122°F	
beer	1.0	1.1 at 40°F	N	black liquor tar		2000 at 300°F	
brewers concentrated yeast, 80% solids		16,000 at 40°F	T	paper coating, 35%		400	
wine	1.0			sulfide, 6%		1600	
chemicals, miscellaneous							
glycols	1.1	35 at range		petroleum and petroleum products			
confectionary				asphalt, unblended	1.3	500–2500	
caramel	1.2	400 at 140°F		gasoline	0.7	0.8 at 60°F	N
chocolate	1.1	17,000 at 120°F	T	kerosene	0.8	3 at 68°F	N
fudge, hot	1.1	36,000	T	fuel oil no. 6	0.9	660 at 122°F	N
toffee	1.2	87,000	T	auto lube oil SAE 40	0.9	200 at 100°F	N
cosmetics, soaps				auto trans oil SAE 90	0.9	320 at 100°F	N
face cream		10,000	T	propane	0.46	0.2 at 100°F	N
gel, hair	1.4	5000	T	tars	1.2	wide range	
shampoo		5000	T	pharmaceuticals			
toothpaste		20,000	T	castor oil	0.96	350	N
hand cleaner		2000	T	cough syrup	1.0	190	N
dairy				"stomach" remedy slurries		1500	T
cottage cheese	1.08	225	T	pill pastes		5000±	T
cream	1.02	20 at 40°F	N	plastics*, resins			
milk	1.03	1.2 at 60°F	N	butadiene	0.94	0.17 at 40°F	
cheese, processed		30,000 at 160°F	T	polyester resin (typ)	1.4	3000	
yogurt		1100	T	PVA resin (typ)	1.3	65,000	T
detergents				starches, gums			
detergent concentrate		10	N	cornstarch sol 22°B	1.18	32	T
dyes and inks				cornstarch sol 25°B	1.21	300	T
ink, printers	1–1.38	10,000	T	sugar, syrups, molasses			
dye	1.1	10	N	corn syrup 41°Be	1.39	15,000 at 60°F	N
gum		5000	T	corn syrup 45°Be	1.45	12,000 at 130°F	N
fats and oils				glucose	1.42	10,000 at 100°F	
corn oils	0.92	30	N	molasses			
lard	0.96	60 at 100°F	N	A	1.4–1.47	280–5000 at 100°F	
linseed oil	0.93	30 at 100°F	N				

APPENDIX 14.H *(continued)*

Properties of Uncommon Fluids

peanut oil	0.92	42 at 100°F	N		B	1.43–1.48	1400–13,000 at 100°F	
soybean oil	0.95	36 at 100°F	N					
vegetable oil	0.92	3 at 300°F	N		C	1.46–1.49	2600–5000 at 100°F	
foods, miscellaneous								
black bean paste		10,000	T					
cream style corn		130 at 190°F	T		sugar syrups			
catsup	1.11	560 at 145°F	T		60 brix	1.29	75 at 60°F	N
pablum		4500	T		68 brix	1.34	360 at 60°F	N
pear pulp		4000 at 160°F	T		76 brix	1.39	4000 at 60°F	N
potato, mashed	1.0	20,000	T					
potato skins and caustic		20,000 at 100°F	T		water and waste treatment			
prune juice	1.0	60 at 120°F	T		clarified sewage sludge	1.1	2000 range	
orange juice concentrate	1.1	5000 at 38°F	T					
tapioca pudding	0.7	1000 at 235°F	T					
mayonnaise	1.0	5000 at 75°F	T					
tomato paste, 33%	1.14	7000	T					
honey	1.5	1500 at 100°F	T					

*A wide variety of plastics can be pumped; viscosity varies greatly.

APPENDIX 14.I
Vapor Pressure of Various Hydrocarbons and Water
(Cox Chart)

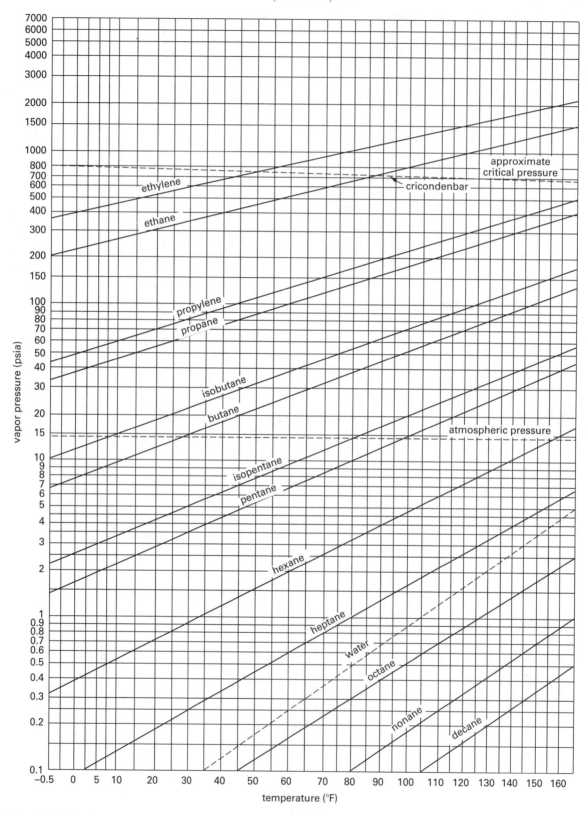

From *Hydraulic Handbook*, copyright © 1988, by Fairbanks Morse Pump. Reproduced with permission.

APPENDIX 14.J
Specific Gravity of Hydrocarbons

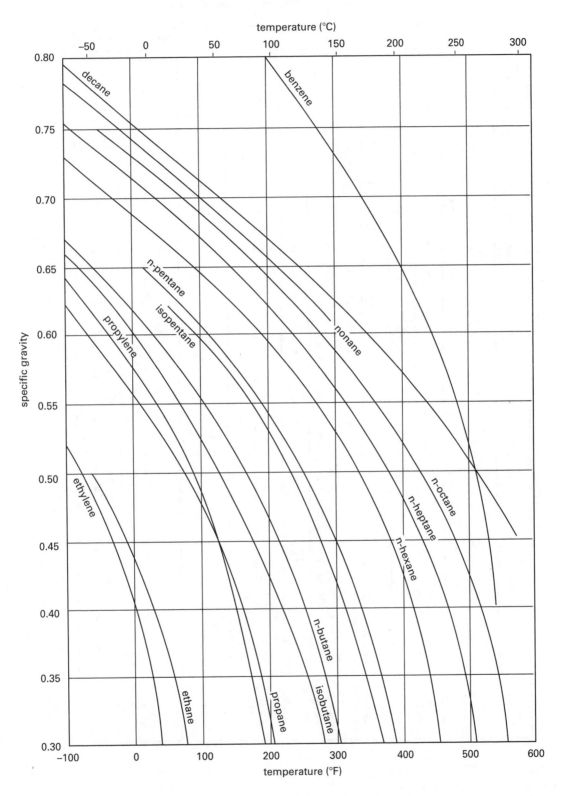

Appendices

APPENDIX 14.K
Viscosity Conversion Chart
(Approximate Conversions for Newtonian Fluids)

kinematic viscosity scales	absolute viscosity scales

Scales (kinematic viscosity):
- centistokes: 0, 100, 200, 300, 400, 500, 600, 700, 800, 900, 1000, 1100, 1200, 1300, 1400, 1500
- Mobilometer 100 g 10 cm seconds: 10, 20, 30, 40, 50, 60, 70
- Engler degrees: 25, 50, 75, 100, 125, 150, 175
- Ford 4 seconds: 25, 50, 75, 100, 125, 150, 175, 200, 225, 250, 275, 300, 325, 350, 375
- Ford 3 seconds: 25, 50, 75, 100, 125, 150, 175, 200, 225, 250, 275, 300, 325, 350, 375, 400, 425, 450, 475, 500, 550, 600, 650
- Saybolt Universal seconds: 500, 1000, 1500, 2000, 2500, 3000, 3500, 4000, 4500, 5000, 5500, 6000, 6500
- Saybolt Furol seconds: 50, 100, 150, 200, 250, 300, 350, 400, 450, 500, 550, 600, 650
- Redwood 1 Standard seconds: 250, 500, 750, 1000, 1250, 1500, 2000, 2500, 3000, 3500, 4000, 4500, 5000, 5500, 6000
- Ubbelohde cSt: 200, 300, 400, 500, 600, 700, 800, 900, 1000
- Gardner Holdt: A2, A4, A, C, E, G, H, I, J, K, L, M, N, O, P, Q, R, S, T, U, V, W, X, A3, A1, B, D, F
- Zahn 5 seconds: 13, 20, 30, 40, 50, 60
- Zahn 3 seconds: 23, 30, 40, 50, 60

Scales (absolute viscosity):
- Kreb Stormer 200 g K.U.: 50, 60, 70, 80, 90
- Stormer Cylinder 150 g seconds: 16, 27, 50, 115, 223
- centipoise: 0, 100, 200, 300, 400, 500, 600, 700, 800, 900, 1000, 1100, 1200, 1300, 1400, 1500

(Multiply centistokes by 1.0764×10^{-5} to get $\mathrm{ft}^2/\mathrm{sec}$.)
(Multiply centistokes by 1.000×10^{-6} to get m^2/s.)

APPENDIX 14.L
Viscosity Index Chart: 0–100 VI

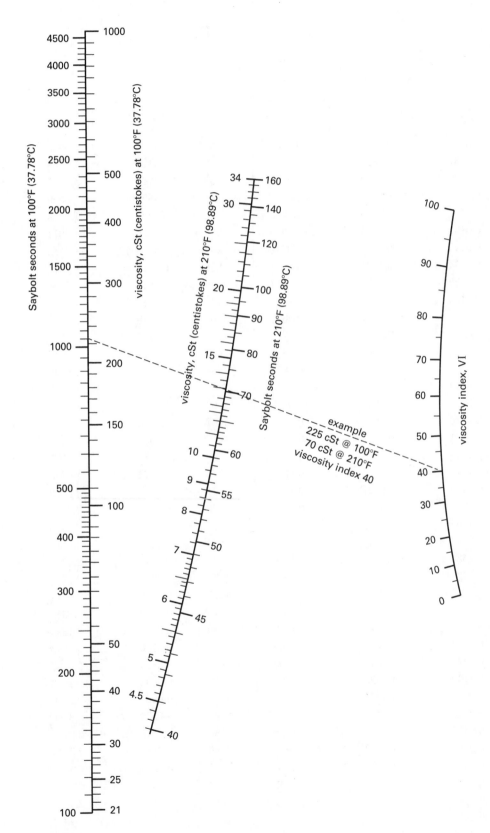

Based on correlations presented in ASTM D2270.

Appendices

APPENDIX 16.A
Area, Wetted Perimeter, and Hydraulic Radius of Partially Filled Circular Pipes

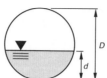

$\dfrac{d}{D}$	$\dfrac{\text{area}}{D^2}$	$\dfrac{\text{wetted perimeter}}{D}$	$\dfrac{r_h}{D}$	$\dfrac{d}{D}$	$\dfrac{\text{area}}{D^2}$	$\dfrac{\text{wetted perimeter}}{D}$	$\dfrac{r_h}{D}$
0.01	0.0013	0.2003	0.0066	0.51	0.4027	1.5908	0.2531
0.02	0.0037	0.2838	0.0132	0.52	0.4127	1.6108	0.2561
0.03	0.0069	0.3482	0.0197	0.53	0.4227	1.6308	0.2591
0.04	0.0105	0.4027	0.0262	0.54	0.4327	1.6509	0.2620
0.05	0.0147	0.4510	0.0326	0.55	0.4426	1.6710	0.2649
0.06	0.0192	0.4949	0.0389	0.56	0.4526	1.6911	0.2676
0.07	0.0242	0.5355	0.0451	0.57	0.4625	1.7113	0.2703
0.08	0.0294	0.5735	0.0513	0.58	0.4723	1.7315	0.2728
0.09	0.0350	0.6094	0.0574	0.59	0.4822	1.7518	0.2753
0.10	0.0409	0.6435	0.0635	0.60	0.4920	1.7722	0.2776
0.11	0.0470	0.6761	0.0695	0.61	0.5018	1.7926	0.2797
0.12	0.0534	0.7075	0.0754	0.62	0.5115	1.8132	0.2818
0.13	0.0600	0.7377	0.0813	0.63	0.5212	1.8338	0.2839
0.14	0.0688	0.7670	0.0871	0.64	0.5308	1.8546	0.2860
0.15	0.0739	0.7954	0.0929	0.65	0.5404	1.8755	0.2881
0.16	0.0811	0.8230	0.0986	0.66	0.5499	1.8965	0.2899
0.17	0.0885	0.8500	0.1042	0.67	0.5594	1.9177	0.2917
0.18	0.0961	0.8763	0.1097	0.68	0.5687	1.9391	0.2935
0.19	0.1039	0.9020	0.1152	0.69	0.5780	1.9606	0.2950
0.20	0.1118	0.9273	0.1206	0.70	0.5872	1.9823	0.2962
0.21	0.1199	0.9521	0.1259	0.71	0.5964	2.0042	0.2973
0.22	0.1281	0.9764	0.1312	0.72	0.6054	2.0264	0.2984
0.23	0.1365	1.0003	0.1364	0.73	0.6143	2.0488	0.2995
0.24	0.1449	1.0239	0.1416	0.74	0.6231	2.0714	0.3006
0.25	0.1535	1.0472	0.1466	0.75	0.6318	2.0944	0.3017
0.26	0.1623	1.0701	0.1516	0.76	0.6404	2.1176	0.3025
0.27	0.1711	1.0928	0.1566	0.77	0.6489	2.1412	0.3032
0.28	0.1800	1.1152	0.1614	0.78	0.6573	2.1652	0.3037
0.29	0.1890	1.1373	0.1662	0.79	0.6655	2.1895	0.3040
0.30	0.1982	1.1593	0.1709	0.80	0.6736	2.2143	0.3042
0.31	0.2074	1.1810	0.1755	0.81	0.6815	2.2395	0.3044
0.32	0.2167	1.2025	0.1801	0.82	0.6893	2.2653	0.3043
0.33	0.2260	1.2239	0.1848	0.83	0.6969	2.2916	0.3041
0.34	0.2355	1.2451	0.1891	0.84	0.7043	2.3186	0.3038
0.35	0.2450	1.2661	0.1935	0.85	0.7115	2.3462	0.3033
0.36	0.2546	1.2870	0.1978	0.86	0.7186	2.3746	0.3026
0.37	0.2642	1.3078	0.2020	0.87	0.7254	2.4038	0.3017
0.38	0.2739	1.3284	0.2061	0.88	0.7320	2.4341	0.3008
0.39	0.2836	1.3490	0.2102	0.89	0.7384	2.4655	0.2995
0.40	0.2934	1.3694	0.2142	0.90	0.7445	2.4981	0.2980
0.41	0.3032	1.3898	0.2181	0.91	0.7504	2.5322	0.2963

APPENDIX 16.A *(continued)*
Area, Wetted Perimeter, and Hydraulic Radius of Partially Filled Circular Pipes

$\dfrac{d}{D}$	$\dfrac{\text{area}}{D^2}$	$\dfrac{\text{wetted perimeter}}{D}$	$\dfrac{r_h}{D}$	$\dfrac{d}{D}$	$\dfrac{\text{area}}{D^2}$	$\dfrac{\text{wetted perimeter}}{D}$	$\dfrac{r_h}{D}$
0.42	0.3130	1.4101	0.2220	0.92	0.7560	2.5681	0.2944
0.43	0.3229	1.4303	0.2257	0.93	0.7612	2.6061	0.2922
0.44	0.3328	1.4505	0.2294	0.94	0.7662	2.6467	0.2896
0.45	0.3428	1.4706	0.2331	0.95	0.7707	2.6906	0.2864
0.46	0.3527	1.4907	0.2366	0.96	0.7749	2.7389	0.2830
0.47	0.3627	1.5108	0.2400	0.97	0.7785	2.7934	0.2787
0.48	0.3727	1.5308	0.2434	0.98	0.7816	2.8578	0.2735
0.49	0.3827	1.5508	0.2467	0.99	0.7841	2.9412	0.2665
0.50	0.3927	1.5708	0.2500	1.00	0.7854	3.1416	0.2500

APPENDIX 16.B
Dimensions of Welded and Seamless Steel Pipe[a,b]
(selected sizes)[c] (customary U.S. units)

nominal diameter (in)	schedule	outside diameter (in)	wall thickness (in)	internal diameter (in)	internal area (in²)	internal diameter (ft)	internal area (ft²)
$\frac{1}{8}$	40 (S)	0.405	0.068	0.269	0.0568	0.0224	0.00039
	80 (X)		0.095	0.215	0.0363	0.0179	0.00025
$\frac{1}{4}$	40 (S)	0.540	0.088	0.364	0.1041	0.0303	0.00072
	80 (X)		0.119	0.302	0.0716	0.0252	0.00050
$\frac{3}{8}$	40 (S)	0.675	0.091	0.493	0.1909	0.0411	0.00133
	80 (X)		0.126	0.423	0.1405	0.0353	0.00098
$\frac{1}{2}$	40 (S)	0.840	0.109	0.622	0.3039	0.0518	0.00211
	80 (X)		0.147	0.546	0.2341	0.0455	0.00163
	160		0.187	0.466	0.1706	0.0388	0.00118
	(XX)		0.294	0.252	0.0499	0.0210	0.00035
$\frac{3}{4}$	40 (S)	1.050	0.113	0.824	0.5333	0.0687	0.00370
	80 (X)		0.154	0.742	0.4324	0.0618	0.00300
	160		0.218	0.614	0.2961	0.0512	0.00206
	(XX)		0.308	0.434	0.1479	0.0362	0.00103
1	40 (S)	1.315	0.133	1.049	0.8643	0.0874	0.00600
	80 (X)		0.179	0.957	0.7193	0.0798	0.00500
	160		0.250	0.815	0.5217	0.0679	0.00362
	(XX)		0.358	0.599	0.2818	0.0499	0.00196
$1\frac{1}{4}$	40 (S)	1.660	0.140	1.380	1.496	0.1150	0.01039
	80 (X)		0.191	1.278	1.283	0.1065	0.00890
	160		0.250	1.160	1.057	0.0967	0.00734
	(XX)		0.382	0.896	0.6305	0.0747	0.00438
$1\frac{1}{2}$	40 (S)	1.900	0.145	1.610	2.036	0.1342	0.01414
	80 (X)		0.200	1.500	1.767	0.1250	0.01227
	160		0.281	1.338	1.406	0.1115	0.00976
	(XX)		0.400	1.100	0.9503	0.0917	0.00660
2	40 (S)	2.375	0.154	2.067	3.356	0.1723	0.02330
	80 (X)		0.218	1.939	2.953	0.1616	0.02051
	160		0.343	1.689	2.240	0.1408	0.01556
	(XX)		0.436	1.503	1.774	0.1253	0.01232
$2\frac{1}{2}$	40 (S)	2.875	0.203	2.469	4.788	0.2058	0.03325
	80 (X)		0.276	2.323	4.238	0.1936	0.02943
	160		0.375	2.125	3.547	0.1771	0.02463

APPENDIX 16.B *(continued)*
Dimensions of Welded and Seamless Steel Pipe[a,b]
(selected sizes)[c] (customary U.S. units)

nominal diameter (in)	schedule	outside diameter (in)	wall thickness (in)	internal diameter (in)	internal area (in²)	internal diameter (ft)	internal area (ft²)
	(XX)		0.552	1.771	2.464	0.1476	0.01711
3	40 (S)	3.500	0.216	3.068	7.393	0.2557	0.05134
	80 (X)		0.300	2.900	6.605	0.2417	0.04587
	160		0.437	2.626	5.416	0.2188	0.03761
	(XX)		0.600	2.300	4.155	0.1917	0.02885
$3\frac{1}{2}$	40 (S)	4.000	0.226	3.548	9.887	0.2957	0.06866
	80 (X)		0.318	3.364	8.888	0.2803	0.06172
	(XX)		0.636	2.728	5.845	0.2273	0.04059
4	40 (S)	4.500	0.237	4.026	12.73	0.3355	0.08841
	80 (X)		0.337	3.826	11.50	0.3188	0.07984
	120		0.437	3.626	10.33	0.3022	0.07171
	160		0.531	3.438	9.283	0.2865	0.06447
	(XX)		0.674	3.152	7.803	0.2627	0.05419
5	40 (S)	5.563	0.258	5.047	20.01	0.4206	0.1389
	80 (X)		0.375	4.813	18.19	0.4011	0.1263
	120		0.500	4.563	16.35	0.3803	0.1136
	160		0.625	4.313	14.61	0.3594	0.1015
	(XX)		0.750	4.063	12.97	0.3386	0.09004
6	40 (S)	6.625	0.280	6.065	28.89	0.5054	0.2006
	80 (X)		0.432	5.761	26.07	0.4801	0.1810
	120		0.562	5.501	23.77	0.4584	0.1650
	160		0.718	5.189	21.15	0.4324	0.1469
	(XX)		0.864	4.897	18.83	0.4081	0.1308
8	20	8.625	0.250	8.125	51.85	0.6771	0.3601
	30		0.277	8.071	51.16	0.6726	0.3553
	40 (S)		0.322	7.981	50.03	0.6651	0.3474
	60		0.406	7.813	47.94	0.6511	0.3329
	80 (X)		0.500	7.625	45.66	0.6354	0.3171
	100		0.593	7.439	43.46	0.6199	0.3018
	120		0.718	7.189	40.59	0.5990	0.2819
	140		0.812	7.001	38.50	0.5834	0.2673
	(XX)		0.875	6.875	37.12	0.5729	0.2578
	160		0.906	6.813	36.46	0.5678	0.2532
10	20	10.75	0.250	10.250	82.52	0.85417	0.5730
	30		0.307	10.136	80.69	0.84467	0.5604
	40 (S)		0.365	10.020	78.85	0.83500	0.5476
	60 (X)		0.500	9.750	74.66	0.8125	0.5185
	80		0.593	9.564	71.84	0.7970	0.4989
	100		0.718	9.314	68.13	0.7762	0.4732
	120		0.843	9.064	64.53	0.7553	0.4481

APPENDIX 16.B *(continued)*
Dimensions of Welded and Seamless Steel Pipe[a,b]
(selected sizes)[c] (customary U.S. units)

nominal diameter (in)	schedule	outside diameter (in)	wall thickness (in)	internal diameter (in)	internal area (in²)	internal diameter (ft)	internal area (ft²)
10 *(continued)*	140 (XX)		1.000	8.750	60.13	0.7292	0.4176
	160		1.125	8.500	56.75	0.7083	0.3941
12	20	12.75	0.250	12.250	117.86	1.0208	0.8185
	30		0.330	12.090	114.80	1.0075	0.7972
	(S)		0.375	12.000	113.10	1.0000	0.7854
	40		0.406	11.938	111.93	0.99483	0.7773
	(X)		0.500	11.750	108.43	0.97917	0.7530
	60		0.562	11.626	106.16	0.96883	0.7372
	80		0.687	11.376	101.64	0.94800	0.7058
	100		0.843	11.064	96.14	0.92200	0.6677
12	120 (XX)		1.000	10.750	90.76	0.89583	0.6303
	140		1.125	10.500	86.59	0.87500	0.6013
	160		1.312	10.126	80.53	0.84383	0.5592
14 O.D.	10	14.00	0.250	13.500	143.14	1.1250	0.9940
	20		0.312	13.376	140.52	1.1147	0.9758
	30 (S)		0.375	13.250	137.89	1.1042	0.9575
	40		0.437	13.126	135.32	1.0938	0.9397
	(X)		0.500	13.000	132.67	1.0833	0.9213
	60		0.593	12.814	128.96	1.0679	0.8956
	80		0.750	12.500	122.72	1.0417	0.8522
	100		0.937	12.126	115.48	1.0105	0.8020
	120		1.093	11.814	109.62	0.98450	0.7612
	140		1.250	11.500	103.87	0.95833	0.7213
	160		1.406	11.188	98.31	0.93233	0.6827
16 O.D.	10	16.00	0.250	15.500	188.69	1.2917	1.3104
	20		0.312	15.376	185.69	1.2813	1.2895
	30 (S)		0.375	15.250	182.65	1.2708	1.2684
	40 (X)		0.500	15.000	176.72	1.2500	1.2272
	60		0.656	14.688	169.44	1.2240	1.1767
	80		0.843	14.314	160.92	1.1928	1.1175
	100		1.031	13.938	152.58	1.1615	1.0596
	120		1.218	13.564	144.50	1.1303	1.0035
	140		1.437	13.126	135.32	1.0938	0.9397
	160		1.593	12.814	128.96	1.0678	0.8956
18 O.D.	10	18.00	0.250	17.500	240.53	1.4583	1.6703
	20		0.312	17.376	237.13	1.4480	1.6467
	(S)		0.375	17.250	233.71	1.4375	1.6230
	30		0.437	17.126	230.36	1.4272	1.5997
	(X)		0.500	17.000	226.98	1.4167	1.5762
	40		0.562	16.876	223.68	1.4063	1.5533
	60		0.750	16.500	213.83	1.3750	1.4849
	80		0.937	16.126	204.24	1.3438	1.4183
	100		1.156	15.688	193.30	1.3073	1.3423
	120		1.375	15.250	182.65	1.2708	1.2684
	140		1.562	14.876	173.81	1.2397	1.2070
	160		1.781	14.438	163.72	1.2032	1.1370

APPENDIX 16.B *(continued)*
Dimensions of Welded and Seamless Steel Pipe[a,b]
(selected sizes)[c] (customary U.S. units)

nominal diameter (in)	schedule	outside diameter (in)	wall thickness (in)	internal diameter (in)	internal area (in²)	internal diameter (ft)	internal area (ft²)
20 O.D.	10	20.00	0.250	19.500	298.65	1.6250	2.0739
	20 (S)		0.375	19.250	291.04	1.6042	2.0211
	30 (X)		0.500	19.000	283.53	1.5833	1.9689
	40		0.593	18.814	278.00	1.5678	1.9306
	60		0.812	18.376	265.21	1.5313	1.8417
	80		1.031	17.938	252.72	1.4948	1.7550
	100		1.281	17.438	238.83	1.4532	1.6585
	120		1.500	17.000	226.98	1.4167	1.5762
	140		1.750	16.500	213.83	1.3750	1.4849
	160		1.968	16.064	202.67	1.3387	1.4075
24 O.D.	10	24.00	0.250	23.500	433.74	1.9583	3.0121
	20 (S)		0.375	23.250	424.56	1.9375	2.9483
	(X)		0.500	23.000	415.48	1.9167	2.8852
	30		0.562	22.876	411.01	1.9063	2.8542
	40		0.687	22.626	402.07	1.8855	2.7922
	60		0.968	22.060	382.20	1.8383	2.6542
	80		1.218	21.564	365.21	1.7970	2.5362
	100		1.531	20.938	344.32	1.7448	2.3911
	120		1.812	20.376	326.92	1.6980	2.2645
	140		2.062	19.876	310.28	1.6563	2.1547
	160		2.343	19.310	292.87	1.6092	2.0337
30 O.D.	10	30.00	0.312	29.376	677.76	2.4480	4.7067
	(S)		0.375	29.250	671.62	2.4375	4.6640
	20 (X)		0.500	29.000	660.52	2.4167	4.5869
	30		0.625	28.750	649.18	2.3958	4.5082

(Multiply in by 25.4 to obtain mm.)

(Multiply in² by 645 to obtain mm².)

[a]Designations are per ANSI B36.10.

[b]The "S" wall thickness was formerly designated as "standard weight." Standard weight and schedule-40 are the same for all diameters through 10 in. For diameters between 12 in and 24 in, standard weight pipe has a wall thickness of 0.375 in. The "X" (or "XS") wall thickness was formerly designated as "extra strong." Extra strong weight and schedule-80 are the same for all diameters through 8 in. For diameters between 10 in and 24 in, extra strong weight pipe has a wall thickness of 0.50 in. The "XX" (or "XXS") wall thickness was formerly designed as "double extra strong." Double extra strong weight pipe does not have a corresponding schedule number.

[c]Pipe sizes and weights in most common usage are listed. Other weights and sizes exist.

Appendices

APPENDIX 16.C
Dimensions of Welded and Seamless Steel Pipe
(SI units)

nominal diameter (in)	(mm)	outside diameter (mm)	schedule		wall thickness (mm)	internal diameter (mm)	internal area (cm²)
1/8	6	10.3	10S		1.245	7.811	0.479
			(S)	40	1.727	6.846	0.368
			(X)	80	2.413	5.474	0.235
1/4	8	13.7	10S		1.651	10.398	0.846
			(S)	40	2.235	9.23	0.669
			(X)	80	3.023	7.654	0.460
3/8	10	17.145	10S		1.651	13.843	1.505
			(S)	40	2.311	12.523	1.232
			(X)	80	3.2	10.745	0.907
1/2	15	21.336	5S		1.651	18.034	2.554
			10S		2.108	17.12	2.302
			(S)	40	2.769	15.798	1.960
			(X)	80	3.734	13.868	1.510
				160	4.75	11.836	1.100
			(XX)		7.468	6.4	0.322
3/4	20	26.67	5S		1.651	23.368	4.289
			10S		2.108	22.454	3.960
			(S)	40	2.87	20.93	3.441
			(X)	80	3.912	18.846	2.790
				160	5.537	15.596	1.910
			(XX)		7.823	11.024	0.954
1	25	33.401	5S		1.651	30.099	7.115
			10S		2.769	27.863	6.097
			(S)	40	3.378	26.645	5.576
			(X)	80	4.547	24.307	4.640
				160	6.35	20.701	3.366
			(XX)		9.093	15.215	1.818
1 1/4	32	42.164	5S		1.651	38.862	11.862
			10S		2.769	36.626	10.563
			(S)	40	3.556	35.052	9.650
			(X)	80	4.851	32.462	8.276
				160	6.35	29.464	6.818
			(XX)		9.703	22.758	4.068
1 1/2	40	48.26	5S		1.651	44.958	15.875
			10S		2.769	42.722	14.335
			(S)	40	3.683	40.894	13.134
			(X)	80	5.08	38.1	11.401
				160	7.137	33.986	9.072
			(XX)		10.16	27.94	6.131
					13.335	21.59	3.661
					15.875	16.51	2.141
2	50	60.325	5S		1.651	57.023	25.538
			10S		2.769	54.787	23.575
			(S)	40	3.912	52.501	21.648

APPENDIX 16.C *(continued)*
Dimensions of Welded and Seamless Steel Pipe
(SI units)

nominal diameter (in)	nominal diameter (mm)	outside diameter (mm)	schedule		wall thickness (mm)	internal diameter (mm)	internal area (cm²)
			(X)	80	5.537	49.251	19.051
				160	8.712	42.901	14.455
			(XX)		11.074	38.177	11.447
					14.275	31.775	7.930
					17.45	25.425	5.077
$2\frac{1}{2}$	65	73.025	5S		2.108	68.809	37.186
			10S		3.048	66.929	35.182
			(S)	40	5.156	62.713	30.889
			(X)	80	7.01	59.005	27.344
				160	9.525	53.975	22.881
			(XX)		14.021	44.983	15.892
					17.145	38.735	11.784
					20.32	32.385	8.237
3	80	88.9	5S		2.108	84.684	56.324
			10S		3.048	82.804	53.851
			(S)	40	5.486	77.928	47.696
			(X)	80	7.62	73.66	42.614
				160	11.1	66.7	34.942
			(XX)		15.24	58.42	26.805
					18.415	52.07	21.294
					21.59	45.72	16.417
$3\frac{1}{2}$	90	101.6	5S		2.108	97.384	74.485
			10S	40	3.048	95.504	71.636
			(S)	80	5.74	90.12	63.787
			(X)		8.077	85.446	57.342
			(XX)		16.154	69.292	37.710
4	100	114.3	5S		2.108	110.084	95.179
			10S		3.048	108.204	91.955
					4.775	104.75	86.179
			(S)	40	6.02	102.26	82.130
			(X)	80	8.56	97.18	74.173
				120	11.1	92.1	66.621
					12.7	88.9	62.072
				160	13.487	87.326	59.893
			(XX)		17.12	80.06	50.341
					20.32	73.66	42.614
					23.495	67.31	35.584
5	125	141.3	5S		2.769	135.762	144.76
			10S		3.404	134.492	142.06
			(S)	40	6.553	128.194	129.07
			(X)	80	9.525	122.25	117.38
				120	12.7	115.9	105.50
				160	15.875	109.55	94.254
			(XX)		19.05	103.2	83.647
					22.225	96.85	73.670
					25.4	90.5	64.326

APPENDIX 16.C *(continued)*
Dimensions of Welded and Seamless Steel Pipe
(SI units)

nominal diameter (in)	(mm)	outside diameter (mm)	schedule		wall thickness (mm)	internal diameter (mm)	internal area (cm²)
			5S		2.769	162.737	208.00
			10S		3.404	161.467	204.77
					5.563	157.149	193.96
			(S)	40	7.112	154.051	186.39
			(X)	80	10.973	146.329	168.17
6	150	168.275		120	14.275	139.725	153.33
				160	18.237	131.801	136.44
			(XX)		21.946	124.383	121.51
					25.4	117.475	108.39
					28.575	111.125	96.987
			5S		2.769	213.537	358.13
			10S		3.759	211.557	351.52
					5.563	207.949	339.63
				20	6.35	206.375	334.51
				30	7.036	205.003	330.07
			(S)	40	8.179	202.717	322.75
8	200	219.075		60	10.312	198.451	309.31
			(X)	80	12.7	193.675	294.60
				100	15.062	188.951	280.41
				120	18.237	182.601	261.88
				140	20.625	177.825	248.36
				160	23.012	173.051	235.20
					25.4	168.275	222.40
					28.575	161.925	205.93

(Multiply in by 25.4 to obtain mm.)
(Multiply lbf/in² by 6.895 to obtain kPa.)

APPENDIX 16.D
Dimensions of Rigid PVC and CPVC Pipe
(customary U.S. units)

nominal diameter (in)	outside diameter (in)	schedule-40 ASTM D1785			schedule-80 ASTM D1785			class 200 ASTM D2241		
		internal diameter (in)	wall thickness (in)	pressure rating* (lbf/in²)	internal diameter (in)	wall thickness (in)	pressure rating* (lbf/in²)	internal diameter (in)	wall thickness (in)	pressure rating* (lbf/in²)
$\frac{1}{8}$	0.405	0.249	0.068	810	0.195	0.095	1230	–	–	–
$\frac{1}{4}$	0.540	0.344	0.088	780	0.282	0.119	1130	–	–	–
$\frac{3}{8}$	0.675	0.473	0.091	620	0.403	0.126	920	–	–	–
$\frac{1}{2}$	0.840	0.622	0.109	600	0.546	0.147	850	0.716	0.062	200
$\frac{3}{4}$	1.050	0.824	0.113	480	0.742	0.154	690	0.930	0.060	200
1	1.315	1.049	0.133	450	0.957	0.179	630	1.189	0.063	200
$1\frac{1}{4}$	1.660	1.380	0.140	370	1.278	0.191	520	1.502	0.079	200
$1\frac{1}{2}$	1.900	1.610	0.145	330	1.500	0.200	470	1.720	0.090	200
2	2.375	2.067	0.154	280	1.939	0.218	400	2.149	0.113	200
$2\frac{1}{2}$	2.875	2.469	0.203	300	2.323	0.276	420	2.601	0.137	200
3	3.500	3.068	0.216	260	2.900	0.300	370	3.166	0.167	200
4	4.500	4.026	0.237	220	3.826	0.337	320	4.072	0.214	200
5	5.563	5.047	0.258	190	4.768	0.375	290	–	–	–
6	6.625	6.065	0.280	180	5.761	0.432	280	5.993	0.316	200
8	8.625	7.961	0.332	160	7.565	0.500	250	7.740	0.410	200
10	10.750	9.976	0.365	140	9.492	0.593	230	9.650	0.511	200
12	12.750	11.890	0.406	130	11.294	0.687	230	11.450	0.606	200
14	14.000	13.073	0.447	130	12.410	0.750	220	–	–	–
16	16.000	14.940	0.500	130	14.213	0.843	220	–	–	–
18	18.000	16.809	0.562	130	16.014	0.937	220	–	–	–
20	20.000	18.743	0.5937	130	17.814	1.031	220	–	–	–
24	24.000	22.554	0.687	120	21.418	1.218	210	–	–	–

(Multiply in by 25.4 to obtain mm.)

(Multiply lbf/in² by 6.895 to obtain kPa.)

*Pressure ratings are for a pipe temperature of 68°F (20°C) and are subject to the following temperature derating factors. Operation above 140°F (60°C) is not permitted.

pipe operating temperature °F (°C)	73 (23)	80 (27)	90 (32)	100 (38)	110 (43)	120 (49)	130 (54)	140 (60)
derating factor	1.0	0.88	0.75	0.62	0.51	0.40	0.31	0.22

Appendices

APPENDIX 16.E
Dimensions of Large Diameter, Nonpressure, PVC Sewer and Water Pipe
(customary U.S. units)

nominal size (in)	nominal size (mm)	designations and dimensional controls	minimum wall thickness (in)	outside diameter (inside diameter for profile wall pipes) (in)
ASTM F679 – Gravity Sewer Pipe (solid wall)				
18	450	PS-46, T-2	0.499	18.701
		PS-46, T-1	0.536	
		PS-115, T-2	0.671	
		PS-115, T-1	0.720	
21	525	PS-46, T-2	0.588	22.047
		PS-46, T-1	0.632	
		PS-115, T-2	0.791	
		PS-115, T-1	0.849	
24	600	PS-46, T-2	0.661	24.803
		PS-46, T-1	0.709	
		PS-115, T-2	0.889	
		PS-115, T-1	0.954	
27	675	PS-46, T-2	0.745	27.953
		PS-46, T-1	0.745	
		PS-115, T-2	1.002	
		PS-115, T-1	1.075	
30	750	PS-46, T-2	0.853	32.00
		PS-46, T-1	0.914	
		PS-115, T-2	1.148	
		PS-115, T-1	1.231	
36	900	PS-46, T-2	1.021	38.30
		PS-46, T-1	1.094	
		PS-115, T-2	1.373	
		PS-115, T-1	1.473	
42	1050	PS-46, T-2	1.186	44.50
		PS-46, T-1	1.271	
		PS-115, T-2	1.596	
		PS-115, T-1	1.712	
48	1200	PS-46, T-2	1.354	50.80
		PS-46, T-1	1.451	
		PS-115, T-2	1.822	
		PS-115, T-1	1.954	
ASTM F758 – PVC Highway Underdrain Pipe (perforated)				
4	100		0.120	4.215
6	150		0.180	6.275
8	200		0.240	8.40
ASTM F789 – PVC Sewage Pipe (solid wall)				
available in sizes		available in two stiffnesses and three material grades:		
4–18 in		PS-46; T-1, T-2, T-3		
(100–450 mm)		PS-115; T-1, T-2, T-3		
(obsolete)				
ASTM F794 – Open Profile Wall Pipe				
available in sizes		PS-46		internal diameter controlled
4–48 in				
(100–1200 mm)				

APPENDIX 16.E *(continued)*
Dimensions of Large Diameter, Nonpressure, PVC Sewer and Water Pipe
(customary U.S. units)

nominal size (in)	nominal size (mm)	designations and dimensional controls	minimum wall thickness (in)	outside diameter (inside diameter for profile wall pipes) (in)
AWWA C900 – Water and Wastewater Pressure Pipe (solid wall)				
4	100	DR-25 (PC-100,		4.80
6	150	PR-165, CL100);		6.90
8	200	DR-18 (PC-150,	$t_{min} = D_o/\text{DR}$	9.05
10	250	PR-235, CL150);		11.10
12	300	DR-14 (PC-200, PC-305, CL200)		13.20
AWWA C905 – Water and Wastewater Pipe (solid wall)				
14	350	DR-51 (PR-80);		15.30
16	400	DR-41 (PR-100);		17.40
18	450	DR-32.5 (PR-125);		19.50
20	500	DR-25 (PR-165);	$t_{min} = D_o/\text{DR}$	21.60
24	600	DR-21 (PR-200);		25.8
30	750	DR-18 (PR-235);		32.0
36	900	DR-14 (PR-305)		38.3

ASTM F949 – Open Profile Wall Pipe

available in sizes   PS-46   internal diameter
4–36 in   controlled
(100–900 mm)

ASTM D1785
See App. 16.D.

ASTM F1803 – Closed Profile Wall "Truss" Pipe (DWCP)

available in sizes   PS-46   internal diameter
18–60 in   controlled
(450–1500 mm)

ASTM D2241 – PVC Sewer Pipe (solid wall)

nominal size (in)	nominal size (mm)	designations	minimum wall thickness	outside diameter (in)
1	25			1.315
$1\frac{1}{4}$	31			1.660
$1\frac{1}{2}$	37			1.900
2	50	SDR-41		2.375
$2\frac{1}{2}$	62	SDR-32.5 SDR-26		2.875
3	75	SDR-21	$t_{min} = D_o/\text{SDR}$	3.500
$3\frac{1}{2}$	87	SDR-17		4.000
4	100	SDR-13.5		4.500
6	150			6.625
8	200			8.625
10	250			10.750
12	300			12.750

ASTM D2729 – PVC Drainage Pipe (solid wall; perforated)

3		0.070	3.250
4		0.075	4.215
6		0.100	6.275

APPENDIX 16.E *(continued)*
Dimensions of Large Diameter, Nonpressure, PVC Sewer and Water Pipe
(customary U.S. units)

nominal size (in)	(mm)	designations and dimensional controls	minimum wall thickness (in)	outside diameter (inside diameter for profile wall pipes) (in)
		**ASTM D3034 – PVC Gravity Sewer Pipe and Perforated Drain Pipe (solid wall)**		
4	100	SDR-35 & PS-46 (regular); SDR-26 & PS-115 (HW); SDR-23.5 & PS-135	$t_{min} = D_o/SDR$	4.215
5	135			5.640
6	150			6.275
8	200	SDR-41 & PS-28; SDR-35 & PS-46 (regular); SDR-26 & PS-115 (HW)		8.400
10	250			10.50
12	300			12.50
15	375			15.30

abbreviations

CL – pressure class (same as PC)

DR – dimensional ratio (constant ratio of outside diameter/wall thickness)

DWCP – double wall corrugated pipe

HW – heavy weight

PC – pressure class (same as CL)

PS – pipe stiffness (resistance to vertical diametral deflection due to compression from burial soil and surface loading, in lbf/in²; calculated uncorrected as vertical load in lbf per inch of pipe length divided by deflection in inches)

SDR – standard dimension ratio (constant ratio of outside diameter/wall thickness)

(Multiply in by 25.4 to obtain mm.)

(Multiply lbf/in² by 6.895 to obtain kPa.)

APPENDIX 16.F
Dimensions and Weights of Concrete Sewer Pipe
(customary U.S. units)

ASTM C14 – Nonreinforced Round Sewer and Culvert Pipe for Nonpressure (Gravity Flow) Applications

nominal size and internal diameter		minimum wall thickness and required ASTM C14 $D_{0.01}$-load rating[a] (in and lbf/ft)		
(in)	(mm)	class 1[b]	class 2[b]	class 3[b]
4	100	0.625 (1500)	0.75 (2000)	0.875 (2400)
6	150	0.625 (1500)	0.75 (2000)	1.0 (2400)
8	200	0.75 (1500)	0.875 (2000)	1.125 (2400)
10	250	0.875 (1600)	1.0 (2000)	1.25 (2400)
12	300	1.0 (1800)	1.375 (2250)	1.75 (2500)
15	375	1.25 (2000)	1.625 (2500)	1.875 (2900)
18	450	1.5 (2200)	2.0 (3000)	2.25 (3300)
21	525	1.75 (2400)	2.25 (3300)	2.75 (3850)
24	600	2.125 (2600)	3.0 (3600)	3.375 (4400)
27	675	3.25 (2800)	3.75 (3950)	3.75 (4600)
30	750	3.5 (3000)	4.25 (4300)	4.25 (4750)
33	825	3.75 (3150)	4.5 (4400)	4.5 (4850)
36	900	4.0 (3300)	4.75 (4500)	4.75 (5000)

(Multiply in by 25.4 to obtain mm.)
(Multiply lbf/ft by 14.594 to obtain kN/m.)
[a]For C14 pipe, D-loads are stated in lbf/ft (pound per foot length of pipe).
[b]As defined by the strength (D-load) requirements of ASTM C14.

ASTM C76 – Reinforced Round Concrete Culvert, Storm Drain, and Sewer Pipe for Nonpressure (Gravity Flow) Applications

nominal size and internal diameter		minimum wall thickness and maximum ASTM C76 $D_{0.01}$-load rating (in and lbf/ft^2)		
(in)	(mm)	wall A[a]	wall B[a]	wall C[a]
8[b]	200		2.0	
10[b]	250		2.0	
12	300	1.75 (2000)	2.0 (3000)	2.75 (3000)
15	375	1.875 (2000)	2.25 (3000)	3.0 (3000)
18	450	2.0 (2000)	2.5 (3000)	3.25 (3000)
21	525	2.25 (2000)	2.75 (3000)	3.50 (3000)
24	600	2.5 (2000)	3.0 (3000)	3.75 (3000)
27	675	2.625 (2000)	3.25 (3000)	4.0 (3000)
30	750	2.75 (2000)	3.5 (3000)	4.25 (3000)
33	825	2.875 (1350)	3.75 (3000)	4.5 (3000)
36	900	3.0 (1350)	4.0 (3000)	4.75 (3000)
42	1050	3.5 (1350)	4.5 (3000)	5.25 (3000)
48	1200	4.0 (1350)	5.0 (3000)	5.75 (3000)
54	1350	4.5 (1350)	5.5 (2000)	6.25 (3000)
60	1500	5.0 (1350)	6.0 (2000)	6.75 (3000)
66	1650	5.5 (1350)	6.5 (2000)	7.25 (3000)
72	1800	6.0 (1350)	7.0 (2000)	7.75 (3000)
78	1950	6.5 (1350)	7.5 (1350)	8.25 (2000)
84	2100	7.0 (1350)	8.0 (1350)	8.75 (1350)
90	2250	7.5 (1350)	8.5 (1350)	9.25 (1350)
96	2400	8.0 (1350)	9.0 (1350)	9.75 (1350)
102	2550	8.5 (1350)	9.5 (1350)	10.25 (1350)
108	2700	9.0 (1350)	10.0 (1350)	10.75 (1350)

(Multiply in by 25.4 to obtain mm.)
(Multiply lbf/ft^2 by 0.04788 to obtain kN/m^2 (kPa).)
[a]wall A thickness in inches = diameter in feet; wall B thickness in inches = diameter in feet + 1 in; wall C thickness in inches = diameter in feet + 1.75 in
[b]Although not specifically called out in ASTM C76, 8 in and 10 in diameter circular concrete pipes are routinely manufactured.

APPENDIX 16.F *(continued)*
Dimensions and Weights of Concrete Sewer Pipe
(customary U.S. units)

ASTM C76 – Large Sizes of Pipe for Nonpressure (Gravity Flow) Applications

nominal size and internal diameter		minimum wall thickness (in)
(in)	(mm)	wall A[*]
114	2850	9.5
120	3000	10.0
126	3150	10.5
132	3300	11.0
138	3450	11.5
144	3600	12.0
150	3750	12.5
156	3900	13.0
162	4050	13.5
168	4200	14.0
174	4350	14.5
180	4500	15.0

(Multiply in by 25.4 to obtain mm.)

[*]wall A thickness in inches = diameter in feet

ASTM C361 – RCPP: Reinforced Concrete Low Head Pressure Pipe

"Low head" means 125 ft (54 psi; 375 kPa) or less. Standard pipe with 12 in (300 mm) to 108 in (2700 mm) diameters are available. ASTM C361 limits tensile stress (and, therefore, strain) and flexural deformation. Any reinforcement design that meets these limitations is permitted. Internal and external dimensions typically correspond to C76 (wall B) pipe, but wall C may also be used.

ASTM C655 – Reinforced Concrete D-Load Culvert, Storm Drain, and Sewer Pipe

Pipes smaller than 72 in (1800 mm) are often specified by D-load. Pipe manufactured to satisfy ASTM C655 will support a specified concentrated vertical load (applied in three-point loading) in pounds per ft of length per ft of diameter. Some standard C76 pipe meets this standard. In order to use C76 (and other) pipes, D-loads are mapped onto C76 pipe classes. In some cases, special pipe designs are required. Pipe selection by D-load is accomplished by first determining the size of pipe required to carry the flow, then choosing the appropriate class of pipe according to the required D-load.

$D_{0.01}$-load range	C76 pipe class
1–800	class 1 (I)
801–1000	class 2 (II)
1001–1350	class 3 (III)
1351–2000	class 4 (IV)
2000–3000	class 5 (V)
>3000	no equivalent class pipe

(Multiply in by 25.4 to obtain mm.)
(Multiply lbf/ft² by 0.04788 to obtain kN/m² (kPa).)

APPENDIX 16.G
Dimensions of Cast-Iron and Ductile Iron Pipe Standard Pressure Classes
(customary U.S. units)

ANSI/AWWA C106 Gray Cast-Iron Pipe

The ANSI/AWWA C106 standard is obsolete. The outside diameters of 4–48 in (100–1200 mm) diameter gray cast-iron pipes are the same as for C150 pipes. Ductile iron pipe is interchangeable with respect to joining diameters, accessories, and fittings. However, inside diameters of cast-iron pipe are substantially greater than C150 pipe. Outside diameters and thicknesses of AWWA 1908 standard cast-iron pipe are substantially different from C150 values.

ANSI/AWWA C150/A21.50 and ANSI/AWWA C151/A21.51

calculated minimum wall thickness[a], t

(in)

pressure class and head

nominal size (in)	outside diameter, D_o (in)	(lbf/in^2 and ft)					casting tolerance (in)	inside diameter
		150 (346)	200 (462)	250 (577)	300	350		
3	3.96					0.25[b]	0.05[c]	
4	4.80					0.25[b]	0.05[c]	
6	6.90					0.25[b]	0.05[c]	
8	9.05					0.25[b]	0.05[c]	
10	11.10					0.26	0.06	
12	13.20					0.28	0.06	
14	15.30			0.28	0.30	0.31	0.07	
16	17.40			0.30	0.32	0.34	0.07	
18	19.50			0.31	0.34	0.36	0.07	$D_i = D_o - 2t$
20	21.60			0.33	0.36	0.38	0.07	
24	25.80		0.33	0.37	0.40	0.43	0.07	
30	32.00	0.34	0.38	0.42	0.45	0.49	0.07	
36	38.30	0.38	0.42	0.47	0.51	0.56	0.07	
42	44.50	0.41	0.47	0.52	0.57	0.63	0.07	
48	50.80	0.46	0.52	0.58	0.64	0.70	0.08	
54	57.56	0.51	0.58	0.65	0.72	0.79	0.09	
60	61.61	0.54	0.61	0.68	0.76	0.83	0.09	
64	65.67	0.56	0.64	0.72	0.80	0.87	0.09	

(Multiply in by 25.4 to obtain mm.)

(Multiply lbf/in^2 by 6.895 to obtain kPa.)

[a]Per ANSI/AWWA C150/A21.50, the tabulated minimum wall thicknesses include a 0.08 in (2 mm) service allowance and the appropriate casting tolerance. Listed thicknesses are adequate for the rated water working pressure plus a surge allowance of 100 lbf/in^2 (690 kPa). Values are based on a yield strength of 42,000 lbf/in^2 (290 MPa), the sum of the working pressure and 100 lbf/in^2 (690 kPa) surge allowance, and a safety factor of 2.0.

[b]Pressure class is defined as the rated gage water pressure of the pipe in lbf/in^2.

[c]Limited by manufacturing. Calculated required thickness is less.

APPENDIX 16.H
Dimensions of Ductile Iron Pipe Special Pressure Classes
(customary U.S. units)

ANSI/AWWA C150/A21.50 and ANSI/AWWA C151/A21.51

nominal size[a] (in)	outside diameter (in)	wall thickness (in) special pressure class[b]						
		50	51	52	53	54	55	56
4	4.80	–	0.26	0.29	0.32	0.35	0.38	0.41
6	6.90	0.25	0.28	0.31	0.34	0.37	0.40	0.43
8	9.05	0.27	0.30	0.33	0.36	0.39	0.42	0.45
10	11.10	0.29	0.32	0.35	0.38	0.41	0.44	0.47
12	13.20	0.31	0.34	0.37	0.40	0.43	0.46	0.49
14	15.30	0.33	0.36	0.39	0.42	0.45	0.48	0.51
16	17.40	0.34	0.37	0.40	0.43	0.46	0.49	0.52
18	19.50	0.35	0.38	0.41	0.44	0.47	0.50	0.53
20	21.60	0.36	0.39	0.42	0.45	0.48	0.51	0.54
24	25.80	0.38	0.41	0.44	0.47	0.50	0.53	0.56
30	32.00	0.39	0.43	0.47	0.51	0.55	0.59	0.63
36	38.30	0.43	0.48	0.53	0.58	0.63	0.68	0.73
42	44.50	0.47	0.53	0.59	0.65	0.71	0.77	0.83
48	50.80	0.51	0.58	0.65	0.72	0.79	0.86	0.93
54	57.56	0.57	0.65	0.73	0.81	0.89	0.97	1.05

(Multiply in by 25.4 to obtain mm.)

(Multiply lbf/in^2 by 6.895 to obtain kPa.)

[a]Formerly designated "standard thickness classes." These special pressure classes are as shown in AWWA C150 and C151. Special classes are most appropriate for some threaded, grooved, or ball and socket pipes, or for extraordinary design conditions. They are generally less available than standard pressure class pipe.

[b]60 in (1500 mm) and 64 in (1600 mm) pipe sizes are not available in special pressure classes.

ASTM A746 Cement-Lined Gravity Sewer Pipe

The thicknesses for cement-lined pipe are the same as those in AWWA C150 and C151.

ASTM A746 Flexible Lining Gravity Sewer Pipe

The thicknesses for flexible lining pipe are the same as those in AWWA C150 and C151.

APPENDIX 16.I
Standard ASME/ANSI Z32.2.3 Piping Symbols*

	flanged	screwed	bell and spigot	welded	soldered
joint					
elbow—90°					
elbow—45°					
elbow—turned up					
elbow—turned down					
elbow—long radius					
reducing elbow					
tee					
tee—outlet up					
tee—outlet down					
side outlet tee—outlet up					
cross					
reducer—concentric					
reducer—eccentric					
lateral					
gate valve					
globe valve					
check valve					
stop cock					
safety valve					
expansion joint					
union					
sleeve					
bushing					

*Similar to MIL-STD-17B.

Appendices

APPENDIX 16.J
Dimensions of Copper Water Tubing
(customary U.S. units)

classification	nominal tube size (in)	outside diameter (in)	wall thickness (in)	inside diameter (in)	transverse area (in²)	safe working pressure (psi)
hard	1/4	3/8	0.025	0.325	0.083	1000
	3/8	1/2	0.025	0.450	0.159	1000
	1/2	5/8	0.028	0.569	0.254	890
	3/4	7/8	0.032	0.811	0.516	710
	1	1 1/8	0.035	1.055	0.874	600
	1 1/4	1 3/8	0.042	1.291	1.309	590
type "M" 250 psi working pressure	1 1/2	1 5/8	0.049	1.527	1.831	580
	2	2 1/8	0.058	2.009	3.17	520
	2 1/2	2 5/8	0.065	2.495	4.89	470
	3	3 1/8	0.072	2.981	6.98	440
	3 1/2	3 5/8	0.083	3.459	9.40	430
	4	4 1/8	0.095	3.935	12.16	430
	5	5 1/8	0.109	4.907	18.91	400
	6	6 1/8	0.122	5.881	27.16	375
	8	8 1/8	0.170	7.785	47.6	375
hard	3/8	1/2	0.035	0.430	0.146	1000
	1/2	5/8	0.040	0.545	0.233	1000
	3/4	7/8	0.045	0.785	0.484	1000
	1	1 1/8	0.050	1.025	0.825	880
	1 1/4	1 3/8	0.055	1.265	1.256	780
	1 1/2	1 5/8	0.060	1.505	1.78	720
type "L" 250 psi working pressure	2	2 1/8	0.070	1.985	3.094	640
	2 1/2	2 5/8	0.080	2.465	4.77	580
	3	3 1/8	0.090	2.945	6.812	550
	3 1/2	3 5/8	0.100	3.425	9.213	530
	4	4 1/8	0.110	3.905	11.97	510
	5	5 1/8	0.125	4.875	18.67	460
	6	6 1/8	0.140	5.845	26.83	430
hard	1/4	3/8	0.032	0.311	0.076	1000
	3/8	1/2	0.049	0.402	0.127	1000
	1/2	5/8	0.049	0.527	0.218	1000
	3/4	7/8	0.065	0.745	0.436	1000
	1	1 1/8	0.065	0.995	0.778	780
	1 1/4	1 3/8	0.065	1.245	1.217	630
type "K" 400 psi working pressure	1 1/2	1 5/8	0.072	1.481	1.722	580
	2	2 1/8	0.083	1.959	3.014	510
	2 1/2	2 5/8	0.095	2.435	4.656	470
	3	3 1/8	0.109	2.907	6.637	450
	3 1/2	3 5/8	0.120	3.385	8.999	430
	4	4 1/8	0.134	3.857	11.68	420
	5	5 1/8	0.160	4.805	18.13	400
	6	6 1/8	0.192	5.741	25.88	400
soft	1/4	3/8	0.032	0.311	0.076	1000
	3/8	1/2	0.049	0.402	0.127	1000
	1/2	5/8	0.049	0.527	0.218	1000
	3/4	7/8	0.065	0.745	0.436	1000
	1	1 1/8	0.065	0.995	0.778	780
	1 1/4	1 3/8	0.065	1.245	1.217	630
type "K" 250 psi working pressure	1 1/2	1 5/8	0.072	1.481	1.722	580
	2	2 1/8	0.083	1.959	3.014	510
	2 1/2	2 5/8	0.095	2.435	4.656	470
	3	3 1/8	0.109	2.907	6.637	450
	1 1/2	2 5/8	0.120	3.385	8.999	430
	4	4 1/8	0.134	3.857	11.68	420
	5	5 2/8	0.160	4.805	18.13	400
	6	6 1/8	0.192	5.741	25.88	400

(Multiply in by 25.4 to obtain mm.)
(Multiply in² by 645 to obtain mm².)

APPENDIX 16.K
Dimensions of Brass and Copper Tubing
(customary U.S. units)

regular

pipe size (in)	nominal dimensions (in)			cross-sectional area of bore (in²)	lbm/ft	
	O.D.	I.D.	wall		red brass	copper
1/8	0.405	0.281	0.062	0.062	0.253	0.259
1/4	0.540	0.376	0.082	0.110	0.447	0.457
3/8	0.675	0.495	0.090	0.192	0.627	0.641
1/2	0.840	0.626	0.107	0.307	0.934	0.955
3/4	1.050	0.822	0.114	0.531	1.270	1.300
1	1.315	1.063	0.126	0.887	1.780	1.820
1 1/4	1.660	1.368	0.146	1.470	2.630	2.690
1 1/2	1.900	1.600	0.150	2.010	3.130	3.200
2	2.375	2.063	0.156	3.340	4.120	4.220
2 1/2	2.875	2.501	0.187	4.910	5.990	6.120
3	3.500	3.062	0.219	7.370	8.560	8.750
3 1/2	4.000	3.500	0.250	9.620	11.200	11.400
4	4.500	4.000	0.250	12.600	12.700	12.900
5	5.562	5.062	0.250	20.100	15.800	16.200
6	6.625	6.125	0.250	29.500	19.000	19.400
8	8.625	8.001	0.312	50.300	30.900	31.600
10	10.750	10.020	0.365	78.800	45.200	46.200
12	12.750	12.000	0.375	113.000	55.300	56.500

extra strong

pipe size (in)	nominal dimensions (in)			cross-sectional area of bore (in²)	lbm/ft	
	O.D.	I.D.	wall		red brass	copper
1/8	0.405	0.205	0.100	0.033	0.363	0.371
1/4	0.540	0.294	0.123	0.068	0.611	0.625
3/8	0.675	0.421	0.127	0.139	0.829	0.847
1/2	0.840	0.542	0.149	0.231	1.230	1.250
3/4	1.050	0.736	0.157	0.425	1.670	1.710
1	1.315	0.951	0.182	0.710	2.460	2.510
1 1/4	1.660	1.272	0.194	1.270	3.390	3.460
1 1/2	1.990	1.494	0.203	1.750	4.100	4.190
2	2.375	1.933	0.221	2.94	5.670	5.800
2 1/2	2.875	2.315	0.280	4.21	8.660	8.850
3	3.500	2.892	0.304	6.57	11.600	11.800
3 1/2	4.000	3.358	0.321	8.86	14.100	14.400
4	4.500	3.818	0.341	11.50	16.900	17.300
5	5.562	4.812	0.375	18.20	23.200	23.700
6	6.625	5.751	0.437	26.00	32.200	32.900
8	8.625	7.625	0.500	45.70	48.400	49.500
10	10.750	9.750	0.500	74.70	61.100	62.400

(Multiply in by 25.4 to obtain mm.)
(Multiply in² by 645 to obtain mm².)

APPENDIX 16.L
Dimensions of Seamless Steel Boiler (BWG) Tubing[a,b,c] (customary U.S. units)

O.D. (in)	BWG	wall thickness (in)	O.D. (in)	BWG	wall thickness (in)
1	13	0.095	3	12	0.109
	12	0.109		11	0.120
	11	0.120		10	0.134
	10	0.134		9	0.148
$1^1/_4$	13	0.095	$3^1/_4$	11	0.120
	12	0.109		10	0.134
	11	0.120		9	0.148
	10	0.134		8	0.165
$1^1/_2$	13	0.095	$3^1/_2$	11	0.120
	12	0.109		10	0.134
	11	0.120		9	0.148
	10	0.134		8	0.165
$1^3/_4$	13	0.095	4	10	0.134
	12	0.109		9	0.148
	11	0.120		8	0.165
	10	0.134		7	0.180
2	13	0.095	$4^1/_2$	10	0.134
	12	0.109		9	0.148
	11	0.120		8	0.165
	10	0.134		7	0.180
$2^1/_4$	13	0.095	5	9	0.148
	12	0.109		8	0.165
	11	0.120		7	0.180
	10	0.134		6	0.203
$2^1/_2$	12	0.109	$5^1/_2$	9	0.148
	11	0.120		8	0.165
	10	0.134		7	0.180
	9	0.148		6	0.203
$2^3/_4$	12	0.109	6	7	0.180
	11	0.120		6	0.203
	10	0.134		5	0.220
	9	0.148		4	0.238

(Multiply in by 25.4 to obtain mm.)
(Multiply in^2 by 645 to obtain mm^2.)
[a]Abstracted from information provided by the United States Steel Corporation.
[b]Values in this table are not to be used for tubes in condensers and heat-exchangers unless those tubes are specified by BWG.
[c]Birmingham wire gauge, commonly used for ferrous tubing, is identical to Stubs iron-wire gauge.

APPENDIX 17.A
Specific Roughness and Hazen-Williams Constants for Various Water Pipe Materials[a]
(Multiply ft by 0.3048 to obtain m.)

type of pipe or surface	ϵ (ft) range	ϵ (ft) design	C range	C clean	C design[b]
steel					
welded and seamless	0.0001–0.0003	0.0002	150–80	140	100
interior riveted, no projecting rivets				139	100
projecting girth rivets				130	100
projecting girth and horizontal rivets				115	100
vitrified, spiral-riveted, flow with lap				110	100
vitrified, spiral-riveted, flow against lap				100	90
corrugated			80–40	80	60
mineral					
concrete	0.001–0.01	0.004	150–60	120	100
cement-asbestos			160–140	150	140
vitrified clays					110
brick sewer					100
iron					
cast, plain	0.0004–0.002	0.0008	150–80	130	100
cast, tar (asphalt) coated	0.0002–0.0006	0.0004	145–50	130	100
cast, cement lined		0.00001		150	140
cast, bituminous lined		0.00001	160–130	148	140
cast, centrifugally spun	0.00001	0.00001			
ductile iron	0.0004–0.002	0.0008	150–100	150	140
cement lined		0.00001	150–120	150	140
asphalt coated	0.0002–0.0006	0.0004	145–50	130	160
galvanized, plain	0.0002–0.0008	0.0005			
wrought, plain	0.0001–0.0003	0.0002	150–80	130	100
miscellaneous					
aluminum, irrigation pipe			135–100	135	130
copper and brass	0.000005	0.000005	150–120	140	130
wood stave	0.0006–0.003	0.002	145–110	120	110
transite	0.000008	0.000008			
lead, tin, glass		0.000005	150–120	140	130
plastic (PVC, ABS, and HDPE)		0.000005	150–120	155	150
fiberglass	0.000017	0.000017	160–150	155	150

[a]C values for sludge pipes are 20% to 40% less than the corresponding water pipe values.

[b]The following guidelines are provided for selecting Hazen-Williams coefficients for cast-iron pipes of different ages. Values for welded steel pipe are similar to those of cast-iron pipe five years older. New pipe, all sizes: $C = 130$. 5 yr old pipe: $C = 120$ ($d < 24$ in); $C = 115$ ($d \geq 24$ in). 10 yr old pipe: $C = 105$ ($d = 4$ in); $C = 110$ ($d = 12$ in); $C = 85$ ($d \geq 30$ in). 40 yr old pipe: $C = 65$ ($d = 4$ in); $C = 80$ ($d = 16$ in).

APPENDIX 17.B
Darcy Friction Factors (turbulent flow)

relative roughness, ϵ/D

Reynolds no.	0.00000	0.000001	0.0000015	0.00001	0.00002	0.00004	0.00005	0.00006	0.00008
2×10^3	0.0495	0.0495	0.0495	0.0495	0.0495	0.0495	0.0495	0.0495	0.0495
2.5×10^3	0.0461	0.0461	0.0461	0.0461	0.0461	0.0461	0.0461	0.0461	0.0461
3×10^3	0.0435	0.0435	0.0435	0.0435	0.0435	0.0436	0.0436	0.0436	0.0436
4×10^3	0.0399	0.0399	0.0399	0.0399	0.0399	0.0399	0.0400	0.0400	0.0400
5×10^3	0.0374	0.0374	0.0374	0.0374	0.0374	0.0374	0.0374	0.0375	0.0375
6×10^3	0.0355	0.0355	0.0355	0.0355	0.0355	0.0356	0.0356	0.0356	0.0356
7×10^3	0.0340	0.0340	0.0340	0.0340	0.0340	0.0341	0.0341	0.0341	0.0341
8×10^3	0.0328	0.0328	0.0328	0.0328	0.0328	0.0328	0.0329	0.0329	0.0329
9×10^3	0.0318	0.0318	0.0318	0.0318	0.0318	0.0318	0.0318	0.0319	0.0319
1×10^4	0.0309	0.0309	0.0309	0.0309	0.0309	0.0309	0.0310	0.0310	0.0310
1.5×10^4	0.0278	0.0278	0.0278	0.0278	0.0278	0.0279	0.0279	0.0279	0.0280
2×10^4	0.0259	0.0259	0.0259	0.0259	0.0259	0.0260	0.0260	0.0260	0.0261
2.5×10^4	0.0245	0.0245	0.0245	0.0245	0.0246	0.0246	0.0246	0.0247	0.0247
3×10^4	0.0235	0.0235	0.0235	0.0235	0.0235	0.0236	0.0236	0.0236	0.0237
4×10^4	0.0220	0.0220	0.0220	0.0220	0.0220	0.0221	0.0221	0.0222	0.0222
5×10^4	0.0209	0.0209	0.0209	0.0209	0.0210	0.0210	0.0211	0.0211	0.0212
6×10^4	0.0201	0.0201	0.0201	0.0201	0.0201	0.0202	0.0203	0.0203	0.0204
7×10^4	0.0194	0.0194	0.0194	0.0194	0.0195	0.0196	0.0196	0.0197	0.0197
8×10^4	0.0189	0.0189	0.0189	0.0189	0.0190	0.0190	0.0191	0.0191	0.0192
9×10^4	0.0184	0.0184	0.0184	0.0184	0.0185	0.0186	0.0186	0.0187	0.0188
1×10^5	0.0180	0.0180	0.0180	0.0180	0.0181	0.0182	0.0183	0.0183	0.0184
1.5×10^5	0.0166	0.0166	0.0166	0.0166	0.0167	0.0168	0.0169	0.0170	0.0171
2×10^5	0.0156	0.0156	0.0156	0.0157	0.0158	0.0160	0.0160	0.0161	0.0163
2.5×10^5	0.0150	0.0150	0.0150	0.0151	0.0152	0.0153	0.0154	0.0155	0.0157
3×10^5	0.0145	0.0145	0.0145	0.0146	0.0147	0.0149	0.0150	0.0151	0.0153
4×10^5	0.0137	0.0137	0.0137	0.0138	0.0140	0.0142	0.0143	0.0144	0.0146
5×10^5	0.0132	0.0132	0.0132	0.0133	0.0134	0.0137	0.0138	0.0140	0.0142
6×10^5	0.0127	0.0128	0.0128	0.0129	0.0131	0.0133	0.0135	0.0136	0.0139
7×10^5	0.0124	0.0124	0.0124	0.0126	0.0127	0.0131	0.0132	0.0134	0.0136
8×10^5	0.0121	0.0121	0.0121	0.0123	0.0125	0.0128	0.0130	0.0131	0.0134
9×10^5	0.0119	0.0119	0.0119	0.0121	0.0123	0.0126	0.0128	0.0130	0.0133
1×10^6	0.0116	0.0117	0.0117	0.0119	0.0121	0.0125	0.0126	0.0128	0.0131
1.5×10^6	0.0109	0.0109	0.0109	0.0112	0.0114	0.0119	0.0121	0.0123	0.0127
2×10^6	0.0104	0.0104	0.0104	0.0107	0.0110	0.0116	0.0118	0.0120	0.0124
2.5×10^6	0.0100	0.0100	0.0101	0.0104	0.0108	0.0113	0.0116	0.0118	0.0123
3×10^6	0.0097	0.0098	0.0098	0.0102	0.0105	0.0112	0.0115	0.0117	0.0122
4×10^6	0.0093	0.0094	0.0094	0.0098	0.0103	0.0110	0.0113	0.0115	0.0120
5×10^6	0.0090	0.0091	0.0091	0.0096	0.0101	0.0108	0.0111	0.0114	0.0119
6×10^6	0.0087	0.0088	0.0089	0.0094	0.0099	0.0107	0.0110	0.0113	0.0118
7×10^6	0.0085	0.0086	0.0087	0.0093	0.0098	0.0106	0.0110	0.0113	0.0118

APPENDIX 17.B *(continued)*
Darcy Friction Factors (turbulent flow)

Reynolds no.	relative roughness, ϵ/D								
	0.00000	0.000001	0.0000015	0.00001	0.00002	0.00004	0.00005	0.00006	0.00008
8×10^6	0.0084	0.0085	0.0085	0.0092	0.0097	0.0106	0.0109	0.0112	0.0118
9×10^6	0.0082	0.0083	0.0084	0.0091	0.0097	0.0105	0.0109	0.0112	0.0117
1×10^7	0.0081	0.0082	0.0083	0.0090	0.0096	0.0105	0.0109	0.0112	0.0117
1.5×10^7	0.0076	0.0078	0.0079	0.0087	0.0094	0.0104	0.0108	0.0111	0.0116
2×10^7	0.0073	0.0075	0.0076	0.0086	0.0093	0.0103	0.0107	0.0110	0.0116
2.5×10^7	0.0071	0.0073	0.0074	0.0085	0.0093	0.0103	0.0107	0.0110	0.0116
3×10^7	0.0069	0.0072	0.0073	0.0084	0.0092	0.0103	0.0107	0.0110	0.0116
4×10^7	0.0067	0.0070	0.0071	0.0084	0.0092	0.0102	0.0106	0.0110	0.0115
5×10^7	0.0065	0.0068	0.0070	0.0083	0.0092	0.0102	0.0106	0.0110	0.0115

Reynolds no.	relative roughness, ϵ/D								
	0.0001	0.00015	0.00020	0.00025	0.00030	0.00035	0.0004	0.0006	0.0008
2×10^3	0.0495	0.0496	0.0496	0.0496	0.0497	0.0497	0.0498	0.0499	0.0501
2.5×10^3	0.0461	0.0462	0.0462	0.0463	0.0463	0.0463	0.0464	0.0466	0.0467
3×10^3	0.0436	0.0437	0.0437	0.0437	0.0438	0.0438	0.0439	0.0441	0.0442
4×10^3	0.0400	0.0401	0.0401	0.0402	0.0402	0.0403	0.0403	0.0405	0.0407
5×10^3	0.0375	0.0376	0.0376	0.0377	0.0377	0.0378	0.0378	0.0381	0.0383
6×10^3	0.0356	0.0357	0.0357	0.0358	0.0359	0.0359	0.0360	0.0362	0.0365
7×10^3	0.0341	0.0342	0.0343	0.0343	0.0344	0.0345	0.0345	0.0348	0.0350
8×10^3	0.0329	0.0330	0.0331	0.0331	0.0332	0.0333	0.0333	0.0336	0.0339
9×10^3	0.0319	0.0320	0.0321	0.0321	0.0322	0.0323	0.0323	0.0326	0.0329
1×10^4	0.0310	0.0311	0.0312	0.0313	0.0313	0.0314	0.0315	0.0318	0.0321
1.5×10^4	0.0280	0.0281	0.0282	0.0283	0.0284	0.0285	0.0285	0.0289	0.0293
2×10^4	0.0261	0.0262	0.0263	0.0264	0.0265	0.0266	0.0267	0.0272	0.0276
2.5×10^4	0.0248	0.0249	0.0250	0.0251	0.0252	0.0254	0.0255	0.0259	0.0264
3×10^4	0.0238	0.0239	0.0240	0.0241	0.0243	0.0244	0.0245	0.0250	0.0255
4×10^4	0.0223	0.0224	0.0226	0.0227	0.0229	0.0230	0.0232	0.0237	0.0243
5×10^4	0.0212	0.0214	0.0216	0.0218	0.0219	0.0221	0.0223	0.0229	0.0235
6×10^4	0.0205	0.0207	0.0208	0.0210	0.0212	0.0214	0.0216	0.0222	0.0229
7×10^4	0.0198	0.0200	0.0202	0.0204	0.0206	0.0208	0.0210	0.0217	0.0224
8×10^4	0.0193	0.0195	0.0198	0.0200	0.0202	0.0204	0.0206	0.0213	0.0220
9×10^4	0.0189	0.0191	0.0194	0.0196	0.0198	0.0200	0.0202	0.0210	0.0217
1×10^5	0.0185	0.0188	0.0190	0.0192	0.0195	0.0197	0.0199	0.0207	0.0215
1.5×10^5	0.0172	0.0175	0.0178	0.0181	0.0184	0.0186	0.0189	0.0198	0.0207
2×10^5	0.0164	0.0168	0.0171	0.0174	0.0177	0.0180	0.0183	0.0193	0.0202
2.5×10^5	0.0158	0.0162	0.0166	0.0170	0.0173	0.0176	0.0179	0.0190	0.0199
3×10^5	0.0154	0.0159	0.0163	0.0166	0.0170	0.0173	0.0176	0.0188	0.0197
4×10^5	0.0148	0.0153	0.0158	0.0162	0.0166	0.0169	0.0172	0.0184	0.0195
5×10^5	0.0144	0.0150	0.0154	0.0159	0.0163	0.0167	0.0170	0.0183	0.0193
6×10^5	0.0141	0.0147	0.0152	0.0157	0.0161	0.0165	0.0168	0.0181	0.0192
7×10^5	0.0139	0.0145	0.0150	0.0155	0.0159	0.0163	0.0167	0.0180	0.0191

Appendices

APPENDIX 17.B (continued)
Darcy Friction Factors (turbulent flow)

Reynolds no.	relative roughness, ϵ/D								
	0.0001	0.00015	0.00020	0.00025	0.00030	0.00035	0.0004	0.0006	0.0008
8×10^5	0.0137	0.0143	0.0149	0.0154	0.0158	0.0162	0.0166	0.0180	0.0191
9×10^5	0.0136	0.0142	0.0148	0.0153	0.0157	0.0162	0.0165	0.0179	0.0190
1×10^6	0.0134	0.0141	0.0147	0.0152	0.0157	0.0161	0.0165	0.0178	0.0190
1.5×10^6	0.0130	0.0138	0.0144	0.0149	0.0154	0.0159	0.0163	0.0177	0.0189
2×10^6	0.0128	0.0136	0.0142	0.0148	0.0153	0.0158	0.0162	0.0176	0.0188
2.5×10^6	0.0127	0.0135	0.0141	0.0147	0.0152	0.0157	0.0161	0.0176	0.0188
3×10^6	0.0126	0.0134	0.0141	0.0147	0.0152	0.0157	0.0161	0.0176	0.0187
4×10^6	0.0124	0.0133	0.0140	0.0146	0.0151	0.0156	0.0161	0.0175	0.0187
5×10^6	0.0123	0.0132	0.0139	0.0146	0.0151	0.0156	0.0160	0.0175	0.0187
6×10^6	0.0123	0.0132	0.0139	0.0145	0.0151	0.0156	0.0160	0.0175	0.0187
7×10^6	0.0122	0.0132	0.0139	0.0145	0.0151	0.0155	0.0160	0.0175	0.0187
8×10^6	0.0122	0.0131	0.0139	0.0145	0.0150	0.0155	0.0160	0.0175	0.0187
9×10^6	0.0122	0.0131	0.0139	0.0145	0.0150	0.0155	0.0160	0.0175	0.0187
1×10^7	0.0122	0.0131	0.0138	0.0145	0.0150	0.0155	0.0160	0.0175	0.0186
1.5×10^7	0.0121	0.0131	0.0138	0.0144	0.0150	0.0155	0.0159	0.0174	0.0186
2×10^7	0.0121	0.0130	0.0138	0.0144	0.0150	0.0155	0.0159	0.0174	0.0186
2.5×10^7	0.0121	0.0130	0.0138	0.0144	0.0150	0.0155	0.0159	0.0174	0.0186
3×10^7	0.0120	0.0130	0.0138	0.0144	0.0150	0.0155	0.0159	0.0174	0.0186
4×10^7	0.0120	0.0130	0.0138	0.0144	0.0150	0.0155	0.0159	0.0174	0.0186
5×10^7	0.0120	0.0130	0.0138	0.0144	0.0150	0.0155	0.0159	0.0174	0.0186

Reynolds no.	relative roughness, ϵ/D								
	0.001	0.0015	0.002	0.0025	0.003	0.0035	0.004	0.006	0.008
2×10^3	0.0502	0.0506	0.0510	0.0513	0.0517	0.0521	0.0525	0.0539	0.0554
2.5×10^3	0.0469	0.0473	0.0477	0.0481	0.0485	0.0489	0.0493	0.0509	0.0524
3×10^3	0.0444	0.0449	0.0453	0.0457	0.0462	0.0466	0.0470	0.0487	0.0503
4×10^3	0.0409	0.0414	0.0419	0.0424	0.0429	0.0433	0.0438	0.0456	0.0474
5×10^3	0.0385	0.0390	0.0396	0.0401	0.0406	0.0411	0.0416	0.0436	0.0455
6×10^3	0.0367	0.0373	0.0378	0.0384	0.0390	0.0395	0.0400	0.0421	0.0441
7×10^3	0.0353	0.0359	0.0365	0.0371	0.0377	0.0383	0.0388	0.0410	0.0430
8×10^3	0.0341	0.0348	0.0354	0.0361	0.0367	0.0373	0.0379	0.0401	0.0422
9×10^3	0.0332	0.0339	0.0345	0.0352	0.0358	0.0365	0.0371	0.0394	0.0416
1×10^4	0.0324	0.0331	0.0338	0.0345	0.0351	0.0358	0.0364	0.0388	0.0410
1.5×10^4	0.0296	0.0305	0.0313	0.0320	0.0328	0.0335	0.0342	0.0369	0.0393
2×10^4	0.0279	0.0289	0.0298	0.0306	0.0315	0.0323	0.0330	0.0358	0.0384
2.5×10^4	0.0268	0.0278	0.0288	0.0297	0.0306	0.0314	0.0322	0.0352	0.0378
3×10^4	0.0260	0.0271	0.0281	0.0291	0.0300	0.0308	0.0317	0.0347	0.0374
4×10^4	0.0248	0.0260	0.0271	0.0282	0.0291	0.0301	0.0309	0.0341	0.0369
5×10^4	0.0240	0.0253	0.0265	0.0276	0.0286	0.0296	0.0305	0.0337	0.0365
6×10^4	0.0235	0.0248	0.0261	0.0272	0.0283	0.0292	0.0302	0.0335	0.0363
7×10^4	0.0230	0.0245	0.0257	0.0269	0.0280	0.0290	0.0299	0.0333	0.0362

APPENDIX 17.B *(continued)*
Darcy Friction Factors (turbulent flow)

Reynolds no.	relative roughness, ϵ/D								
	0.001	0.0015	0.002	0.0025	0.003	0.0035	0.004	0.006	0.008
8×10^4	0.0227	0.0242	0.0255	0.0267	0.0278	0.0288	0.0298	0.0331	0.0361
9×10^4	0.0224	0.0239	0.0253	0.0265	0.0276	0.0286	0.0296	0.0330	0.0360
1×10^5	0.0222	0.0237	0.0251	0.0263	0.0275	0.0285	0.0295	0.0329	0.0359
1.5×10^5	0.0214	0.0231	0.0246	0.0259	0.0271	0.0281	0.0292	0.0327	0.0357
2×10^5	0.0210	0.0228	0.0243	0.0256	0.0268	0.0279	0.0290	0.0325	0.0355
2.5×10^5	0.0208	0.0226	0.0241	0.0255	0.0267	0.0278	0.0289	0.0325	0.0355
3×10^5	0.0206	0.0225	0.0240	0.0254	0.0266	0.0277	0.0288	0.0324	0.0354
4×10^5	0.0204	0.0223	0.0239	0.0253	0.0265	0.0276	0.0287	0.0323	0.0354
5×10^5	0.0202	0.0222	0.0238	0.0252	0.0264	0.0276	0.0286	0.0323	0.0353
6×10^5	0.0201	0.0221	0.0237	0.0251	0.0264	0.0275	0.0286	0.0323	0.0353
7×10^5	0.0201	0.0221	0.0237	0.0251	0.0264	0.0275	0.0286	0.0322	0.0353
8×10^5	0.0200	0.0220	0.0237	0.0251	0.0263	0.0275	0.0286	0.0322	0.0353
9×10^5	0.0200	0.0220	0.0236	0.0251	0.0263	0.0275	0.0285	0.0322	0.0353
1×10^6	0.0199	0.0220	0.0236	0.0250	0.0263	0.0275	0.0285	0.0322	0.0353
1.5×10^6	0.0198	0.0219	0.0235	0.0250	0.0263	0.0274	0.0285	0.0322	0.0352
2×10^6	0.0198	0.0218	0.0235	0.0250	0.0262	0.0274	0.0285	0.0322	0.0352
2.5×10^6	0.0198	0.0218	0.0235	0.0249	0.0262	0.0274	0.0285	0.0322	0.0352
3×10^6	0.0197	0.0218	0.0235	0.0249	0.0262	0.0274	0.0285	0.0321	0.0352
4×10^6	0.0197	0.0218	0.0235	0.0249	0.0262	0.0274	0.0284	0.0321	0.0352
5×10^6	0.0197	0.0218	0.0235	0.0249	0.0262	0.0274	0.0284	0.0321	0.0352
6×10^6	0.0197	0.0218	0.0235	0.0249	0.0262	0.0274	0.0284	0.0321	0.0352
7×10^6	0.0197	0.0218	0.0234	0.0249	0.0262	0.0274	0.0284	0.0321	0.0352
8×10^6	0.0197	0.0218	0.0234	0.0249	0.0262	0.0274	0.0284	0.0321	0.0352
9×10^6	0.0197	0.0218	0.0234	0.0249	0.0262	0.0274	0.0284	0.0321	0.0352
1×10^7	0.0197	0.0218	0.0234	0.0249	0.0262	0.0273	0.0284	0.0321	0.0352
1.5×10^7	0.0197	0.0217	0.0234	0.0249	0.0262	0.0273	0.0284	0.0321	0.0352
2×10^7	0.0197	0.0217	0.0234	0.0249	0.0262	0.0273	0.0284	0.0321	0.0352
2.5×10^7	0.0196	0.0217	0.0234	0.0249	0.0262	0.0273	0.0284	0.0321	0.0352
3×10^7	0.0196	0.0217	0.0234	0.0249	0.0262	0.0273	0.0284	0.0321	0.0352
4×10^7	0.0196	0.0217	0.0234	0.0249	0.0262	0.0273	0.0284	0.0321	0.0352
5×10^7	0.0196	0.0217	0.0234	0.0249	0.0262	0.0273	0.0284	0.0321	0.0352

APPENDIX 17.B *(continued)*
Darcy Friction Factors (turbulent flow)

Reynolds no.	relative roughness, ϵ/D								
	0.01	0.015	0.02	0.025	0.03	0.035	0.04	0.045	0.05
2×10^3	0.0568	0.0602	0.0635	0.0668	0.0699	0.0730	0.0760	0.0790	0.0819
2.5×10^3	0.0539	0.0576	0.0610	0.0644	0.0677	0.0709	0.0740	0.0770	0.0800
3×10^3	0.0519	0.0557	0.0593	0.0628	0.0661	0.0694	0.0725	0.0756	0.0787
4×10^3	0.0491	0.0531	0.0570	0.0606	0.0641	0.0674	0.0707	0.0739	0.0770
5×10^3	0.0473	0.0515	0.0555	0.0592	0.0628	0.0662	0.0696	0.0728	0.0759
6×10^3	0.0460	0.0504	0.0544	0.0583	0.0619	0.0654	0.0688	0.0721	0.0752
7×10^3	0.0450	0.0495	0.0537	0.0576	0.0613	0.0648	0.0682	0.0715	0.0747
8×10^3	0.0442	0.0489	0.0531	0.0571	0.0608	0.0644	0.0678	0.0711	0.0743
9×10^3	0.0436	0.0484	0.0526	0.0566	0.0604	0.0640	0.0675	0.0708	0.0740
1×10^4	0.0431	0.0479	0.0523	0.0563	0.0601	0.0637	0.0672	0.0705	0.0738
1.5×10^4	0.0415	0.0466	0.0511	0.0553	0.0592	0.0628	0.0664	0.0698	0.0731
2×10^4	0.0407	0.0459	0.0505	0.0547	0.0587	0.0624	0.0660	0.0694	0.0727
2.5×10^4	0.0402	0.0455	0.0502	0.0544	0.0584	0.0621	0.0657	0.0691	0.0725
3×10^4	0.0398	0.0452	0.0499	0.0542	0.0582	0.0619	0.0655	0.0690	0.0723
4×10^4	0.0394	0.0448	0.0496	0.0539	0.0579	0.0617	0.0653	0.0688	0.0721
5×10^4	0.0391	0.0446	0.0494	0.0538	0.0578	0.0616	0.0652	0.0687	0.0720
6×10^4	0.0389	0.0445	0.0493	0.0536	0.0577	0.0615	0.0651	0.0686	0.0719
7×10^4	0.0388	0.0443	0.0492	0.0536	0.0576	0.0614	0.0650	0.0685	0.0719
8×10^4	0.0387	0.0443	0.0491	0.0535	0.0576	0.0614	0.0650	0.0685	0.0718
9×10^4	0.0386	0.0442	0.0491	0.0535	0.0575	0.0613	0.0650	0.0684	0.0718
1×10^5	0.0385	0.0442	0.0490	0.0534	0.0575	0.0613	0.0649	0.0684	0.0718
1.5×10^5	0.0383	0.0440	0.0489	0.0533	0.0574	0.0612	0.0648	0.0683	0.0717
2×10^5	0.0382	0.0439	0.0488	0.0532	0.0573	0.0612	0.0648	0.0683	0.0717
2.5×10^5	0.0381	0.0439	0.0488	0.0532	0.0573	0.0611	0.0648	0.0683	0.0716
3×10^5	0.0381	0.0438	0.0488	0.0532	0.0573	0.0611	0.0648	0.0683	0.0716
4×10^5	0.0381	0.0438	0.0487	0.0532	0.0573	0.0611	0.0647	0.0682	0.0716
5×10^5	0.0380	0.0438	0.0487	0.0531	0.0572	0.0611	0.0647	0.0682	0.0716
6×10^5	0.0380	0.0438	0.0487	0.0531	0.0572	0.0611	0.0647	0.0682	0.0716
7×10^5	0.0380	0.0438	0.0487	0.0531	0.0572	0.0611	0.0647	0.0682	0.0716
8×10^5	0.0380	0.0437	0.0487	0.0531	0.0572	0.0611	0.0647	0.0682	0.0716
9×10^5	0.0380	0.0437	0.0487	0.0531	0.0572	0.0610	0.0647	0.0682	0.0716
1×10^6	0.0380	0.0437	0.0487	0.0531	0.0572	0.0610	0.0647	0.0682	0.0716
1.5×10^6	0.0379	0.0437	0.0487	0.0531	0.0572	0.0610	0.0647	0.0682	0.0716
2×10^6	0.0379	0.0437	0.0487	0.0531	0.0572	0.0610	0.0647	0.0682	0.0716
2.5×10^6	0.0379	0.0437	0.0487	0.0531	0.0572	0.0610	0.0647	0.0682	0.0716
3×10^6	0.0379	0.0437	0.0487	0.0531	0.0572	0.0610	0.0647	0.0682	0.0716
4×10^6	0.0379	0.0437	0.0486	0.0531	0.0572	0.0610	0.0647	0.0682	0.0716
5×10^6	0.0379	0.0437	0.0486	0.0531	0.0572	0.0610	0.0647	0.0682	0.0716
6×10^6	0.0379	0.0437	0.0486	0.0531	0.0572	0.0610	0.0647	0.0682	0.0716
7×10^6	0.0379	0.0437	0.0486	0.0531	0.0572	0.0610	0.0647	0.0682	0.0716

APPENDIX 17.B *(continued)*
Darcy Friction Factors (turbulent flow)

Reynolds no.	relative roughness, ϵ/D								
	0.01	0.015	0.02	0.025	0.03	0.035	0.04	0.045	0.05
8×10^6	0.0379	0.0437	0.0486	0.0531	0.0572	0.0610	0.0647	0.0682	0.0716
9×10^6	0.0379	0.0437	0.0486	0.0531	0.0572	0.0610	0.0647	0.0682	0.0716
1×10^7	0.0379	0.0437	0.0486	0.0531	0.0572	0.0610	0.0647	0.0682	0.0716
1.5×10^7	0.0379	0.0437	0.0486	0.0531	0.0572	0.0610	0.0647	0.0682	0.0716
2×10^7	0.0379	0.0437	0.0486	0.0531	0.0572	0.0610	0.0647	0.0682	0.0716
2.5×10^7	0.0379	0.0437	0.0486	0.0531	0.0572	0.0610	0.0647	0.0682	0.0716
3×10^7	0.0379	0.0437	0.0486	0.0531	0.0572	0.0610	0.0647	0.0682	0.0716
4×10^7	0.0379	0.0437	0.0486	0.0531	0.0572	0.0610	0.0647	0.0682	0.0716
5×10^7	0.0379	0.0437	0.0486	0.0531	0.0572	0.0610	0.0647	0.0682	0.0716

Appendices

APPENDIX 17.C
Water Pressure Drop in Schedule-40 Steel Pipe

pressure drop per 1000 ft of schedule-40 steel pipe (lbf/in²)

(Pipe sizes are staggered across the nine velocity/pressure-drop column pairs. Column 1 = 1 in; successive sizes shift one pair to the right. Beginning at 90 gal/min the columns are reused for the larger sizes: col 1 = 6 in, col 2 = 8 in, col 3 = 10 in, col 4 = 12 in, col 5 = 14 in, col 6 = 16 in, col 7 = 18 in, col 8 = 20 in, col 9 = 24 in. Pipe-size labels below mark where each size begins.)

discharge (gal/min)	velocity (ft/sec)	pressure drop	velocity (ft/sec)	pressure drop	velocity (ft/sec)	pressure drop	velocity (ft/sec)	pressure drop	velocity (ft/sec)	pressure drop	velocity (ft/sec)	pressure drop	velocity (ft/sec)	pressure drop	velocity (ft/sec)	pressure drop	velocity (ft/sec)	pressure drop
	1 in																	
1	0.37	0.49	*1¼ in*															
2	0.74	1.70	0.43	0.45	*1½ in*													
3	1.12	3.53	0.64	0.94	0.47	0.44												
4	1.49	5.94	0.86	1.55	0.63	0.74												
5	1.86	9.02	1.07	2.36	0.79	1.12	*2 in*											
6	2.24	12.25	1.28	3.30	0.95	1.53	0.57	0.46										
8	2.98	21.1	1.72	5.52	1.26	2.63	0.76	0.75	*2½ in*									
10	3.72	30.8	2.14	8.34	1.57	3.86	0.96	1.14	0.67	0.48								
15	5.60	64.6	3.21	17.6	2.36	8.13	1.43	2.33	1.00	0.99	*3 in*							
20	7.44	110.5	4.29	29.1	3.15	13.5	1.91	3.86	1.34	1.64	0.87	0.59	*3½ in*					
25			5.36	43.7	3.94	20.2	2.39	5.81	1.68	2.48	1.08	0.67	0.81	0.42				
30			6.43	62.9	4.72	29.1	2.87	8.04	2.01	3.43	1.30	1.21	0.97	0.60	*4 in*			
35			7.51	82.5	5.51	38.2	3.35	10.95	2.35	4.49	1.52	1.58	1.14	0.79	0.88	0.42		
40					6.30	47.8	3.82	13.7	2.68	5.88	1.74	2.06	1.30	1.00	1.01	0.53		
45					7.08	60.6	4.30	17.4	3.00	7.14	1.95	2.51	1.46	1.21	1.13	0.67		
50					7.87	74.7	4.78	20.6	3.35	8.82	2.17	3.10	1.62	1.44	1.26	0.80		
60							5.74	29.6	4.02	12.2	2.60	4.29	1.95	2.07	1.51	1.10	*5 in*	
70							6.69	38.6	4.69	15.3	3.04	5.84	2.27	2.71	1.76	1.50	1.12	0.48
80							7.65	50.3	5.37	21.7	3.48	7.62	2.59	3.53	2.01	1.87	1.28	0.63
90	*6 in*						8.60	63.6	6.04	26.1	3.91	9.22	2.92	4.46	2.26	2.37	1.44	0.80
100	1.11	0.39					9.56	75.1	6.71	32.3	4.34	11.4	3.24	5.27	2.52	2.81	1.60	0.95
125	1.39	0.56							8.38	48.2	5.42	17.1	4.05	7.86	3.15	4.38	2.00	1.48
150	1.67	0.78							10.06	60.4	6.51	23.5	4.86	11.3	3.78	6.02	2.41	2.04
175	1.94	1.06							11.73	90.0	7.59	32.0	5.67	14.7	4.41	8.20	2.81	2.78
200	2.22	1.32	*8 in*								8.68	39.7	6.48	19.2	5.04	10.2	3.21	3.46
225	2.50	1.66	1.44	0.44							9.77	50.2	7.29	23.1	5.67	12.9	3.61	4.37
250	2.78	2.05	1.60	0.55							10.85	61.9	8.10	28.5	6.30	15.9	4.01	5.14
275	3.06	2.36	1.76	0.63							11.94	75.0	8.91	34.4	6.93	18.3	4.41	6.22
300	3.33	2.80	1.92	0.75							13.02	84.7	9.72	40.9	7.56	21.8	4.81	7.41
325	3.61	3.29	2.08	0.88									10.53	45.5	8.18	25.5	5.21	8.25
350	3.89	3.62	2.24	0.97									11.35	52.7	8.82	29.7	5.61	9.57
375	4.16	4.16	2.40	1.11									12.17	60.7	9.45	32.3	6.01	11.0
400	4.44	4.72	2.56	1.27									12.97	68.9	10.08	36.7	6.41	12.5
425	4.72	5.34	2.72	1.43									13.78	77.8	10.70	41.5	6.82	14.1
450	5.00	5.96	2.88	1.60	*10 in*								14.59	87.3	11.33	46.5	7.22	15.0
475	5.27	6.66	3.04	1.69	1.93	0.30									11.96	51.7	7.62	16.7
500	5.55	7.39	3.20	1.87	2.04	0.63									12.59	57.3	8.02	18.5
550	6.11	8.94	3.53	2.26	2.24	0.70									13.84	69.3	8.82	22.4
600	6.66	10.6	3.85	2.70	2.44	0.86									15.10	82.5	9.62	26.7
650	7.21	11.8	4.17	3.16	2.65	1.01	*12 in*										10.42	31.3
700	7.77	13.7	4.49	3.69	2.85	1.18	2.01	0.48									11.22	36.3
750	8.32	15.7	4.81	4.21	3.05	1.35	2.15	0.55									12.02	41.6
800	8.88	17.8	5.13	4.79	3.26	1.54	2.29	0.62	*14 in*								12.82	44.7
850	9.44	20.2	5.45	5.11	3.46	1.74	2.44	0.70	2.02	0.43							13.62	50.5
900	10.00	22.6	5.77	5.73	3.66	1.94	2.58	0.79	2.14	0.48							14.42	56.6
950	10.55	23.7	6.09	6.38	3.87	2.23	2.72	0.88	2.25	0.53							15.22	63.1
1000	11.10	26.3	6.41	7.08	4.07	2.40	2.87	0.98	2.38	0.59							16.02	70.0
1100	12.22	31.8	7.05	8.56	4.48	2.74	3.16	1.18	2.61	0.68	*16 in*						17.63	84.6
1200	13.32	37.8	7.69	10.2	4.88	3.27	3.45	1.40	2.85	0.81	2.18	0.40						
1300	14.43	44.4	8.33	11.3	5.29	3.86	3.73	1.56	3.09	0.95	2.36	0.47						
1400	15.54	51.5	8.97	13.0	5.70	4.44	4.02	1.80	3.32	1.10	2.54	0.54						
1500	16.65	55.5	9.62	15.0	6.10	5.11	4.30	2.07	3.55	1.19	2.73	0.62						
1600	17.76	63.1	10.26	17.0	6.51	5.16	4.59	2.36	3.80	1.35	2.91	0.71						
1800	19.98	79.8	11.54	21.6	7.32	6.91	5.16	2.98	4.27	1.71	3.27	0.85	*18 in*					
2000	22.20	98.5	12.83	25.0	8.13	8.54	5.73	3.47	4.74	2.11	3.63	1.05	2.58	0.48				
2500			16.03	39.0	10.18	12.5	7.17	5.41	5.92	3.09	4.54	1.63	3.59	0.88	*20 in*			
3000			19.24	52.4	12.21	18.0	8.60	7.31	7.12	4.45	5.45	2.21	4.31	1.27	3.45	0.73		
3500			22.43	71.4	14.25	22.9	10.03	9.95	8.32	6.18	6.35	3.00	5.03	1.52	4.03	0.94	*24 in*	
4000			25.65	93.3	16.28	29.9	11.48	13.0	9.49	7.92	7.25	3.92	5.74	2.12	4.61	1.22	3.19	0.51
4500					18.31	37.8	12.90	15.4	10.67	9.36	8.17	4.97	6.47	2.50	5.19	1.55	3.59	0.60
5000					20.35	46.7	14.34	18.9	11.84	11.6	9.08	5.72	7.17	3.08	5.76	1.78	3.99	0.74
6000					24.42	67.2	17.21	27.3	14.32	15.4	10.88	8.24	8.62	4.45	6.92	2.57	4.80	1.00
7000					28.50	85.1	20.08	37.2	16.60	21.0	12.69	12.2	10.04	6.06	8.06	3.50	5.68	1.36
8000							22.95	45.1	18.98	27.4	14.52	13.6	11.48	7.34	9.23	4.57	6.38	1.78
9000							25.80	57.0	21.35	34.7	16.32	17.2	12.92	9.20	10.37	5.36	7.19	2.25
10,000							28.63	70.4	23.75	42.9	18.16	21.2	14.37	11.5	11.53	6.63	7.96	2.78
12,000							34.38	93.6	28.50	61.8	21.80	30.9	17.23	16.5	13.83	9.54	9.57	3.71
14,000									33.20	84.0	25.42	41.6	20.10	20.7	16.14	12.0	11.18	5.05
16,000											29.05	54.4	22.96	27.1	18.43	15.7	12.77	6.60

(Multiply gal/min by 0.0631 to obtain L/s.)

(Multiply ft/sec by 0.3048 to obtain m/s.)

(Multiply in by 25.4 to obtain mm.)

(Multiply lbf/in²-1000 ft by 2.3 to obtain kPa/100 m.)

APPENDIX 17.D
Equivalent Length of Straight Pipe for Various (generic) Fittings
(in feet, turbulent flow only, for any fluid)

fittings			1/4	3/8	1/2	3/4	1	1 1/4	1 1/2	2	2 1/2	3	4	5	6	8	10	12	14	16	18	20	24
regular 90° ell	screwed	steel	2.3	3.1	3.6	4.4	5.2	6.6	7.4	8.5	9.3	11.0	13.0										
		cast iron										9.0	11.0										
	flanged	steel			0.92	1.2	1.6	2.1	2.4	3.1	3.6	4.4	5.9	7.3	8.9	12.0	14.0	17.0	18.0	21.0	23.0	25.0	30.0
		cast iron										3.6	4.8		7.2	9.8	12.0	15.0	17.0	19.0	22.0	24.0	28.0
long radius 90° ell	screwed	steel	1.5	2.0	2.2	2.3	2.7	3.2	3.4	3.6	3.6	4.0	4.6										
		cast iron										3.3	3.7										
	flanged	steel			1.1	1.3	1.6	2.0	2.3	2.7	2.9	3.4	4.2	5.0	5.7	7.0	8.0	9.0	9.4	10.0	11.0	12.0	14.0
		cast iron										2.8	3.4		4.7	5.7	6.8	7.8	8.6	9.6	11.0	11.0	13.0
regular 45° ell	screwed	steel	0.34	0.52	0.71	0.92	1.3	1.7	2.1	2.7	3.2	4.0	5.5										
		cast iron										3.3	4.5										
	flanged	steel			0.45	0.59	0.81	1.1	1.3	1.7	2.0	2.6	3.5	4.5	5.6	7.7	9.0	11.0	13.0	15.0	16.0	18.0	22.0
		cast iron										2.1	2.9		4.5	6.3	8.1	9.7	12.0	13.0	15.0	17.0	20.0
tee-line flow	screwed	steel	0.79	1.2	1.7	2.4	3.2	4.6	5.6	7.7	9.3	12.0	17.0										
		cast iron										9.9	14.0										
	flanged	steel			0.69	0.82	1.0	1.3	1.5	1.8	1.9	2.2	2.8	3.3	3.8	4.7	5.2	6.0	6.4	7.2	7.6	8.2	9.6
		cast iron										1.9	2.2		3.1	3.9	4.6	5.2	5.9	6.5	7.2	7.7	8.8
tee-branch flow	screwed	steel	2.4	3.5	4.2	5.3	6.6	8.7	9.9	12.0	13.0	17.0	21.0										
		cast iron										14.0	17.0										
	flanged	steel			2.0	2.6	3.3	4.4	5.2	6.6	7.5	9.4	12.0	15.0	18.0	24.0	30.0	34.0	37.0	43.0	47.0	52.0	62.0
		cast iron										7.7	10.0		15.0	20.0	25.0	30.0	35.0	39.0	44.0	49.0	57.0
180° return bend	regular screwed	steel	2.3	3.1	3.6	4.4	5.2	6.6	7.4	8.5	9.3	11.0	13.0										
		cast iron										9.0	11.0										
	regular flanged	steel			0.92	1.2	1.6	2.1	2.4	3.1	3.6	4.4	5.9	7.3	8.9	12.0	14.0	17.0	18.0	21.0	23.0	25.0	30.0
		cast iron										3.6	4.8		7.2	9.8	12.0	15.0	17.0	19.0	22.0	24.0	28.0
	long radius flanged	steel			1.1	1.3	1.6	2.0	2.3	2.7	2.9	3.4	4.2	5.0	5.7	7.0	8.0	9.0	9.4	10.0	11.0	12.0	14.0
		cast iron										2.8	3.4		4.7	5.7	6.8	7.8	8.6	9.6	11.0	11.0	13.0
globe valve	screwed	steel	21.0	22.0	22.0	24.0	29.0	37.0	42.0	54.0	62.0	79.0	110.0										
		cast iron										65.0	86.0										
	flanged	steel			38.0	40.0	45.0	54.0	59.0	70.0	77.0	94.0	120.0	150.0	190.0	260.0	310.0	390.0					
		cast iron										77.0	99.0		150.0	210.0	270.0	330.0					
gate valve	screwed	steel	0.32	0.45	0.56	0.67	0.84	1.1	1.2	1.5	1.7	1.9	2.5										
		cast iron										1.6	2.0										
	flanged	steel								2.6	2.7	2.8	2.9	3.1	3.2	3.2	3.2	3.2	3.2	3.2	3.2	3.2
		cast iron										2.3	2.4		2.6	2.7	2.8	2.9	2.9	3.0	3.0	3.0	3.0
angle valve	screwed	steel	12.8	15.0	15.0	15.0	17.0	18.0	18.0	18.0	18.0	18.0	18.0										
		cast iron										15.0	15.0										
	flanged	steel			15.0	15.0	17.0	18.0	18.0	21.0	22.0	28.0	38.0	50.0	63.0	90.0	120.0	140.0	160.0	190.0	210.0	240.0	300.0
		cast iron										23.0	31.0		52.0	74.0	98.0	120.0	150.0	170.0	200.0	230.0	280.0
swing check valve	screwed	steel	7.2	7.3	8.0	8.8	11.0	13.0	15.0	19.0	22.0	27.0	38.0										
		cast iron										22.0	31.0										
	flanged	steel			3.8	5.3	7.2	10.0	12.0	17.0	21.0	27.0	38.0	50.0	63.0	90.0	120.0	140.0					
		cast iron										22.0	31.0		52.0	74.0	98.0	120.0					
coupling or union	screwed	steel	0.14	0.18	0.21	0.24	0.29	0.36	0.39	0.45	0.47	0.53	0.65										
		cast iron										0.44	0.52										
inlet	bell mouth inlet	steel	0.04	0.07	0.10	0.13	0.18	0.26	0.31	0.43	0.52	0.67	0.95	1.3	1.6	2.3	2.9	3.5	4.0	4.7	5.3	6.1	7.6
		cast iron										0.55	0.77		1.3	1.9	2.4	3.0	3.6	4.3	5.0	5.7	7.0
	square mouth inlet	steel	0.44	0.68	0.96	1.3	1.8	2.6	3.1	4.3	5.2	6.7	9.5	13.0	16.0	23.0	29.0	35.0	40.0	47.0	53.0	61.0	76.0
		cast iron										5.5	7.7		13.0	19.0	24.0	30.0	36.0	43.0	50.0	57.0	70.0
	re-entrant pipe	steel	0.88	1.4	1.9	2.6	3.6	5.1	6.2	8.5	10.0	13.0	19.0	25.0	32.0	45.0	58.0	70.0	80.0	95.0	110.0	120.0	150.0
		cast iron										11.0	15.0		26.0	37.0	49.0	61.0	73.0	86.0	100.0	110.0	140.0

(Multiply in by 25.4 to obtain mm.)
(Multiply ft by 0.3048 to obtain m.)

From *Engineering Data Book*, Second Edition, copyright © 1990, by the Hydraulic Institute. Reproduced with permission.

APPENDIX 17.E
Hazen-Williams Nomograph
($C = 100$)

Quantity (i.e., flow rate) and velocity are proportional to the C-value. For values of C other than 100, the quantity and velocity must be converted according to $\dot{V}_{actual} = \dot{V}_{chart}\, C_{actual}/100$. When quantity is the unknown, use the chart with known values of diameter, slope, or velocity to find $\dot{V}_{chart}$, and then convert to $\dot{V}_{actual}$. When velocity is the unknown, use the chart with the known values of diameter, slope, or quantity to find $\dot{V}_{chart}$, then convert to $\dot{V}_{actual}$. If $\dot{V}_{actual}$ is known, it must be converted to $\dot{V}_{chart}$ before this nomograph can be used. In that case, the diameter, loss, and quantity are as read from this chart.

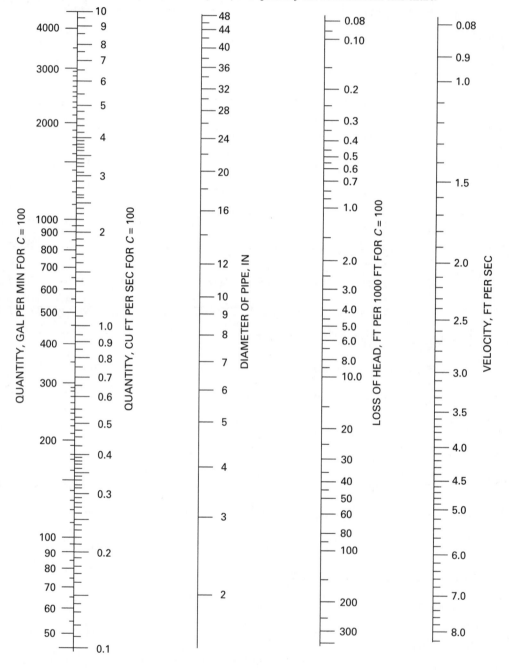

(Multiply gal/min by 0.0631 to obtain L/s.)
(Multiply ft³/sec by 28.3 to obtain L/s.)
(Multiply in by 25.4 to obtain mm.)
(Multiply ft/1000 ft by 0.1 to obtain m/100 m.)
(Multiply ft/sec by 0.3048 to obtain m/s.)

APPENDIX 17.F
Corrugated Metal Pipe

Corrugated metal pipe (CMP, also known as *corrugated steel pipe*) is frequently used for culverts. Pipe is made from corrugated sheets of galvanized steel that are rolled and riveted together along a longitudinal seam. Aluminized steel may also be used in certain ranges of soil pH. Standard round pipe diameters range from 8 in to 96 in (200 mm to 2450 mm). Metric dimensions of standard diameters are usually rounded to the nearest 25 mm or 50 mm (e.g., a 42 in culvert would be specified as a 1050 mm culvert, not 1067 mm).

Larger and noncircular culverts can be created out of curved steel plate. Standard section lengths are 10 ft to 20 ft (3 m to 6 m). Though most corrugations are transverse (i.e., annular), helical corrugations are also used. Metal gages of 8, 10, 12, 14, and 16 are commonly used, depending on the depth of burial.

The most common corrugated steel pipe has transverse corrugations that are $\frac{1}{2}$ in (13 mm) deep and $2\frac{2}{3}$ in (68 mm) from crest to crest. These are referred to as "$2\frac{1}{2}$ inch" or "68×13" corrugations. For larger culverts, corrugations with a 2 in (25 mm) depth and 3, 5, and 6 in (76, 125, or 152 mm) pitches are used. Plate-based products using 6 in by 2 in (152 mm by 51 mm) corrugations are known as *structural plate corrugated steel pipe* (SPCSP) and *multiplate* after the trade-named product "Multi-Plate™."

The flow area for circular culverts is based on the nominal culvert diameter, regardless of the gage of the plate metal used to construct the pipe. Flow area is calculated to (at most) three significant digits.

A Hazen-Williams coefficient, C, of 60 is typically used with all sizes of corrugated pipe. Values of C and Manning's constant, n, for corrugated pipe are generally not affected by age. *Design Charts for Open Channel Flow* (U.S. Department of Transportation, 1979) recommends a Manning constant of $n = 0.024$ for all cases. The U.S. Department of the Interior recommends the following values. For standard ($2\frac{2}{3}$ in by $\frac{1}{2}$ in or 68 mm by 13 mm) corrugated pipe with the diameters given: 12 in (457 mm), 0.027; 24 in (610 mm), 0.025; 36 in to 48 in (914 mm to 1219 mm), 0.024; 60 in to 84 in (1524 mm to 2134 mm), 0.023; 96 in (2438 mm), 0.022. For (6 in by 2 in or 152 mm by 51 mm) multiplate construction with the diameters given: 5 ft to 6 ft (1.5 m to 1.8 m), 0.034; 7 ft to 8 ft (2.1 m to 2.4 m), 0.033; 9 ft to 11 ft (2.7 m to 3.3 m), 0.032; 12 ft to 13 ft (3.6 m to 3.9 m), 0.031; 14 ft to 15 ft (4.2 m to 4.5 m), 0.030; 16 ft to 18 ft (4.8 m to 5.4 m), 0.029; 19 ft to 20 ft (5.8 m to 6.0 m), 0.028; 21 ft to 22 ft (6.3 m to 6.6 m), 0.027.

If the inside of the corrugated pipe has been asphalted completely smooth 360° circumferentially, Manning's n ranges from 0.009 to 0.011. For culverts with 40% asphalted inverts, $n = 0.019$. For other percentages of paved invert, the resulting value is proportional to the percentage and the values normally corresponding to that diameter pipe. For field-bolted corrugated metal pipe arches, $n = 0.025$.

It is also possible to calculate the Darcy friction loss if the corrugation depth, 0.5 in (13 mm) for standard corrugations and 2.0 in (51 mm) for multiplate, is taken as the specific roughness.

Appendices

APPENDIX 18.A
International Standard Atmosphere

customary U.S. units		
altitude (ft)	temperature (°R)	pressure (psia)
0	518.7	14.696
1000	515.1	14.175
2000	511.6	13.664
3000	508.0	13.168
4000	504.4	12.692
5000	500.9	12.225
6000	497.3	11.778
7000	493.7	11.341
8000	490.2	10.914
9000	486.6	10.501
10,000	483.0	10.108
11,000	479.5	9.720
12,000	475.9	9.347
13,000	472.3	8.983
14,000	468.8	8.630
15,000	465.2	8.291
16,000	461.6	7.962
17,000	458.1	7.642
18,000	454.5	7.338
19,000	450.9	7.038
20,000	447.4	6.753
21,000	443.8	6.473
22,000	440.2	6.203
23,000	436.7	5.943
24,000	433.1	5.693
25,000	429.5	5.452
26,000	426.0	5.216
27,000	422.4	4.990
28,000	418.8	4.774
29,000	415.3	4.563
30,000	411.7	4.362
31,000	408.1	4.165
32,000	404.6	3.978
33,000	401.0	3.797
34,000	397.5	3.625
35,000	393.9	3.458
36,000	392.7	3.296
37,000	392.7	3.143
38,000	392.7	2.996
39,000	392.7	2.854
40,000	392.7	2.721
41,000	392.7	2.593
42,000	392.7	2.475
43,000	392.7	2.358
44,000	392.7	2.250
45,000	392.7	2.141
46,000	392.7	2.043
47,000	392.7	1.950
48,000	392.7	1.857
49,000	392.7	1.768
50,000	392.7	1.690
51,000	392.7	1.611
52,000	392.7	1.532
53,000	392.7	1.464
54,000	392.7	1.395
55,000	392.7	1.331
56,000	392.7	1.267
57,000	392.7	1.208
58,000	392.7	1.154
59,000	392.7	1.100
60,000	392.7	1.046
61,000	392.7	0.997
62,000	392.7	0.953
63,000	392.7	0.909
64,000	392.7	0.864
65,000	392.7	0.825

troposphere (0 to ~34,000 ft); tropopause; stratosphere (to approximately 160,000 ft)

SI units		
altitude (m)	temperature (K)	pressure (bar)
0	288.15	1.01325
500	284.9	0.9546
1000	281.7	0.8988
1500	278.4	0.8456
2000	275.2	0.7950
2500	271.9	0.7469
3000	268.7	0.7012
3500	265.4	0.6578
4000	262.2	0.6166
4500	258.9	0.5775
5000	255.7	0.5405
5500	252.4	0.5054
6000	249.2	0.4722
6500	245.9	0.4408
7000	242.7	0.4111
7500	239.5	0.3830
8000	236.2	0.3565
8500	233.0	0.3315
9000	229.7	0.3080
9500	226.5	0.2858
10 000	223.3	0.2650
10 500	220.0	0.2454
11 000	216.8	0.2270
11 500	216.7	0.2098
12 000	216.7	0.1940
12 500	216.7	0.1793
13 000	216.7	0.1658
13 500	216.7	0.1533
14 000	216.7	0.1417
14 500	216.7	0.1310
15 000	216.7	0.1211
15 500	216.7	0.1120
16 000	216.7	0.1035
16 500	216.7	0.09572
17 000	216.7	0.08850
17 500	216.7	0.08182
18 000	216.7	0.07565
18 500	216.7	0.06995
19 000	216.7	0.06467
19 500	216.7	0.05980
20 000	216.7	0.05529
22 000	218.6	0.04047
24 000	220.6	0.02972
26 000	222.5	0.02188
28 000	224.5	0.01616
30 000	226.5	0.01197
32 000	228.5	0.00889

troposphere (0 to ~10 000 m); tropopause; stratosphere (to approximately 50 000 m)

APPENDIX 18.B
Properties of Saturated Steam by Temperature
(customary U.S. units)

temp. (°F)	absolute pressure (psia)	specific volume (ft³/lbm)		internal energy (Btu/lbm)		enthalpy (Btu/lbm)			entropy (Btu/lbm-°R)		temp. (°F)
		sat. liquid, v_f	sat. vapor, v_g	sat. liquid, u_f	sat. vapor, u_g	sat. liquid, h_f	evap., h_{fg}	sat. vapor, h_g	sat. liquid, s_f	sat. vapor, s_g	
32	0.0886	0.01602	3302	−0.02	1021.0	−0.02	1075.2	1075.2	−0.0004	2.1868	32
34	0.0961	0.01602	3059	2.00	1021.7	2.00	1074.1	1076.1	0.00405	2.1797	34
36	0.1040	0.01602	2836	4.01	1022.3	4.01	1072.9	1076.9	0.00812	2.1727	36
38	0.1126	0.01602	2632	6.02	1023.0	6.02	1071.8	1077.8	0.01217	2.1658	38
40	0.1217	0.01602	2443	8.03	1023.7	8.03	1070.7	1078.7	0.01620	2.1589	40
42	0.1316	0.01602	2270	10.04	1024.3	10.04	1069.5	1079.6	0.02022	2.1522	42
44	0.1421	0.01602	2111	12.05	1025.0	12.05	1068.4	1080.4	0.02421	2.1454	44
46	0.1533	0.01602	1964	14.06	1025.6	14.06	1067.3	1081.3	0.02819	2.1388	46
48	0.1653	0.01602	1828	16.06	1026.3	16.06	1066.1	1082.2	0.03215	2.1322	48
50	0.1781	0.01602	1703	18.07	1026.9	18.07	1065.0	1083.1	0.03609	2.1257	50
52	0.1918	0.01603	1587	20.07	1027.6	20.07	1063.8	1083.9	0.04001	2.1192	52
54	0.2065	0.01603	1481	22.07	1028.2	22.07	1062.7	1084.8	0.04392	2.1128	54
56	0.2221	0.01603	1382	24.07	1028.9	24.08	1061.6	1085.7	0.04781	2.1065	56
58	0.2387	0.01603	1291	26.08	1029.6	26.08	1060.5	1086.6	0.05168	2.1002	58
60	0.2564	0.01604	1206	28.08	1030.2	28.08	1059.3	1087.4	0.05554	2.0940	60
62	0.2752	0.01604	1128	30.08	1030.9	30.08	1058.2	1088.3	0.05938	2.0879	62
64	0.2953	0.01604	1055	32.08	1031.5	32.08	1057.1	1089.2	0.06321	2.0818	64
66	0.3166	0.01604	987.7	34.08	1032.2	34.08	1055.9	1090.0	0.06702	2.0758	66
68	0.3393	0.01605	925.2	36.08	1032.8	36.08	1054.8	1090.9	0.07081	2.0698	68
70	0.3634	0.01605	867.1	38.07	1033.5	38.08	1053.7	1091.8	0.07459	2.0639	70
72	0.3889	0.01606	813.2	40.07	1034.1	40.07	1052.5	1092.6	0.07836	2.0581	72
74	0.4160	0.01606	763.0	42.07	1034.8	42.07	1051.4	1093.5	0.08211	2.0523	74
76	0.4448	0.01606	716.3	44.07	1035.4	44.07	1050.3	1094.4	0.08585	2.0466	76
78	0.4752	0.01607	672.9	46.07	1036.1	46.07	1049.1	1095.2	0.08957	2.0409	78
80	0.5075	0.01607	632.4	48.06	1036.7	48.07	1048.0	1096.1	0.09328	2.0353	80
82	0.5416	0.01608	594.7	50.06	1037.4	50.06	1046.9	1097.0	0.09697	2.0297	82
84	0.5778	0.01608	559.5	52.06	1038.0	52.06	1045.7	1097.8	0.1007	2.0242	84
86	0.6160	0.01609	526.7	54.05	1038.7	54.06	1044.6	1098.7	0.1043	2.0187	86
88	0.6564	0.01610	496.0	56.05	1039.3	56.05	1043.4	1099.5	0.1080	2.0133	88
90	0.6990	0.01610	467.4	58.05	1039.9	58.05	1042.4	1100.4	0.1116	2.0079	90
92	0.7441	0.01611	440.7	60.04	1040.6	60.05	1041.3	1101.3	0.1152	2.0026	92
94	0.7917	0.01611	415.6	62.04	1041.2	62.04	1040.1	1102.1	0.1189	1.9974	94
96	0.8418	0.01612	392.3	64.04	1041.9	64.04	1039.0	1103.0	0.1225	1.9922	96
98	0.8947	0.01613	370.3	66.03	1042.5	66.04	1037.8	1103.8	0.1260	1.9870	98
100	0.9505	0.01613	349.8	68.03	1043.2	68.03	1036.7	1104.7	0.1296	1.9819	100
105	1.103	0.01615	304.0	73.02	1044.8	73.03	1033.8	1106.8	0.1385	1.9693	105
110	1.277	0.01617	265.0	78.01	1046.4	78.02	1031.0	1109.0	0.1473	1.9570	110
115	1.473	0.01619	231.6	83.01	1048.0	83.01	1028.1	1111.1	0.1560	1.9450	115
120	1.695	0.01621	203.0	88.00	1049.6	88.00	1025.2	1113.2	0.1647	1.9333	120
125	1.945	0.01623	178.3	92.99	1051.1	93.00	1022.3	1115.3	0.1732	1.9218	125
130	2.226	0.01625	157.1	97.99	1052.7	97.99	1019.4	1117.4	0.1818	1.9106	130
135	2.541	0.01627	138.7	102.98	1054.3	102.99	1016.5	1119.5	0.1902	1.8996	135

APPENDIX 18.B *(continued)*
Properties of Saturated Steam by Temperature
(customary U.S. units)

temp. (°F)	absolute pressure (psia)	specific volume (ft³/lbm)		internal energy (Btu/lbm)		enthalpy (Btu/lbm)			entropy (Btu/lbm-°R)		temp. (°F)
		sat. liquid, v_f	sat. vapor, v_g	sat. liquid, u_f	sat. vapor, u_g	sat. liquid, h_f	evap., h_{fg}	sat. vapor, h_g	sat. liquid, s_f	sat. vapor, s_g	
140	2.893	0.01629	122.8	107.98	1055.8	107.99	1013.6	1121.6	0.1986	1.8888	140
145	3.286	0.01632	109.0	112.98	1057.4	112.99	1010.7	1123.7	0.2069	1.8783	145
150	3.723	0.01634	96.93	117.98	1059.0	117.99	1007.7	1125.7	0.2151	1.8680	150
155	4.209	0.01637	86.40	122.98	1060.5	122.99	1004.8	1127.8	0.2233	1.8580	155
160	4.747	0.01639	77.18	127.98	1062.0	128.00	1001.8	1129.8	0.2314	1.8481	160
165	5.343	0.01642	69.09	132.99	1063.6	133.01	998.9	1131.9	0.2394	1.8384	165
170	6.000	0.01645	61.98	138.00	1065.1	138.01	995.9	1133.9	0.2474	1.8290	170
175	6.724	0.01648	55.71	143.01	1066.6	143.03	992.9	1135.9	0.2553	1.8197	175
180	7.520	0.01651	50.17	148.02	1068.1	148.04	989.9	1137.9	0.2632	1.8106	180
185	8.393	0.01654	45.27	153.03	1069.6	153.06	986.8	1139.9	0.2710	1.8017	185
190	9.350	0.01657	40.92	158.05	1071.0	158.08	983.7	1141.8	0.2788	1.7930	190
195	10.396	0.01660	37.05	163.07	1072.5	163.10	980.7	1143.8	0.2865	1.7844	195
200	11.538	0.01663	33.61	168.09	1074.0	168.13	977.6	1145.7	0.2941	1.7760	200
205	12.782	0.01667	30.54	173.12	1075.4	173.16	974.4	1147.6	0.3017	1.7678	205
210	14.14	0.01670	27.79	178.2	1076.8	178.2	971.3	1149.5	0.3092	1.7597	210
212	14.71	0.01672	26.78	180.2	1077.4	180.2	970.1	1150.3	0.3122	1.7565	212
220	17.20	0.01677	23.13	188.2	1079.6	188.3	965.0	1153.3	0.3242	1.7440	220
230	20.80	0.01685	19.37	198.3	1082.4	198.4	958.5	1156.9	0.3389	1.7288	230
240	24.99	0.01692	16.31	208.4	1085.1	208.5	952.0	1160.5	0.3534	1.7141	240
250	29.84	0.01700	13.82	218.5	1087.7	218.6	945.4	1164.0	0.3678	1.7000	250
260	35.45	0.01708	11.76	228.7	1090.3	228.8	938.6	1167.4	0.3820	1.6863	260
270	41.88	0.01717	10.06	238.9	1092.8	239.0	931.7	1170.7	0.3960	1.6730	270
280	49.22	0.01726	8.64	249.0	1095.2	249.2	924.7	1173.9	0.4099	1.6601	280
290	57.57	0.01735	7.46	259.3	1097.5	259.5	917.6	1177.0	0.4236	1.6476	290
300	67.03	0.01745	6.466	269.5	1099.8	269.7	910.3	1180.0	0.4372	1.6354	300
310	77.69	0.01755	5.626	279.8	1101.9	280.1	902.8	1182.8	0.4507	1.6236	310
320	89.67	0.01765	4.914	290.1	1103.9	290.4	895.1	1185.5	0.4640	1.6120	320
330	103.07	0.01776	4.308	300.5	1105.9	300.8	887.2	1188.0	0.4772	1.6007	330
340	118.02	0.01787	3.788	310.9	1107.7	311.2	879.3	1190.5	0.4903	1.5897	340
350	134.63	0.01799	3.343	321.3	1109.4	321.7	871.0	1192.7	0.5032	1.5789	350
360	153.03	0.01811	2.958	331.8	1111.0	332.3	862.5	1194.8	0.5161	1.5684	360
370	173.36	0.01823	2.625	342.3	1112.5	342.9	853.8	1196.7	0.5289	1.5580	370
380	195.74	0.01836	2.336	352.9	1113.9	353.5	845.0	1198.5	0.5415	1.5478	380
390	220.3	0.01850	2.084	363.5	1115.1	364.3	835.8	1200.1	0.5541	1.5378	390
400	247.3	0.01864	1.864	374.2	1116.2	375.1	826.4	1201.4	0.5667	1.5279	400
410	276.7	0.01879	1.671	385.0	1117.1	385.9	816.7	1202.6	0.5791	1.5182	410
420	308.8	0.01894	1.501	395.8	1117.8	396.9	806.8	1203.6	0.5915	1.5085	420
430	343.6	0.01910	1.351	406.7	1118.4	407.9	796.4	1204.3	0.6038	1.4990	430
440	381.5	0.01926	1.218	417.6	1118.9	419.0	785.8	1204.8	0.6161	1.4895	440
450	422.5	0.01944	1.0999	428.7	1119.1	430.2	774.9	1205.1	0.6283	1.4802	450
460	466.8	0.01962	0.9951	439.8	1119.2	441.5	763.6	1205.1	0.6405	1.4708	460
470	514.5	0.01981	0.9015	451.1	1119.0	452.9	752.0	1204.9	0.6527	1.4615	470

APPENDIX 18.B *(continued)*
Properties of Saturated Steam by Temperature
(customary U.S. units)

temp. (°F)	absolute pressure (psia)	specific volume (ft³/lbm)		internal energy (Btu/lbm)		enthalpy (Btu/lbm)			entropy (Btu/lbm-°R)		temp. (°F)
		sat. liquid, v_f	sat. vapor, v_g	sat. liquid, u_f	sat. vapor, u_g	sat. liquid, h_f	evap., h_{fg}	sat. vapor, h_g	sat. liquid, s_f	sat. vapor, s_g	
480	566.0	0.02001	0.8179	462.4	1118.7	464.5	739.8	1204.3	0.6648	1.4522	480
490	621.2	0.02022	0.7429	473.8	1118.1	476.1	727.4	1203.5	0.6769	1.4428	490
500	680.6	0.02044	0.6756	485.4	1117.2	487.9	714.4	1202.3	0.6891	1.4335	500
510	744.1	0.02068	0.6149	497.1	1116.2	499.9	700.9	1200.8	0.7012	1.4240	510
520	812.1	0.02092	0.5601	508.9	1114.8	512.0	686.9	1198.9	0.7134	1.4146	520
530	884.7	0.02119	0.5105	520.8	1113.1	524.3	672.4	1196.7	0.7256	1.4050	530
540	962.2	0.02146	0.4655	533.0	1111.1	536.8	657.2	1194.0	0.7378	1.3952	540
550	1044.8	0.02176	0.4247	545.3	1108.8	549.5	641.4	1190.9	0.7501	1.3853	550
560	1132.7	0.02208	0.3874	557.8	1106.0	562.4	624.8	1187.2	0.7625	1.3753	560
570	1226.2	0.02242	0.3534	570.5	1102.8	575.6	607.4	1183.0	0.7750	1.3649	570
580	1325.5	0.02279	0.3223	583.5	1099.2	589.1	589.3	1178.3	0.7876	1.3543	580
590	1430.8	0.02319	0.2937	596.7	1095.0	602.8	570.0	1172.8	0.8004	1.3434	590
600	1542.5	0.02363	0.2674	610.3	1090.3	617.0	549.6	1166.6	0.8133	1.3320	600
610	1660.9	0.02411	0.2431	624.2	1084.8	631.6	528.0	1159.6	0.8266	1.3201	610
620	1786.2	0.02465	0.2206	638.5	1078.6	646.7	504.8	1151.5	0.8401	1.3077	620
630	1918.9	0.02525	0.1997	653.4	1071.5	662.4	480.1	1142.4	0.8539	1.2945	630
640	2059.2	0.02593	0.1802	668.9	1063.2	678.7	453.1	1131.8	0.8683	1.2803	640
650	2207.8	0.02672	0.1618	685.1	1053.6	696.0	423.7	1119.7	0.8833	1.2651	650
660	2364.9	0.02766	0.1444	702.4	1042.2	714.5	390.8	1105.3	0.8991	1.2482	660
670	2531.2	0.02883	0.1277	721.1	1028.4	734.6	353.6	1088.2	0.9163	1.2292	670
680	2707.3	0.03036	0.1113	742.2	1011.1	757.4	309.4	1066.8	0.9355	1.2070	680
690	2894	0.03258	0.0946	767.1	987.7	784.5	253.8	1038.3	0.9582	1.1790	690
700	3093	0.03665	0.0748	801.5	948.2	822.5	168.5	991.0	0.9900	1.1353	700
705.103	3200.11	0.04975	0.04975	866.6	866.6	896.1	0	896.1	1.0526	1.0526	705.103

Values in this table were calculated from *NIST Standard Reference Database 10*, "NIST/ASME Steam Properties," Ver. 2.11, National Institute of Standards and Technology, U.S. Department of Commerce, Gaithersburg, MD, 1997, which has been licensed to PPI.

APPENDIX 18.C
Properties of Saturated Steam by Pressure
(customary U.S. units)

absolute press (psia)	temp. (°F)	specific volume (ft³/lbm)		internal energy (Btu/lbm)		enthalpy (Btu/lbm)			entropy (Btu/lbm-°R)			absolute press (psia)
		sat. liquid, v_f	sat. vapor, v_g	sat. liquid, u_f	sat. vapor, u_g	sat. liquid, h_f	evap., h_{fg}	sat. vapor, h_g	sat. liquid, s_f	evap., s_{fg}	sat. vapor, s_g	
0.4	72.83	0.01606	791.8	40.91	1034.4	40.91	1052.1	1093.0	0.0799	1.9758	2.0557	0.4
0.6	85.18	0.01609	539.9	53.23	1038.4	53.24	1045.1	1098.3	0.1028	1.9182	2.0210	0.6
0.8	94.34	0.01611	411.6	62.38	1041.3	62.38	1039.9	1102.3	0.1195	1.8770	1.9965	0.8
1.0	101.69	0.01614	333.5	69.72	1043.7	69.72	1035.7	1105.4	0.1326	1.8450	1.9776	1.0
1.2	107.87	0.01616	280.9	75.88	1045.7	75.89	1032.2	1108.1	0.1435	1.8187	1.9622	1.2
1.5	115.64	0.01619	227.7	83.64	1048.2	83.65	1027.8	1111.4	0.1571	1.7864	1.9435	1.5
2.0	126.03	0.01623	173.7	94.02	1051.5	94.02	1021.8	1115.8	0.1750	1.7445	1.9195	2.0
3.0	141.42	0.01630	118.7	109.4	1056.3	109.4	1012.8	1122.2	0.2009	1.6849	1.8858	3.0
4.0	152.91	0.01636	90.63	120.9	1059.9	120.9	1006.0	1126.9	0.2199	1.6423	1.8621	4.0
5.0	162.18	0.01641	73.52	130.2	1062.7	130.2	1000.5	1130.7	0.2349	1.6089	1.8438	5.0
6.0	170.00	0.01645	61.98	138.0	1065.1	138.0	995.9	1133.9	0.2474	1.5816	1.8290	6.0
7.0	176.79	0.01649	53.65	144.8	1067.1	144.8	991.8	1136.6	0.2581	1.5583	1.8164	7.0
8.0	182.81	0.01652	47.34	150.8	1068.9	150.9	988.1	1139.0	0.2676	1.5380	1.8056	8.0
9.0	188.22	0.01656	42.40	156.3	1070.5	156.3	984.8	1141.1	0.2760	1.5201	1.7961	9.0
10	193.16	0.01659	38.42	161.2	1072.0	161.3	981.9	1143.1	0.2836	1.5040	1.7876	10
14.696	211.95	0.01671	26.80	180.1	1077.4	180.2	970.1	1150.3	0.3122	1.4445	1.7566	14.696
15	212.99	0.01672	26.29	181.2	1077.7	181.2	969.5	1150.7	0.3137	1.4412	1.7549	15
20	227.92	0.01683	20.09	196.2	1081.8	196.3	959.9	1156.2	0.3358	1.3961	1.7319	20
25	240.03	0.01692	16.31	208.4	1085.1	208.5	952.0	1160.5	0.3535	1.3606	1.7141	25
30	250.30	0.01700	13.75	218.8	1087.8	218.9	945.2	1164.1	0.3682	1.3313	1.6995	30
35	259.25	0.01708	11.90	227.9	1090.1	228.0	939.2	1167.2	0.3809	1.3064	1.6873	35
40	267.22	0.01715	10.50	236.0	1092.1	236.1	933.7	1169.8	0.3921	1.2845	1.6766	40
45	274.41	0.01721	9.402	243.3	1093.9	243.5	928.7	1172.2	0.4022	1.2650	1.6672	45
50	280.99	0.01727	8.517	250.1	1095.4	250.2	924.0	1174.2	0.4113	1.2475	1.6588	50
55	287.05	0.01732	7.788	256.2	1096.8	256.4	919.7	1176.1	0.4196	1.2316	1.6512	55
60	292.68	0.01738	7.176	262.0	1098.1	262.2	915.6	1177.8	0.4273	1.2170	1.6443	60
65	297.95	0.01743	6.656	267.4	1099.3	267.6	911.8	1179.4	0.4344	1.2035	1.6379	65
70	302.91	0.01748	6.207	272.5	1100.4	272.7	908.1	1180.8	0.4411	1.1908	1.6319	70

APPENDIX 18.C *(continued)*
Properties of Saturated Steam by Pressure
(customary U.S. units)

absolute press (psia)	temp. (°F)	specific volume (ft³/lbm)		internal energy (Btu/lbm)		enthalpy (Btu/lbm)			entropy (Btu/lbm-°R)			absolute press (psia)
		sat. liquid, v_f	sat. vapor, v_g	sat. liquid, u_f	sat. vapor, u_g	sat. liquid, h_f	evap., h_{fg}	sat. vapor, h_g	sat. liquid, s_f	evap., s_{fg}	sat. vapor, s_g	
75	307.58	0.01752	5.816	277.3	1101.4	277.6	904.6	1182.1	0.4474	1.1790	1.6264	75
80	312.02	0.01757	5.473	281.9	1102.3	282.1	901.2	1183.3	0.4534	1.1679	1.6212	80
85	316.24	0.01761	5.169	286.2	1103.2	286.5	898.0	1184.5	0.4590	1.1573	1.6163	85
90	320.26	0.01765	4.897	290.4	1104.0	290.7	894.9	1185.6	0.4643	1.1474	1.6117	90
95	324.11	0.01770	4.653	294.4	1104.8	294.7	891.9	1186.6	0.4694	1.1379	1.6073	95
100	327.81	0.01774	4.433	298.2	1105.5	298.5	889.0	1187.5	0.4743	1.1289	1.6032	100
110	334.77	0.01781	4.050	305.4	1106.8	305.8	883.4	1189.2	0.4834	1.1120	1.5954	110
120	341.25	0.01789	3.729	312.2	1107.9	312.6	878.2	1190.7	0.4919	1.0965	1.5884	120
130	347.32	0.01796	3.456	318.5	1109.0	318.9	873.2	1192.1	0.4998	1.0821	1.5818	130
140	353.03	0.01802	3.220	324.5	1109.9	324.9	868.5	1193.4	0.5071	1.0686	1.5757	140
150	358.42	0.01809	3.015	330.1	1110.8	330.6	863.9	1194.5	0.5141	1.0559	1.5700	150
160	363.54	0.01815	2.835	335.5	1111.6	336.0	859.5	1195.5	0.5206	1.0441	1.5647	160
170	368.41	0.01821	2.675	340.6	1112.3	341.2	855.2	1196.4	0.5268	1.0328	1.5596	170
180	373.07	0.01827	2.532	345.5	1112.9	346.1	851.2	1197.3	0.5328	1.0222	1.5549	180
190	377.52	0.01833	2.404	350.3	1113.5	350.9	847.2	1198.1	0.5384	1.0119	1.5503	190
200	381.80	0.01839	2.288	354.8	1114.1	355.5	843.3	1198.8	0.5438	1.0022	1.5460	200
250	400.97	0.01865	1.844	375.2	1116.2	376.1	825.5	1201.6	0.5679	0.9591	1.5270	250
300	417.35	0.01890	1.544	392.9	1117.7	394.0	809.4	1203.3	0.5882	0.9229	1.5111	300
350	431.74	0.01913	1.326	408.6	1118.5	409.8	794.6	1204.4	0.6059	0.8915	1.4974	350
400	444.62	0.01934	1.162	422.7	1119.0	424.2	780.9	1205.0	0.6217	0.8635	1.4852	400
450	456.31	0.01955	1.032	435.7	1119.2	437.3	767.8	1205.1	0.6360	0.8382	1.4742	450
500	467.04	0.01975	0.928	447.7	1119.1	449.5	755.5	1205.0	0.6491	0.8152	1.4642	500
550	476.98	0.01995	0.842	458.9	1118.8	461.0	743.5	1204.5	0.6611	0.7939	1.4550	550
600	486.24	0.02014	0.770	469.5	1118.3	471.7	732.1	1203.8	0.6724	0.7739	1.4463	600
700	503.13	0.02051	0.656	489.0	1116.9	491.7	710.2	1201.9	0.6929	0.7376	1.4305	700
800	518.27	0.02088	0.569	506.8	1115.0	509.9	689.4	1199.3	0.7113	0.7050	1.4162	800
900	532.02	0.02124	0.501	523.3	1112.7	526.8	669.4	1196.2	0.7280	0.6750	1.4030	900

APPENDIX 18.C *(continued)*
Properties of Saturated Steam by Pressure
(customary U.S. units)

absolute press (psia)	temp. (°F)	specific volume (ft³/lbm) sat. liquid, v_f	sat. vapor, v_g	internal energy (Btu/lbm) sat. liquid, u_f	sat. vapor, u_g	enthalpy (Btu/lbm) sat. liquid, h_f	evap., h_{fg}	sat. vapor, h_g	entropy (Btu/lbm-°R) sat. liquid, s_f	evap., s_{fg}	sat. vapor, s_g	absolute press (psia)
1000	544.65	0.02160	0.446	538.7	1110.1	542.7	649.9	1192.6	0.7435	0.6472	1.3907	1000
1100	556.35	0.02196	0.401	553.2	1107.1	557.7	631.0	1188.6	0.7580	0.6211	1.3790	1100
1200	567.26	0.02233	0.362	567.0	1103.8	571.9	612.3	1184.2	0.7715	0.5963	1.3678	1200
1300	577.49	0.02270	0.330	580.2	1100.2	585.6	593.9	1179.5	0.7844	0.5726	1.3570	1300
1400	587.14	0.02307	0.302	592.9	1096.3	598.9	575.5	1174.4	0.7967	0.5498	1.3465	1400
1500	596.26	0.02346	0.277	605.2	1092.1	611.7	557.3	1169.0	0.8085	0.5278	1.3363	1500
1600	604.93	0.02386	0.255	617.1	1087.7	624.1	539.1	1163.2	0.8198	0.5064	1.3262	1600
1700	613.18	0.02428	0.236	628.7	1082.9	636.3	520.8	1157.1	0.8308	0.4854	1.3162	1700
1800	621.07	0.02471	0.218	640.1	1077.9	648.3	502.3	1150.6	0.8415	0.4648	1.3063	1800
1900	628.61	0.02516	0.203	651.3	1072.5	660.1	483.6	1143.7	0.8520	0.4443	1.2963	1900
2000	635.85	0.02564	0.188	662.4	1066.8	671.8	464.6	1136.4	0.8623	0.4240	1.2863	2000
2250	652.74	0.02696	0.157	689.7	1050.6	701.0	415.1	1116.0	0.8875	0.3731	1.2606	2250
2500	668.17	0.02859	0.131	717.6	1031.1	730.8	360.8	1091.6	0.9130	0.3199	1.2329	2500
2750	682.34	0.03080	0.108	747.6	1006.3	763.2	297.8	1061.0	0.9404	0.2607	1.2011	2750
3000	695.41	0.03434	0.085	783.4	969.9	802.5	214.4	1016.9	0.9733	0.1856	1.1589	3000
3200.11	705.1028	0.04975	0.04975	866.6	866.6	896.1	0	896.1	1.0526	0	1.0526	3200.11

Values in this table were calculated from *NIST Standard Reference Database 10*, "NIST/ASME Steam Properties," Ver. 2.11, National Institute of Standards and Technology, U.S. Department of Commerce, Gaithersburg, MD, 1997, which has been licensed to PPI.

APPENDIX 18.D

Properties of Superheated Steam (customary U.S. units)

specific volume, v, in ft³/lbm; enthalpy, h, in Btu/lbm; entropy, s, in Btu/lbm-°R

absolute pressure (psia) (sat. temp., °F)		temperature (°F)								
		200	300	400	500	600	700	800	900	1000
1.0	v	392.5	452.3	511.9	571.5	631.1	690.7	750.3	809.9	869.5
(101.69)	h	1150.1	1195.7	1241.8	1288.6	1336.2	1384.6	1433.9	1484.1	1535.1
	s	2.0510	2.1153	2.1723	2.2238	2.2710	2.3146	2.3554	2.3937	2.4299
5.0	v	78.15	90.25	102.25	114.21	126.15	138.09	150.02	161.94	173.86
(162.18)	h	1148.5	1194.8	1241.3	1288.2	1335.9	1384.4	1433.7	1483.9	1535.0
	s	1.8716	1.9370	1.9944	2.0461	2.0934	2.1371	2.1779	2.2162	2.2525
10.0	v	38.85	44.99	51.04	57.04	63.03	69.01	74.98	80.95	86.91
(193.16)	h	1146.4	1193.8	1240.6	1287.8	1335.6	1384.2	1433.5	1483.8	1534.9
	s	1.7926	1.8595	1.9174	1.9693	2.0167	2.0605	2.1014	2.1397	2.1760
14.696	v	....	30.53	34.67	38.77	42.86	46.93	51.00	55.07	59.13
(211.95)	h	....	1192.7	1240.0	1287.4	1335.3	1383.9	1433.3	1483.6	1534.8
	s	....	1.8160	1.8744	1.9266	1.9741	2.0179	2.0588	2.0972	2.1335
20.0	v	....	22.36	25.43	28.46	31.47	34.47	37.46	40.45	43.44
(227.92)	h	....	1191.6	1239.3	1286.9	1334.9	1383.6	1433.1	1483.4	1534.6
	s	....	1.7808	1.8398	1.8922	1.9399	1.9838	2.0247	2.0631	2.0994
60.0	v	....	7.260	8.355	9.400	10.426	11.440	12.448	13.453	14.454
(292.68)	h	....	1181.9	1233.7	1283.1	1332.2	1381.6	1431.5	1482.1	1533.5
	s	....	1.6496	1.7138	1.7682	1.8168	1.8613	1.9026	1.9413	1.9778
100.0	v	....	....	4.936	5.588	6.217	6.834	7.446	8.053	8.658
(327.81)	h	....	....	1227.7	1279.3	1329.4	1379.5	1429.8	1480.7	1532.4
	s	....	....	1.6521	1.7089	1.7586	1.8037	1.8453	1.8842	1.9209
150.0	v	....	....	3.222	3.680	4.112	4.531	4.944	5.353	5.759
(358.42)	h	....	....	1219.7	1274.3	1325.9	1376.8	1427.7	1479.0	1531.0
	s	....	....	1.6001	1.6602	1.7114	1.7573	1.7994	1.8386	1.8755
200.0	v	....	....	2.362	2.725	3.059	3.380	3.693	4.003	4.310
(381.80)	h	....	....	1210.9	1269.0	1322.3	1374.1	1425.6	1477.3	1529.6
	s	....	....	1.5602	1.6243	1.6771	1.7238	1.7664	1.8060	1.8430
250.0	v	....	....	....	2.151	2.426	2.688	2.943	3.193	3.440
(400.97)	h	....	....	....	1263.6	1318.6	1371.3	1423.5	1475.6	1528.1
	s	....	....	....	1.5953	1.6499	1.6975	1.7406	1.7804	1.8177
300.0	v	....	....	....	1.767	2.005	2.227	2.442	2.653	2.861
(417.35)	h	....	....	....	1257.9	1314.8	1368.6	1421.3	1473.9	1526.7
	s	....	....	....	1.5706	1.6271	1.6756	1.7192	1.7594	1.7969
400.0	v	....	....	....	1.285	1.477	1.651	1.817	1.978	2.136
(444.62)	h	....	....	....	1245.6	1306.9	1362.9	1416.9	1470.4	1523.9
	s	....	....	....	1.5288	1.5897	1.6402	1.6849	1.7257	1.7637

APPENDIX 18.D *(continued)*
Properties of Superheated Steam (customary U.S. units)
specific volume, v, in ft^3/lbm; enthalpy, h, in Btu/lbm; entropy, s, in Btu/lbm-°R

absolute pressure (psia) (sat. temp., °F)		temperature (°F)									
		500	600	700	800	900	1000	1100	1200	1400	1600
450.0 (456.31)	v	1.123	1.300	1.458	1.608	1.753	1.894	2.034	2.172	2.445	
	h	1238.9	1302.8	1360.0	1414.7	1468.6	1522.4	1576.5	1631.0	1742.0	
	s	1.5103	1.5737	1.6253	1.6706	1.7118	1.7499	1.7858	1.8196	1.8828	
500.0 (467.04)	v	0.993	1.159	1.304	1.441	1.573	1.701	1.827	1.952	2.199	
	h	1231.9	1298.6	1357.0	1412.5	1466.9	1521.0	1575.3	1630.0	1741.2	
	s	1.4928	1.5591	1.6118	1.6576	1.6992	1.7376	1.7736	1.8076	1.8708	
600.0 (486.24)	v	0.795	0.946	1.073	1.190	1.302	1.411	1.518	1.623	1.830	
	h	1216.5	1289.9	1351.0	1408.0	1463.3	1518.1	1572.8	1627.9	1739.7	
	s	1.4596	1.5326	1.5877	1.6348	1.6771	1.7160	1.7523	1.7865	1.8501	
700.0 (503.13)	v	....	0.793	0.908	1.011	1.109	1.204	1.296	1.387	1.566	
	h	....	1280.7	1344.8	1403.4	1459.7	1515.1	1570.4	1625.9	1738.2	
	s	....	1.5087	1.5666	1.6151	1.6581	1.6975	1.7341	1.7686	1.8324	
800.0 (518.27)	v	....	0.678	0.783	0.877	0.964	1.048	1.130	1.211	1.368	
	h	....	1270.9	1338.4	1398.7	1456.0	1512.2	1568.0	1623.8	1736.7	
	s	....	1.4867	1.5476	1.5975	1.6414	1.6812	1.7182	1.7529	1.8171	
900.0 (532.02)	v	....	0.588	0.686	0.772	0.852	0.928	1.001	1.073	1.214	
	h	....	1260.4	1331.8	1393.9	1452.3	1509.2	1565.5	1621.8	1735.2	
	s	....	1.4658	1.5302	1.5816	1.6263	1.6667	1.7040	1.7389	1.8034	
1000 (544.65)	v	....	0.514	0.608	0.688	0.761	0.831	0.898	0.963	1.091	1.216
	h	....	1249.3	1325.0	1389.0	1448.6	1506.2	1563.0	1619.7	1733.6	1849.6
	s	....	1.4457	1.5140	1.5671	1.6126	1.6535	1.6912	1.7264	1.7912	1.8504
1200 (567.26)	v	....	0.402	0.491	0.562	0.626	0.686	0.743	0.798	0.906	1.012
	h	....	1224.2	1310.6	1379.0	1441.0	1500.1	1558.1	1615.5	1730.6	1847.3
	s	....	1.4061	1.4842	1.5408	1.5882	1.6302	1.6686	1.7043	1.7698	1.8294
1400 (587.14)	v	....	0.318	0.406	0.472	0.529	0.582	0.632	0.681	0.774	0.866
	h	....	1193.8	1295.1	1368.5	1433.1	1494.0	1553.0	1611.3	1727.5	1845.0
	s	....	1.3649	1.4566	1.5174	1.5668	1.6100	1.6491	1.6853	1.7515	1.8115
1600 (604.93)	v	....	....	0.342	0.404	0.456	0.504	0.549	0.592	0.676	0.756
	h	....	....	1278.3	1357.6	1425.1	1487.7	1547.9	1607.1	1724.5	1842.7
	s	....	....	1.4302	1.4959	1.5475	1.5920	1.6319	1.6686	1.7354	1.7958
1800 (621.07)	v	....	....	0.291	0.350	0.399	0.443	0.484	0.524	0.599	0.671
	h	....	....	1260.0	1346.2	1416.9	1481.3	1542.8	1602.8	1721.4	1840.3
	s	....	....	1.4044	1.4759	1.5299	1.5756	1.6164	1.6537	1.7211	1.7819
2000 (635.85)	v	....	....	0.249	0.308	0.354	0.395	0.433	0.469	0.537	0.603
	h	....	....	1239.7	1334.3	1408.5	1474.8	1537.6	1598.5	1718.3	1838.0
	s	....	....	1.3783	1.4568	1.5135	1.5606	1.6022	1.6400	1.7082	1.7693

Values in this table were calculated from *NIST Standard Reference Database 10*, "NIST/ASME Steam Properties," Ver. 2.11, National Institute of Standards and Technology, U.S. Department of Commerce, Gaithersburg, MD, 1997, which has been licensed to PPI.

APPENDIX 18.E
Properties of Saturated Steam by Temperature
(SI units)

temp. (°C)	absolute pressure (bars)	specific volume (cm³/g)		internal energy (kJ/kg)		enthalpy (kJ/kg)			entropy (kJ/kg·K)		temp. (°C)
		sat. liquid, v_f	sat. vapor, v_g	sat. liquid, u_f	sat. vapor, u_g	sat. liquid, h_f	evap., h_{ff}	sat. vapor, h_g	sat. liquid, s_f	sat. vapor, s_g	
0.01	0.006117	1.0002	205991	0.00	2374.9	0.0006	2500.9	2500.9	0.0000	9.1555	0.01
4	0.00814	1.0001	157116	16.81	2380.4	16.81	2491.4	2508.2	0.0611	9.0505	4
5	0.00873	1.0001	147011	21.02	2381.8	21.02	2489.1	2510.1	0.0763	9.0248	5
6	0.00935	1.0001	137633	25.22	2383.2	25.22	2486.7	2511.9	0.0913	8.9993	6
8	0.01073	1.0002	120829	33.63	2385.9	33.63	2482.0	2515.6	0.1213	8.9491	8
10	0.01228	1.0003	106303	42.02	2388.6	42.02	2477.2	2519.2	0.1511	8.8998	10
11	0.01313	1.0004	99787	46.22	2390.0	46.22	2474.8	2521.0	0.1659	8.8754	11
12	0.01403	1.0005	93719	50.41	2391.4	50.41	2472.5	2522.9	0.1806	8.8513	12
13	0.01498	1.0007	88064	54.60	2392.8	54.60	2470.1	2524.7	0.1953	8.8274	13
14	0.01599	1.0008	82793	58.79	2394.1	58.79	2467.7	2526.5	0.2099	8.8037	14
15	0.01706	1.0009	77875	62.98	2395.5	62.98	2465.3	2528.3	0.2245	8.7803	15
16	0.01819	1.0011	73286	67.17	2396.9	67.17	2463.0	2530.2	0.2390	8.7570	16
17	0.01938	1.0013	69001	71.36	2398.2	71.36	2460.6	2532.0	0.2534	8.7339	17
18	0.02065	1.0014	64998	75.54	2399.6	75.54	2458.3	2533.8	0.2678	8.7111	18
19	0.02198	1.0016	61256	79.73	2401.0	79.73	2455.9	2535.6	0.2822	8.6884	19
20	0.02339	1.0018	57757	83.91	2402.3	83.91	2453.5	2537.4	0.2965	8.6660	20
21	0.02488	1.0021	54483	88.10	2403.7	88.10	2451.2	2539.3	0.3107	8.6437	21
22	0.02645	1.0023	51418	92.28	2405.0	92.28	2448.8	2541.1	0.3249	8.6217	22
23	0.02811	1.0025	48548	96.46	2406.4	96.47	2446.4	2542.9	0.3391	8.5998	23
24	0.02986	1.0028	45858	100.64	2407.8	100.65	2444.1	2544.7	0.3532	8.5781	24
25	0.03170	1.0030	43337	104.83	2409.1	104.83	2441.7	2546.5	0.3672	8.5566	25
26	0.03364	1.0033	40973	109.01	2410.5	109.01	2439.3	2548.3	0.3812	8.5353	26
27	0.03568	1.0035	38754	113.19	2411.8	113.19	2436.9	2550.1	0.3952	8.5142	27
28	0.03783	1.0038	36672	117.37	2413.2	117.37	2434.5	2551.9	0.4091	8.4933	28
29	0.04009	1.0041	34716	121.55	2414.6	121.55	2432.2	2553.7	0.4229	8.4725	29
30	0.04247	1.0044	32878	125.73	2415.9	125.73	2429.8	2555.5	0.4368	8.4520	30
31	0.04497	1.0047	31151	129.91	2417.3	129.91	2427.4	2557.3	0.4505	8.4316	31
32	0.04760	1.0050	29526	134.09	2418.6	134.09	2425.1	2559.2	0.4642	8.4113	32
33	0.05035	1.0054	27998	138.27	2420.0	138.27	2422.7	2561.0	0.4779	8.3913	33
34	0.05325	1.0057	26560	142.45	2421.3	142.45	2420.4	2562.8	0.4916	8.3714	34
35	0.05629	1.0060	25205	146.63	2422.7	146.63	2417.9	2564.5	0.5051	8.3517	35
36	0.05948	1.0064	23929	150.81	2424.0	150.81	2415.5	2566.3	0.5187	8.3321	36
38	0.06633	1.0071	21593	159.17	2426.7	159.17	2410.7	2569.9	0.5456	8.2935	38
40	0.07385	1.0079	19515	167.53	2429.4	167.53	2406.0	2573.5	0.5724	8.2555	40
45	0.09595	1.0099	15252	188.43	2436.1	188.43	2394.0	2582.4	0.6386	8.1633	45
50	0.1235	1.0121	12027	209.33	2442.7	209.34	2382.0	2591.3	0.7038	8.0748	50
55	0.1576	1.0146	9564	230.24	2449.3	230.26	2369.8	2600.1	0.7680	7.9898	55
60	0.1995	1.0171	7667	251.16	2455.9	251.18	2357.6	2608.8	0.8313	7.9081	60
65	0.2504	1.0199	6194	272.09	2462.4	272.12	2345.4	2617.5	0.8937	7.8296	65
70	0.3120	1.0228	5040	293.03	2468.9	293.07	2333.0	2626.1	0.9551	7.7540	70
75	0.3860	1.0258	4129	313.99	2475.2	314.03	2320.6	2634.6	1.0158	7.6812	75
80	0.4741	1.0291	3405	334.96	2481.6	335.01	2308.0	2643.0	1.0756	7.6111	80
85	0.5787	1.0324	2826	355.95	2487.8	356.01	2295.3	2651.3	1.1346	7.5434	85
90	0.7018	1.0360	2359	376.97	2494.0	377.04	2282.5	2659.5	1.1929	7.4781	90
95	0.8461	1.0396	1981	398.00	2500.0	398.09	2269.5	2667.6	1.2504	7.4151	95

APPENDIX 18.E *(continued)*
Properties of Saturated Steam by Temperature
(SI units)

temp. (°C)	absolute pressure (bars)	specific volume (cm³/g)		internal energy (kJ/kg)		enthalpy (kJ/kg)			entropy (kJ/kg·K)		temp. (°C)
		sat. liquid, v_f	sat. vapor, v_g	sat. liquid, u_f	sat. vapor, u_g	sat. liquid, h_f	evap., h_{ff}	sat. vapor, h_g	sat. liquid, s_f	sat. vapor, s_g	
100	1.014	1.0435	1672	419.06	2506.0	419.17	2256.4	2675.6	1.3072	7.3541	100
105	1.209	1.0474	1418	440.15	2511.9	440.27	2243.1	2683.4	1.3633	7.2952	105
110	1.434	1.0516	1209	461.26	2517.7	461.42	2229.6	2691.1	1.4188	7.2381	110
115	1.692	1.0559	1036	482.42	2523.3	482.59	2216.0	2698.6	1.4737	7.1828	115
120	1.987	1.0603	891.2	503.60	2528.9	503.81	2202.1	2705.9	1.5279	7.1291	120
125	2.322	1.0649	770.0	524.83	2534.3	525.07	2188.0	2713.1	1.5816	7.0770	125
130	2.703	1.0697	668.0	546.10	2539.5	546.38	2173.7	2720.1	1.6346	7.0264	130
135	3.132	1.0746	581.7	567.41	2544.7	567.74	2159.1	2726.9	1.6872	6.9772	135
140	3.615	1.0798	508.5	588.77	2549.6	589.16	2144.3	2733.4	1.7392	6.9293	140
145	4.157	1.0850	446.0	610.19	2554.4	610.64	2129.2	2739.8	1.7907	6.8826	145
150	4.762	1.0905	392.5	631.66	2559.1	632.18	2113.8	2745.9	1.8418	6.8371	150
155	5.435	1.0962	346.5	653.19	2563.5	653.79	2098.0	2751.8	1.8924	6.7926	155
160	6.182	1.1020	306.8	674.79	2567.8	675.47	2082.0	2757.4	1.9426	6.7491	160
165	7.009	1.1080	272.4	696.46	2571.9	697.24	2065.6	2762.8	1.9923	6.7066	165
170	7.922	1.1143	242.6	718.20	2575.7	719.08	2048.8	2767.9	2.0417	6.6650	170
175	8.926	1.1207	216.6	740.02	2579.4	741.02	2031.7	2772.7	2.0906	6.6241	175
180	10.03	1.1274	193.8	761.92	2582.8	763.05	2014.2	2777.2	2.1392	6.5840	180
185	11.23	1.1343	173.9	783.91	2586.0	785.19	1996.2	2781.4	2.1875	6.5447	185
190	12.55	1.1415	156.4	806.00	2589.0	807.43	1977.9	2785.3	2.2355	6.5059	190
195	13.99	1.1489	140.9	828.18	2591.7	829.79	1959.0	2788.8	2.2832	6.4678	195
200	15.55	1.1565	127.2	850.47	2594.2	852.27	1939.7	2792.0	2.3305	6.4302	200
205	17.24	1.1645	115.1	872.87	2596.4	874.88	1920.0	2794.8	2.3777	6.3930	205
210	19.08	1.1727	104.3	895.39	2598.3	897.63	1899.6	2797.3	2.4245	6.3563	210
215	21.06	1.1813	94.68	918.04	2599.9	920.53	1878.8	2799.3	2.4712	6.3200	215
220	23.20	1.1902	86.09	940.82	2601.3	943.58	1857.4	2801.0	2.5177	6.2840	220
225	25.50	1.1994	78.40	963.74	2602.2	966.80	1835.4	2802.2	2.5640	6.2483	225
230	27.97	1.2090	71.50	986.81	2602.9	990.19	1812.7	2802.9	2.6101	6.2128	230
235	30.63	1.2190	65.30	1010.0	2603.2	1013.8	1789.4	2803.2	2.6561	6.1775	235
240	33.47	1.2295	59.71	1033.4	2603.1	1037.6	1765.4	2803.0	2.7020	6.1423	240
245	36.51	1.2403	54.65	1057.0	2602.7	1061.6	1740.7	2802.2	2.7478	6.1072	245
250	39.76	1.2517	50.08	1080.8	2601.8	1085.8	1715.2	2800.9	2.7935	6.0721	250
255	43.23	1.2636	45.94	1104.8	2600.5	1110.2	1688.8	2799.1	2.8392	6.0369	255
260	46.92	1.2761	42.17	1129.0	2598.7	1135.0	1661.6	2796.6	2.8849	6.0016	260
265	50.85	1.2892	38.75	1153.4	2596.5	1160.0	1633.5	2793.5	2.9307	5.9661	265
270	55.03	1.3030	35.62	1178.1	2593.7	1185.3	1604.4	2789.7	2.9765	5.9304	270
275	59.46	1.3175	32.77	1203.1	2590.3	1210.9	1574.3	2785.2	3.0224	5.8944	275
280	64.17	1.3328	30.15	1228.3	2586.4	1236.9	1543.0	2779.9	3.0685	5.8579	280
285	69.15	1.3491	27.76	1253.9	2581.8	1263.3	1510.5	2773.7	3.1147	5.8209	285
290	74.42	1.3663	25.56	1279.9	2576.5	1290.0	1476.7	2766.7	3.1612	5.7834	290
295	79.99	1.3846	23.53	1306.2	2570.5	1317.3	1441.4	2758.7	3.2080	5.7451	295
300	85.88	1.4042	21.66	1332.9	2563.6	1345.0	1404.6	2749.6	3.2552	5.7059	300
305	92.09	1.4252	19.93	1360.2	2555.9	1373.3	1366.1	2739.4	3.3028	5.6657	305
310	98.65	1.4479	18.34	1387.9	2547.1	1402.2	1325.7	2728.0	3.3510	5.6244	310
315	105.6	1.4724	16.85	1416.3	2537.2	1431.8	1283.2	2715.1	3.3998	5.5816	315
320	112.8	1.4990	15.47	1445.3	2526.0	1462.2	1238.4	2700.6	3.4494	5.5372	320

APPENDIX 18.E *(continued)*
Properties of Saturated Steam by Temperature
(SI units)

temp. (°C)	absolute pressure (bars)	specific volume (cm³/g)		internal energy (kJ/kg)		enthalpy (kJ/kg)			entropy (kJ/kg·K)		temp. (°C)
		sat. liquid, v_f	sat. vapor, v_g	sat. liquid, u_f	sat. vapor, u_g	sat. liquid, h_f	evap., h_{ff}	sat. vapor, h_g	sat. liquid, s_f	sat. vapor, s_g	
325	120.5	1.5283	14.18	1475.1	2513.4	1493.5	1190.8	2684.3	3.5000	5.4908	325
330	128.6	1.5606	12.98	1505.8	2499.2	1525.9	1140.2	2666.0	3.5518	5.4422	330
335	137.1	1.5967	11.85	1537.6	2483.0	1559.5	1085.9	2645.4	3.6050	5.3906	335
340	146.0	1.6376	10.78	1570.6	2464.4	1594.5	1027.3	2621.9	3.6601	5.3356	340
345	155.4	1.6846	9.769	1605.3	2443.1	1631.5	963.4	2594.9	3.7176	5.2762	345
350	165.3	1.7400	8.802	1642.1	2418.1	1670.9	892.8	2563.6	3.7784	5.2110	350
355	175.7	1.8079	7.868	1682.0	2388.4	1713.7	812.9	2526.7	3.8439	5.1380	355
360	186.7	1.8954	6.949	1726.3	2351.8	1761.7	719.8	2481.5	3.9167	5.0536	360
365	198.2	2.0172	6.012	1777.8	2303.8	1817.8	605.2	2423.0	4.0014	4.9497	365
370	210.4	2.2152	4.954	1844.1	2230.3	1890.7	443.8	2334.5	4.1112	4.8012	370
374	220.64000	3.1056	3.1056	2015.70	2015.70	2084.30	0	2084.3	4.4070	4.4070	373.9

(Multiply MPa by 10 to obtain bars.)

Values in this table were calculated from *NIST Standard Reference Database 10*, "NIST/ASME Steam Properties," Ver. 2.11, National Institute of Standards and Technology, U.S. Department of Commerce, Gaithersburg, MD, 1997, which has been licensed to PPI.

APPENDIX 18.F
Properties of Saturated Steam by Pressure
(SI units)

absolute pressure (bars)	temp. (°C)	specific volume (cm³/g)		internal energy (kJ/kg)		enthalpy (kJ/kg)			entropy (kJ/kg·K)		absolute pressure (bars)
		sat. liquid, v_f	sat. vapor, v_g	sat. liquid, u_f	sat. vapor, u_g	sat. liquid, h_f	evap., h_{fg}	sat. vapor, h_g	sat. liquid, s_f	sat. vapor, s_g	
0.04	28.96	1.0041	34791	121.38	2414.5	121.39	2432.3	2553.7	0.4224	8.4734	0.04
0.06	36.16	1.0065	23733	151.47	2424.2	151.48	2415.2	2566.6	0.5208	8.3290	0.06
0.08	41.51	1.0085	18099	173.83	2431.4	173.84	2402.4	2576.2	0.5925	8.2273	0.08
0.10	45.81	1.0103	14670	191.80	2437.2	191.81	2392.1	2583.9	0.6492	8.1488	0.10
0.20	60.06	1.0172	7648	251.40	2456.0	251.42	2357.5	2608.9	0.8320	7.9072	0.20
0.30	69.10	1.0222	5228	289.24	2467.7	289.27	2335.3	2624.5	0.9441	7.7675	0.30
0.40	75.86	1.0264	3993	317.58	2476.3	317.62	2318.4	2636.1	1.0261	7.6690	0.40
0.50	81.32	1.0299	3240	340.49	2483.2	340.54	2304.7	2645.2	1.0912	7.5930	0.50
0.60	85.93	1.0331	2732	359.84	2489.0	359.91	2292.9	2652.9	1.1454	7.5311	0.60
0.70	89.93	1.0359	2365	376.68	2493.9	376.75	2282.7	2659.4	1.1921	7.4790	0.70
0.80	93.49	1.0385	2087	391.63	2498.2	391.71	2273.5	2665.2	1.2330	7.4339	0.80
0.90	96.69	1.0409	1869	405.10	2502.1	405.20	2265.1	2670.3	1.2696	7.3943	0.90
1.00	99.61	1.0432	1694	417.40	2505.6	417.50	2257.4	2674.9	1.3028	7.3588	1.00
1.01325	99.97	1.0434	1673	418.95	2506.0	419.06	2256.5	2675.5	1.3069	7.3544	1.01325
1.50	111.3	1.0527	1159	466.97	2519.2	467.13	2226.0	2693.1	1.4337	7.2230	1.50
2.00	120.2	1.0605	885.7	504.49	2529.1	504.70	2201.5	2706.2	1.5302	7.1269	2.00
2.50	127.4	1.0672	718.7	535.08	2536.8	535.35	2181.1	2716.5	1.6072	7.0524	2.50
3.00	133.5	1.0732	605.8	561.10	2543.2	561.43	2163.5	2724.9	1.6717	6.9916	3.00
3.50	138.9	1.0786	524.2	583.88	2548.5	584.26	2147.7	2732.0	1.7274	6.9401	3.50
4.00	143.6	1.0836	462.4	604.22	2553.1	604.65	2133.4	2738.1	1.7765	6.8955	4.00
4.50	147.9	1.0882	413.9	622.65	2557.1	623.14	2120.2	2743.4	1.8205	6.8560	4.50
5.00	151.8	1.0926	374.8	639.54	2560.7	640.09	2108.0	2748.1	1.8604	6.8207	5.00
6.00	158.8	1.1006	315.6	669.72	2566.8	670.38	2085.8	2756.1	1.9308	6.7592	6.00
7.00	164.9	1.1080	272.8	696.23	2571.8	697.00	2065.8	2762.8	1.9918	6.7071	7.00
8.00	170.4	1.1148	240.3	719.97	2576.0	720.86	2047.4	2768.3	2.0457	6.6616	8.00

APPENDIX 18.F *(continued)*
Properties of Saturated Steam by Pressure
(SI units)

Appendices

absolute pressure (bars)	temp. (°C)	specific volume (cm³/g)		internal energy (kJ/kg)		enthalpy (kJ/kg)			entropy (kJ/kg·K)		absolute pressure (bars)
		sat. liquid, v_f	sat. vapor, v_g	sat. liquid, u_f	sat. vapor, u_g	sat. liquid, h_f	evap., h_{fg}	sat. vapor, h_g	sat. liquid, s_f	sat. vapor, s_g	
9.00	175.4	1.1212	214.9	741.55	2579.6	742.56	2030.5	2773.0	2.0940	6.6213	9.00
10.0	179.9	1.1272	194.4	761.39	2582.7	762.52	2014.6	2777.1	2.1381	6.5850	10.0
15.0	198.3	1.1539	131.7	842.83	2593.4	844.56	1946.4	2791.0	2.3143	6.4430	15.0
20.0	212.4	1.1767	99.59	906.14	2599.1	908.50	1889.8	2798.3	2.4468	6.3390	20.0
25.0	224.0	1.1974	79.95	958.91	2602.1	961.91	1840.0	2801.9	2.5543	6.2558	25.0
30.0	233.9	1.2167	66.66	1004.7	2603.2	1008.3	1794.9	2803.2	2.6455	6.1856	30.0
35.0	242.6	1.2350	57.06	1045.5	2602.9	1049.8	1752.8	2802.6	2.7254	6.1243	35.0
40.0	250.4	1.2526	49.78	1082.5	2601.7	1087.5	1713.3	2800.8	2.7968	6.0696	40.0
45.0	257.4	1.2696	44.06	1116.5	2599.7	1122.3	1675.7	2797.9	2.8615	6.0197	45.0
50.0	263.9	1.2864	39.45	1148.2	2597.0	1154.6	1639.6	2794.2	2.9210	5.9737	50.0
55.0	270.0	1.3029	35.64	1177.9	2593.7	1185.1	1604.6	2789.7	2.9762	5.9307	55.0
60.0	275.6	1.3193	32.45	1206.0	2589.9	1213.9	1570.7	2784.6	3.0278	5.8901	60.0
65.0	280.9	1.3356	29.73	1232.7	2585.7	1241.4	1537.5	2778.9	3.0764	5.8516	65.0
70.0	285.8	1.3519	27.38	1258.2	2581.0	1267.7	1504.9	2772.6	3.1224	5.8148	70.0
75.0	290.5	1.3682	25.33	1282.7	2575.9	1292.9	1473.0	2765.9	3.1662	5.7793	75.0
80.0	295.0	1.3847	23.53	1306.2	2570.5	1317.3	1441.4	2758.7	3.2081	5.7450	80.0
85.0	299.3	1.4013	21.92	1329.0	2564.7	1340.9	1410.1	2751.0	3.2483	5.7117	85.0
90.0	303.3	1.4181	20.49	1351.1	2558.5	1363.9	1379.0	2742.9	3.2870	5.6791	90.0
95.0	307.2	1.4352	19.20	1372.6	2552.0	1386.2	1348.2	2734.4	3.3244	5.6473	95.0
100	311.0	1.4526	18.03	1393.5	2545.2	1408.1	1317.4	2725.5	3.3606	5.6160	100
100	311.0	1.4526	18.03	1393.5	2545.2	1408.1	1317.4	2725.5	3.3606	5.6160	100
110	318.1	1.4885	15.99	1434.1	2530.5	1450.4	1255.9	2706.3	3.4303	5.5545	110
120	324.7	1.5263	14.26	1473.1	2514.3	1491.5	1193.9	2685.4	3.4967	5.4939	120
130	330.9	1.5665	12.78	1511.1	2496.5	1531.5	1131.2	2662.7	3.5608	5.4336	130

APPENDIX 18.F *(continued)*
Properties of Saturated Steam by Pressure
(SI units)

absolute pressure (bars)	temp. (°C)	specific volume (cm³/g)		internal energy (kJ/kg)		enthalpy (kJ/kg)			entropy (kJ/kg·K)		absolute pressure (bars)
		sat. liquid, v_f	sat. vapor, v_g	sat. liquid, u_f	sat. vapor, u_g	sat. liquid, h_f	evap., h_{fg}	sat. vapor, h_g	sat. liquid, s_f	sat. vapor, s_g	
140	336.7	1.6097	11.49	1548.4	2477.1	1571.0	1066.9	2637.9	3.6232	5.3727	140
150	342.2	1.6570	10.34	1585.3	2455.6	1610.2	1000.5	2610.7	3.6846	5.3106	150
160	347.4	1.7094	9.309	1622.3	2431.8	1649.7	931.1	2580.8	3.7457	5.2463	160
170	352.3	1.7693	8.371	1659.9	2405.2	1690.0	857.5	2547.5	3.8077	5.1787	170
180	357.0	1.8398	7.502	1699.0	2374.8	1732.1	777.7	2509.8	3.8718	5.1061	180
190	361.5	1.9268	6.677	1740.5	2339.1	1777.2	688.8	2466.0	3.9401	5.0256	190
200	365.7	2.0400	5.865	1786.4	2295.0	1827.2	585.1	2412.3	4.0156	4.9314	200
210	369.8	2.2055	4.996	1841.2	2233.7	1887.6	451.0	2338.6	4.1064	4.8079	210
220.64	373.95	3.1056	3.1056	2015.7	2015.7	2084.3	0	2084.3	4.4070	4.4070	220.64

(Multiply MPa by 10 to obtain bars.)

Values in this table were calculated from *NIST Standard Reference Database 10*, "NIST/ASME Steam Properties," Ver. 2.11, National Institute of Standards and Technology, U.S. Department of Commerce, Gaithersburg, MD, 1997, which has been licensed to PPI.

APPENDIX 18.G
Properties of Superheated Steam (SI units)
specific volume, v, in m³/kg; enthalpy, h, in kJ/kg; entropy, s, in kJ/kg·K

absolute pressure (kPa)		temperature (°C)							
(sat. temp. °C)		100	150	200	250	300	360	420	500
10	v	17.196	19.513	21.826	24.136	26.446	29.216	31.986	35.680
(45.81)	h	2687.5	2783.0	2879.6	2977.4	3076.7	3197.9	3321.4	3489.7
	s	8.4489	8.6892	8.9049	9.1015	9.2827	9.4837	9.6700	9.8998
50	v	3.419	3.890	4.356	4.821	5.284	5.839	6.394	7.134
(81.32)	h	2682.4	2780.2	2877.8	2976.1	3075.8	3197.2	3320.8	3489.3
	s	7.6953	7.9413	8.1592	8.3568	8.5386	8.7401	8.9266	9.1566
75	v	2.270	2.588	2.900	3.211	3.521	3.891	4.262	4.755
(91.76)	h	2679.2	2778.4	2876.6	2975.3	3075.1	3196.7	3320.4	3489.0
	s	7.5011	7.7509	7.9702	8.1685	8.3507	8.5524	8.7391	8.9692
100	v	1.6959	1.9367	2.172	2.406	2.639	2.917	3.195	3.566
(99.61)	h	2675.8	2776.6	2875.5	2974.5	3074.5	3196.3	3320.1	3488.7
	s	7.3610	7.6148	7.8356	8.0346	8.2172	8.4191	8.6059	8.8361
150	v	......	1.2855	1.4445	1.6013	1.7571	1.9433	2.129	2.376
(111.35)	h	......	2772.9	2873.1	2972.9	3073.3	3195.3	3319.4	3488.2
	s	......	7.4208	7.6447	7.8451	8.0284	8.2309	8.4180	8.6485
400	v	......	0.4709	0.5343	0.5952	0.6549	0.7257	0.7961	0.8894
(143.61)	h	......	2752.8	2860.9	2964.5	3067.1	3190.7	3315.8	3485.5
	s	......	6.9306	7.1723	7.3804	7.5677	7.7728	7.9615	8.1933
700	v	......	......	0.3000	0.3364	0.3714	0.4126	0.4533	0.5070
(164.95)	h	......	......	2845.3	2954.0	3059.4	3185.1	3311.5	3482.3
	s	......	......	6.8884	7.1070	7.2995	7.5080	7.6986	7.9319
1000	v	......	......	0.2060	0.2328	0.2580	0.2874	0.3162	0.3541
(179.88)	h	......	......	2828.3	2943.1	3051.6	3179.4	3307.1	3479.1
	s	......	......	6.6955	6.9265	7.1246	7.3367	7.5294	7.7641
1500	v	......	......	0.13245	0.15201	0.16971	0.18990	0.2095	0.2352
(198.29)	h	......	......	2796.0	2923.9	3038.2	3169.8	3299.8	3473.7
	s	......	......	6.4536	6.7111	6.9198	7.1382	7.3343	7.5718
2000	v	......	......	......	0.11150	0.12551	0.14115	0.15617	0.17568
(212.38)	h	......	......	......	2903.2	3024.2	3159.9	3292.3	3468.2
	s	......	......	......	6.5475	6.7684	6.9937	7.1935	7.4337
2500	v	......	......	......	0.08705	0.09894	0.11188	0.12416	0.13999
(223.95)	h	......	......	......	2880.9	3009.6	3149.8	3284.8	3462.7
	s	......	......	......	6.4107	6.6459	6.8788	7.0824	7.3254
3000	v	......	......	......	0.07063	0.08118	0.09236	0.10281	0.11620
(233.85)	h	......	......	......	2856.5	2994.3	3139.5	3277.1	3457.2
	s	......	......	......	6.2893	6.5412	6.7823	6.9900	7.2359

(Multiply kPa by 0.01 to obtain bars.)

Values in this table were calculated from *NIST Standard Reference Database 10*, "NIST/ASME Steam Properties," Ver. 2.11, National Institute of Standards and Technology, U.S. Department of Commerce, Gaithersburg, MD, 1997, which has been licensed to PPI.

APPENDIX 19.A
Manning's Roughness Coefficient[a,b]
(design use)

channel material	n
plastic (PVC and ABS)	0.009
clean, uncoated cast iron	0.013–0.015
clean, coated cast iron	0.012–0.014
dirty, tuberculated cast iron	0.015–0.035
riveted steel	0.015–0.017
lock-bar and welded steel pipe	0.012–0.013
galvanized iron	0.015–0.017
brass and glass	0.009–0.013
wood stave	
small diameter	0.011–0.012
large diameter	0.012–0.013
concrete	
average value used	0.013
typical commercial, ball and spigot	
rubber gasketed end connections	
– full (pressurized and wet)	0.010
– partially full	0.0085
with rough joints	0.016–0.017
dry mix, rough forms	0.015–0.016
wet mix, steel forms	0.012–0.014
very smooth, finished	0.011–0.012
vitrified sewer	0.013–0.015
common-clay drainage tile	0.012–0.014
asbestos	0.011
planed timber (flume)	0.012 (0.010–0.014)
canvas	0.012
unplaned timber (flume)	0.013 (0.011–0.015)
brick	0.016
rubble masonry	0.017
smooth earth	0.018
firm gravel	0.023
corrugated metal pipe (CMP)	0.022–0.033
natural channels, good condition	0.025
rip rap	0.035
natural channels with stones and weeds	0.035
very poor natural channels	0.060

[a]compiled from various sources

[b]Values outside these ranges have been observed, but these values are typical.

APPENDIX 19.B
Manning Equation Nomograph

$$\left(\text{solves } v = \frac{1.486}{n} R^{2/3}\sqrt{S}\right)$$

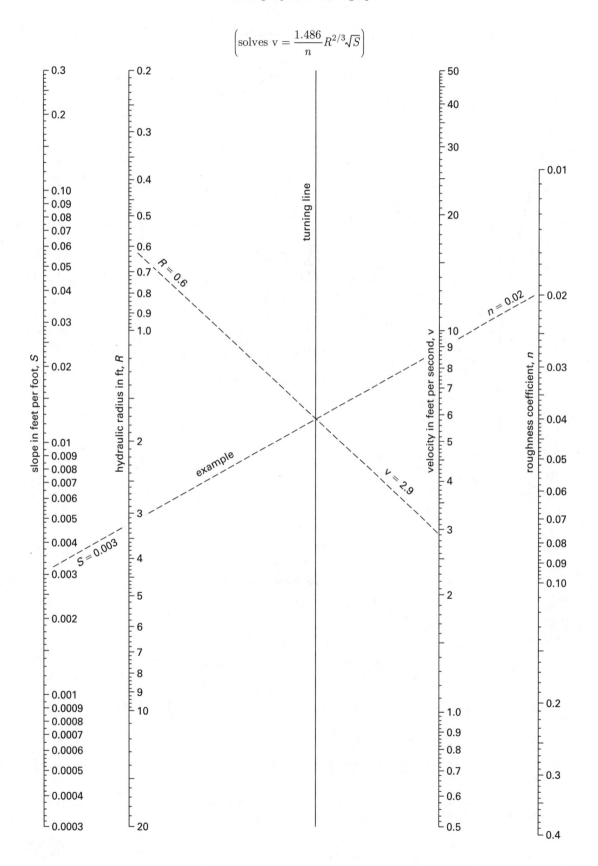

Appendices

APPENDIX 19.C
Circular Channel Ratios*

Experiments have shown that n varies slightly with depth. This figure gives velocity and flow rate ratios for varying n (solid line) and constant n (broken line) assumptions.

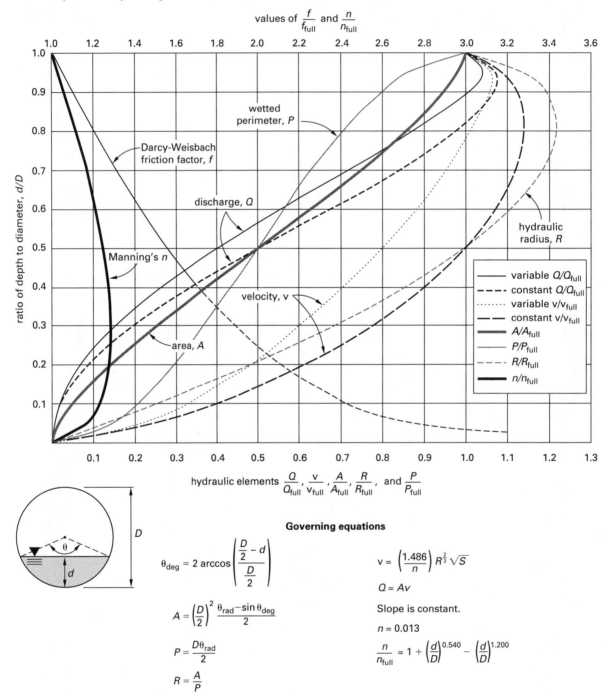

Governing equations

$$\theta_{deg} = 2\arccos\left(\frac{\frac{D}{2}-d}{\frac{D}{2}}\right)$$

$$A = \left(\frac{D}{2}\right)^2 \frac{\theta_{rad}-\sin\theta_{deg}}{2}$$

$$P = \frac{D\theta_{rad}}{2}$$

$$R = \frac{A}{P}$$

$$v = \left(\frac{1.486}{n}\right)R^{\frac{2}{3}}\sqrt{S}$$

$$Q = Av$$

Slope is constant.

$n = 0.013$

$$\frac{n}{n_{full}} = 1 + \left(\frac{d}{D}\right)^{0.540} - \left(\frac{d}{D}\right)^{1.200}$$

*For $n = 0.013$.

Adapted from *Design and Construction of Sanitary and Storm Sewers*, p. 87, ASCE, 1969, as originally presented in "Design of Sewers to Facilitate Flow", Camp, T. R., *Sewage Works Journal*, 18, 3 (1946).

APPENDIX 19.D
Critical Depths in Circular Channels

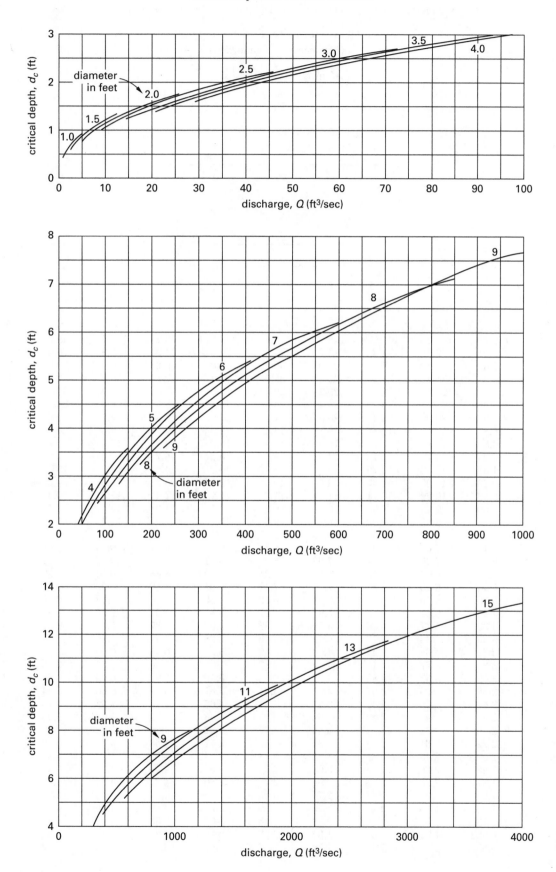

APPENDIX 19.E
Conveyance Factor, K
Symmetrical Rectangular,[a] Trapezoidal, and V-Notch[b] Open Channels

(use for determining Q or b when d is known)
(customary U.S. units[c,d])

$$K \text{ in } Q = K\left(\frac{1}{n}\right)d^{8/3}\sqrt{S}$$

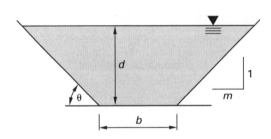

	m and θ									
	0.0	0.25	0.5	0.75	1.0	1.5	2.0	2.5	3.0	4.0
$x = d/b$	90°	76.0°	63.4°	53.1°	45.0°	33.7°	26.6°	21.8°	18.4°	14.0°
0.01	146.7	147.2	147.6	148.0	148.3	148.8	149.2	149.5	149.9	150.5
0.02	72.4	72.9	73.4	73.7	74.0	74.5	74.9	75.3	75.6	76.3
0.03	47.6	48.2	48.6	49.0	49.8	49.8	50.2	50.6	50.9	51.6
0.04	35.3	35.8	36.3	36.6	36.9	37.4	37.8	38.2	38.6	39.3
0.05	27.9	28.4	28.9	29.2	29.5	30.0	30.5	30.9	31.2	32.0
0.06	23.0	23.5	23.9	24.3	24.6	25.1	25.5	26.0	26.3	27.1
0.07	19.5	20.0	20.4	20.8	21.1	21.6	22.0	22.4	22.8	23.6
0.08	16.8	17.3	17.8	18.1	18.4	18.9	19.4	19.8	20.2	21.0
0.09	14.8	15.3	15.7	16.1	16.4	16.9	17.4	17.8	18.2	19.0
0.10	13.2	13.7	14.1	14.4	14.8	15.3	15.7	16.2	16.6	17.4
0.11	11.83	12.33	12.76	13.11	13.42	13.9	14.4	14.9	15.3	16.1
0.12	10.73	11.23	11.65	12.00	12.31	12.8	13.3	13.8	14.2	15.0
0.13	9.80	10.29	10.71	11.06	11.37	11.9	12.4	12.8	13.3	14.1
0.14	9.00	9.49	9.91	10.26	10.57	11.1	11.6	12.0	12.5	13.4
0.15	8.32	8.80	9.22	9.67	9.88	10.4	10.9	11.4	11.8	12.7
0.16	7.72	8.20	8.61	8.96	9.27	9.81	10.29	10.75	11.2	12.1
0.17	7.19	7.67	8.08	8.43	8.74	9.28	9.77	10.23	10.68	11.6
0.18	6.73	7.20	7.61	7.96	8.27	8.81	9.30	9.76	10.21	11.1
0.19	6.31	6.78	7.19	7.54	7.85	8.39	8.88	9.34	9.80	10.7
0.20	5.94	6.40	6.81	7.16	7.47	8.01	8.50	8.97	9.43	10.3
0.22	5.30	5.76	6.16	6.51	6.82	7.36	7.86	8.33	8.79	9.70
0.24	4.77	5.22	5.62	5.96	6.27	6.82	7.32	7.79	8.26	9.18
0.26	4.32	4.77	5.16	5.51	5.82	6.37	6.87	7.35	7.81	8.74
0.28	3.95	4.38	4.77	5.12	5.48	5.98	6.48	6.96	7.43	8.36
0.30	3.62	4.05	4.44	4.78	5.09	5.64	6.15	6.63	7.10	8.04
0.32	3.34	3.77	4.15	4.49	4.80	5.35	5.86	6.34	6.82	7.75
0.34	3.09	3.51	3.89	4.23	4.54	5.10	5.60	6.09	6.56	7.50
0.36	2.88	3.29	3.67	4.01	4.31	4.87	5.38	5.86	6.34	7.28
0.38	2.68	3.09	3.47	3.81	4.11	4.67	5.17	5.66	6.14	7.09
0.40	2.51	2.92	3.29	3.62	3.93	4.48	4.99	5.48	5.96	6.91
0.42	2.36	2.76	3.13	3.46	3.77	4.32	4.83	5.32	5.80	6.75
0.44	2.22	2.61	2.98	3.31	3.62	4.17	4.68	5.17	5.66	6.60
0.46	2.09	2.48	2.85	3.18	3.48	4.04	4.55	5.04	5.52	6.47
0.48	1.98	2.36	2.72	3.06	3.36	3.91	4.43	4.92	5.40	6.35
0.50	1.87	2.26	2.61	2.94	3.25	3.80	4.31	4.81	5.29	6.24
0.55	1.65	2.02	2.37	2.70	3.00	3.55	4.07	4.56	5.05	6.00
0.60	1.46	1.83	2.17	2.50	2.80	3.35	3.86	4.36	4.84	5.80
0.70	1.18	1.53	1.87	2.19	2.48	3.03	3.55	4.04	4.53	5.49
0.80	0.982	1.31	1.64	1.95	2.25	2.80	3.31	3.81	4.30	5.26
0.90	0.831	1.15	1.47	1.78	2.07	2.62	3.13	3.63	4.12	5.08

APPENDIX 19.E *(continued)*
Conveyance Factor, K
Symmetrical Rectangular,[a] Trapezoidal, and V-Notch[b] Open Channels

(use for determining Q or b when d is known)
(customary U.S. units[c,d])

$$K \text{ in } Q = K\left(\frac{1}{n}\right)d^{8/3}\sqrt{S}$$

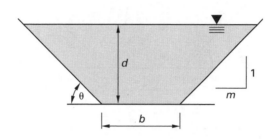

	m and θ									
	0.0	0.25	0.5	0.75	1.0	1.5	2.0	2.5	3.0	4.0
$x = d/b$	90°	76.0°	63.4°	53.1°	45.0°	33.7°	26.6°	21.8°	18.4°	14.0°
1.00	0.714	1.02	1.33	1.64	1.93	2.47	2.99	3.48	3.97	4.93
1.20	0.548	0.836	1.14	1.43	1.72	2.26	2.77	3.27	3.76	4.72
1.40	0.436	0.708	0.998	1.29	1.57	2.11	2.62	3.12	3.60	4.57
1.60	0.357	0.616	0.897	1.18	1.46	2.00	2.51	3.00	3.49	4.45
1.80	0.298	0.546	0.820	1.10	1.38	1.91	2.42	2.91	3.40	4.36
2.00	0.254	0.491	0.760	1.04	1.31	1.84	2.35	2.84	3.33	4.29
2.25	0.212	0.439	0.700	0.973	1.24	1.77	2.28	2.77	3.26	4.22
∞	0.00	0.091	0.274	0.499	0.743	1.24	1.74	2.23	2.71	3.67

[a]For rectangular channels, use the 0.0 (90°, vertical sides) column.

[b]For V-notch triangular channels, use the $d/b = \infty$ row.

[c]Q = flow rate, ft³/sec; d = depth of flow, ft; b = bottom width of channel, ft; S = geometric slope, ft/ft; n = Manning's roughness constant.

[d]For SI units (i.e., Q in m³/s and d and b in m), divide each table value by 1.486.

APPENDIX 19.F
Conveyance Factor, K'
Symmetrical Rectangular,[a] Trapezoidal Open Channels

(use for determining Q or d when b is known)

(customary U.S. units[b,c])

$$K' \text{ in } Q = K'\left(\frac{1}{n}\right)b^{8/3}\sqrt{S}$$

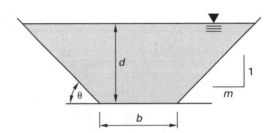

$x = d/b$	m and θ									
	0.0	0.25	0.5	0.75	1.0	1.5	2.0	2.5	3.0	4.0
	90°	76.0°	63.4°	53.1°	45.0°	33.7°	26.6°	21.8°	18.4°	14.0°
0.01	0.00068	0.00068	0.00069	0.00069	0.00069	0.00069	0.00069	0.00069	0.00070	0.00070
0.02	0.00213	0.00215	0.00216	0.00217	0.00218	0.00220	0.00221	0.00222	0.00223	0.00225
0.03	0.00414	0.00419	0.00423	0.00426	0.00428	0.00433	0.00436	0.00439	0.00443	0.00449
0.04	0.00660	0.00670	0.00679	0.00685	0.00691	0.00700	0.00708	0.00716	0.00723	0.00736
0.05	0.00946	0.00964	0.00979	0.00991	0.01002	0.01019	0.01033	0.01047	0.01060	0.01086
0.06	0.0127	0.0130	0.0132	0.0134	0.0136	0.0138	0.0141	0.0148	0.0145	0.0150
0.07	0.0162	0.0166	0.0170	0.0173	0.0175	0.0180	0.0183	0.0187	0.0190	0.0197
0.08	0.0200	0.0206	0.0211	0.0215	0.0219	0.0225	0.0231	0.0236	0.0240	0.0250
0.09	0.0241	0.0249	0.0256	0.0262	0.0267	0.0275	0.0282	0.0289	0.0296	0.0310
0.10	0.0284	0.0294	0.0304	0.0311	0.0318	0.0329	0.0339	0.0348	0.0358	0.0376
0.11	0.0329	0.0343	0.0354	0.0364	0.0373	0.0387	0.0400	0.0413	0.0424	0.0448
0.12	0.0376	0.0393	0.0408	0.0420	0.0431	0.0450	0.0466	0.0482	0.0497	0.0527
0.13	0.0425	0.0446	0.0464	0.0480	0.0493	0.0516	0.0537	0.0556	0.0575	0.0613
0.14	0.0476	0.0502	0.0524	0.0542	0.0559	0.0587	0.0612	0.0636	0.0659	0.0706
0.15	0.0528	0.0559	0.0585	0.0608	0.0627	0.0662	0.0692	0.0721	0.0749	0.0805
0.16	0.0582	0.0619	0.0650	0.0676	0.0700	0.0740	0.0777	0.0811	0.0845	0.0912
0.17	0.0638	0.0680	0.0716	0.0748	0.0775	0.0823	0.0866	0.0907	0.0947	0.1026
0.18	0.0695	0.0744	0.0786	0.0822	0.0854	0.0910	0.0960	0.1008	0.1055	0.1148
0.19	0.0753	0.0809	0.0857	0.0899	0.0936	0.1001	0.1059	0.1115	0.1169	0.1277
0.20	0.0812	0.0876	0.0931	0.0979	0.1021	0.1096	0.1163	0.1227	0.1290	0.1414
0.22	0.0934	0.1015	0.109	0.115	0.120	0.130	0.139	0.147	0.155	0.171
0.24	0.1061	0.1161	0.125	0.133	0.140	0.152	0.163	0.173	0.184	0.204
0.26	0.119	0.131	0.142	0.152	0.160	0.175	0.189	0.202	0.215	0.241
0.28	0.132	0.147	0.160	0.172	0.182	0.201	0.217	0.234	0.249	0.281
0.30	0.146	0.163	0.179	0.193	0.205	0.228	0.248	0.267	0.287	0.324
0.32	0.160	0.180	0.199	0.215	0.230	0.256	0.281	0.304	0.327	0.371
0.34	0.174	0.198	0.219	0.238	0.256	0.287	0.316	0.343	0.370	0.423
0.36	0.189	0.216	0.241	0.263	0.283	0.319	0.353	0.385	0.416	0.478
0.38	0.203	0.234	0.263	0.288	0.312	0.353	0.392	0.429	0.465	0.537
0.40	0.218	0.253	0.286	0.315	0.341	0.389	0.434	0.476	0.518	0.600
0.42	0.233	0.273	0.309	0.342	0.373	0.427	0.478	0.526	0.574	0.668
0.44	0.248	0.293	0.334	0.371	0.405	0.467	0.525	0.580	0.633	0.740
0.46	0.264	0.313	0.359	0.401	0.439	0.509	0.574	0.636	0.696	0.816
0.48	0.279	0.334	0.385	0.432	0.474	0.553	0.625	0.695	0.763	0.897
0.50	0.295	0.355	0.412	0.463	0.511	0.598	0.679	0.757	0.833	0.983
0.55	0.335	0.410	0.482	0.548	0.609	0.722	0.826	0.926	1.025	1.22

APPENDIX 19.F *(continued)*
Conveyance Factor, K'
Symmetrical Rectangular,[a] Trapezoidal Open Channels

(use for determining Q or d when b is known)
(customary U.S. units[b,c])

$$K' \text{ in } Q = K'\left(\frac{1}{n}\right)b^{8/3}\sqrt{S}$$

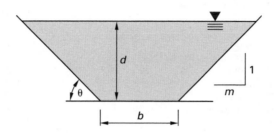

$x = d/b$	m and θ									
	0.0	0.25	0.5	0.75	1.0	1.5	2.0	2.5	3.0	4.0
	90°	76.0°	63.4°	53.1°	45.0°	33.7°	26.6°	21.8°	18.4°	14.0°
0.60	0.375	0.468	0.557	0.640	0.717	0.858	0.990	1.117	1.24	1.49
0.70	0.457	0.592	0.722	0.844	0.959	1.17	1.37	1.56	1.75	2.12
0.80	0.542	0.725	0.906	1.078	1.24	1.54	1.83	2.10	2.37	2.90
0.90	0.628	0.869	1.11	1.34	1.56	1.98	2.36	2.74	3.11	3.83
1.00	0.714	1.022	1.33	1.64	1.93	2.47	2.99	3.48	3.97	4.93
1.20	0.891	1.36	1.85	2.33	2.79	3.67	4.51	5.32	6.11	7.67
1.40	1.07	1.74	2.45	3.16	3.85	5.17	6.42	7.64	8.84	11.2
1.60	1.25	2.16	3.14	4.14	5.12	6.99	8.78	10.52	12.2	15.6
1.80	1.43	2.62	3.93	5.28	6.60	9.15	11.6	14.0	16.3	20.9
2.00	1.61	3.12	4.82	6.58	8.32	11.7	14.9	18.1	21.2	27.3
2.25	1.84	3.81	6.09	8.46	10.8	15.4	19.8	24.1	28.4	36.7

[a]For rectangular channels, use the 0.0 (90°, vertical sides) column.

[b]Q = flow rate, ft³/sec; d = depth of flow, ft; b = bottom width of channel, ft; S = geometric slope, ft/ft; n = Manning's roughness constant.

[c]For SI units (i.e., Q in m³/s, and d and b in m), divide each table value by 1.486.

APPENDIX 20.A
Rational Method Runoff *C*-Coefficients

categorized by surface
forested	0.059–0.2
asphalt	0.7–0.95
brick	0.7–0.85
concrete	0.8–0.95
shingle roof	0.75–0.95
lawns, well-drained (sandy soil)	
up to 2% slope	0.05–0.1
2% to 7% slope	0.10–0.15
over 7% slope	0.15–0.2
lawns, poor drainage (clay soil)	
up to 2% slope	0.13–0.17
2% to 7% slope	0.18–0.22
over 7% slope	0.25–0.35
driveways, walkways	0.75–0.85

categorized by use
farmland	0.05–0.3
pasture	0.05–0.3
unimproved	0.1–0.3
parks	0.1–0.25
cemeteries	0.1–0.25
railroad yards	0.2–0.35
playgrounds (except asphalt or concrete)	0.2–0.35
business districts	
neighborhood	0.5–0.7
city (downtown)	0.7–0.95
residential	
single family	0.3–0.5
multiplexes, detached	0.4–0.6
multiplexes, attached	0.6–0.75
suburban	0.25–0.4
apartments, condominiums	0.5–0.7
industrial	
light	0.5–0.8
heavy	0.6–0.9

APPENDIX 20.B
Random Numbers[*]

78466	83326	96589	88727	72655	49682	82338	28583	01522	11248
78722	47603	03477	29528	63956	01255	29840	32370	18032	82051
06401	87397	72898	32441	88861	71803	55626	77847	29925	76106
04754	14489	39420	94211	58042	43184	60977	74801	05931	73822
97118	06774	87743	60156	38037	16201	35137	54513	68023	34380
71923	49313	59713	95710	05975	64982	79253	93876	33707	84956
78870	77328	09637	67080	49168	75290	50175	34312	82593	76606
61208	17172	33187	92523	69895	28284	77956	45877	08044	58292
05033	24214	74232	33769	06304	54676	70026	41957	40112	66451
95983	13391	30369	51035	17042	11729	88647	70541	36026	23113
19946	55448	75049	24541	43007	11975	31797	05373	45893	25665
03580	67206	09635	84610	62611	86724	77411	99415	58901	86160
56823	49819	20283	22272	00114	92007	24369	00543	05417	92251
87633	31761	99865	31488	49947	06060	32083	47944	00449	06550
95152	10133	52693	22480	50336	49502	06296	76414	18358	05313
05639	24175	79438	92151	57602	03590	25465	54780	79098	73594
65927	55525	67270	22907	55097	63177	34119	94216	84861	10457
59005	29000	38395	80367	34112	41866	30170	84658	84441	03926
06626	42682	91522	45955	23263	09764	26824	82936	16813	13878
11306	02732	34189	04228	58541	72573	89071	58066	67159	29633
45143	56545	94617	42752	31209	14380	81477	36952	44934	97435
97612	87175	22613	84175	96413	83336	12408	89318	41713	90669
97035	62442	06940	45719	39918	60274	54353	54497	29789	82928
62498	00257	19179	06313	07900	46733	21413	63627	48734	92174
80306	19257	18690	54653	07263	19894	89909	76415	57246	02621
84114	84884	50129	68942	93264	72344	98794	16791	83861	32007
58437	88807	92141	88677	02864	02052	62843	21692	21373	29408
15702	53457	54258	47485	23399	71692	56806	70801	41548	94809
59966	41287	87001	26462	94000	28457	09469	80416	05897	87970
43641	05920	81346	02507	25349	93370	02064	62719	45740	62080
25501	50113	44600	87433	00683	79107	22315	42162	25516	98434
98294	08491	25251	26737	00071	45090	68628	64390	42684	94956
52582	89985	37863	60788	27412	47502	71577	13542	31077	13353
26510	83622	12546	00489	89304	15550	09482	07504	64588	92562
24755	71543	31667	83624	27085	65905	32386	30775	19689	41437
38399	88796	58856	18220	51056	04976	54062	49109	95563	48244
18889	87814	52232	58244	95206	05947	26622	01381	28744	38374
51774	89694	02654	63161	54622	31113	51160	29015	64730	07750
88375	37710	61619	69820	13131	90406	45206	06386	06398	68652
10416	70345	93307	87360	53452	61179	46845	91521	32430	74795

[*]To use, enter the table randomly and arbitrarily select any direction (i.e., up, down, to the right, left, or diagonally).

Appendices

APPENDIX 22.A
Atomic Numbers and Weights of the Elements
(referred to carbon-12)

name	symbol	atomic number	atomic weight	name	symbol	atomic number	atomic weight
actinium	Ac	89	–	meitnerium	Mt	109	–
aluminum	Al	13	26.9815	mendelevium	Md	101	–
americium	Am	95	–	mercury	Hg	80	200.59
antimony	Sb	51	121.760	molybdenum	Mo	42	95.96
argon	Ar	18	39.948	neodymium	Nd	60	144.242
arsenic	As	33	74.9216	neon	Ne	10	20.1797
astatine	At	85	–	neptunium	Np	93	237.048
barium	Ba	56	137.327	nickel	Ni	28	58.693
berkelium	Bk	97	–	niobium	Nb	41	92.906
beryllium	Be	4	9.0122	nitrogen	N	7	14.0067
bismuth	Bi	83	208.980	nobelium	No	102	–
bohrium	Bh	107	–	osmium	Os	76	190.23
boron	B	5	10.811	oxygen	O	8	15.9994
bromine	Br	35	79.904	palladium	Pd	46	106.42
cadmium	Cd	48	112.411	phosphorus	P	15	30.9738
calcium	Ca	20	40.078	platinum	Pt	78	195.084
californium	Cf	98	–	plutonium	Pu	94	–
carbon	C	6	12.0107	polonium	Po	84	–
cerium	Ce	58	140.116	potassium	K	19	39.0983
cesium	Cs	55	132.9054	praseodymium	Pr	59	140.9077
chlorine	Cl	17	35.453	promethium	Pm	61	–
chromium	Cr	24	51.996	protactinium	Pa	91	231.0359
cobalt	Co	27	58.9332	radium	Ra	88	–
copernicium	Cn	112	–	radon	Rn	86	226.025
copper	Cu	29	63.546	rhenium	Re	75	186.207
curium	Cm	96	–	rhodium	Rh	45	102.9055
darmstadtium	Ds	110	–	roentgenium	Rg	111	–
dubnium	Db	105	–	rubidium	Rb	37	85.4678
dysprosium	Dy	66	162.50	ruthenium	Ru	44	101.07
einsteinium	Es	99	–	rutherfordium	Rf	104	–
erbium	Er	68	167.259	samarium	Sm	62	150.36
europium	Eu	63	151.964	scandium	Sc	21	44.956
fermium	Fm	100	–	seaborgium	Sg	106	–
fluorine	F	9	18.9984	selenium	Se	34	78.96
francium	Fr	87	–	silicon	Si	14	28.0855
gadolinium	Gd	64	157.25	silver	Ag	47	107.868
gallium	Ga	31	69.723	sodium	Na	11	22.9898
germanium	Ge	32	72.64	strontium	Sr	38	87.62
gold	Au	79	196.9666	sulfur	S	16	32.065
hafnium	Hf	72	178.49	tantalum	Ta	73	180.94788
hassium	Hs	108	–	technetium	Tc	43	–
helium	He	2	4.0026	tellurium	Te	52	127.60
holmium	Ho	67	164.930	terbium	Tb	65	158.925
hydrogen	H	1	1.00794	thallium	Tl	81	204.383
indium	In	49	114.818	thorium	Th	90	232.038
iodine	I	53	126.90447	thulium	Tm	69	168.934
iridium	Ir	77	192.217	tin	Sn	50	118.710
iron	Fe	26	55.845	titanium	Ti	22	47.867
krypton	Kr	36	83.798	tungsten	W	74	183.84
lanthanum	La	57	138.9055	uranium	U	92	238.0289
lawrencium	Lr	103	–	vanadium	V	23	50.942
lead	Pb	82	207.2	xenon	Xe	54	131.293
lithium	Li	3	6.941	ytterbium	Yb	70	173.054
lutetium	Lu	71	174.9668	yttrium	Y	39	88.906
magnesium	Mg	12	24.305	zinc	Zn	30	65.38
manganese	Mn	25	54.9380	zirconium	Zr	40	91.224

APPENDIX 22.B
Periodic Table of the Elements
(referred to carbon-12)

The Periodic Table of Elements (Long Form)

The number of electrons in filled shells is shown in the column at the extreme left; the remaining electrons for each element are shown immediately below the symbol for each element. Atomic numbers are enclosed in brackets. Atomic weights (rounded, based on carbon-12) are shown above the symbols. Atomic weight values in parentheses are those of the isotopes of longest half-life for certain radioactive elements whose atomic weights cannot be precisely quoted without knowledge of origin of the element.

metals / transition metals / nonmetals

periods	I A	II A	III B	IV B	V B	VI B	VII B	VIII			I B	II B	III A	IV A	V A	VI A	VII A	0
1 / 0	1.00794 H[1] 1																	4.00260 He[2] 2
2 / 2	6.941 Li[3] 1	9.01218 Be[4] 2											10.811 B[5] 3	12.0107 C[6] 4	14.0067 N[7] 5	15.9994 O[8] 6	18.9984 F[9] 7	20.1797 Ne[10] 8
3 / 2,8	22.9898 Na[11] 1	24.3050 Mg[12] 2											26.9815 Al[13] 3	28.0855 Si[14] 4	30.9738 P[15] 5	32.065 S[16] 6	35.453 Cl[17] 7	39.948 Ar[18] 8
4 / 2,8	39.0983 K[19] 8,1	40.078 Ca[20] 8,2	44.9559 Sc[21] 9,2	47.867 Ti[22] 10,2	50.9415 V[23] 11,2	51.9961 Cr[24] 13,1	54.9380 Mn[25] 13,2	55.845 Fe[26] 14,2	58.9332 Co[27] 15,2	58.6934 Ni[28] 16,2	63.546 Cu[29] 18,1	65.38 Zn[30] 18,2	69.723 Ga[31] 18,3	72.64 Ge[32] 18,4	74.9216 As[33] 18,5	78.96 Se[34] 18,6	79.904 Br[35] 18,7	83.798 Kr[36] 18,8
5 / 2,8,18	85.4678 Rb[37] 8,1	87.62 Sr[38] 8,2	88.9059 Y[39] 9,2	91.224 Zr[40] 10,2	92.9064 Nb[41] 12,1	95.96 Mo[42] 13,1	(98) Tc[43] 14,1	101.07 Ru[44] 15,1	102.906 Rh[45] 16,1	106.42 Pd[46] 18	107.868 Ag[47] 18,1	112.411 Cd[48] 18,2	114.818 In[49] 18,3	118.710 Sn[50] 18,4	121.760 Sb[51] 18,5	127.60 Te[52] 18,6	126.904 I[53] 18,7	131.293 Xe[54] 18,8
6 / 2,8,18	132.905 Cs[55] 18,8,1	137.327 Ba[56] 18,8,2	* (57–71)	178.49 Hf[72] 32,10,2	180.948 Ta[73] 32,11,2	183.84 W[74] 32,12,2	186.207 Re[75] 32,13,2	190.23 Os[76] 32,14,2	192.217 Ir[77] 32,15,2	195.084 Pt[78] 32,17,1	196.967 Au[79] 32,18,1	200.59 Hg[80] 32,18,2	204.383 Tl[81] 32,18,3	207.2 Pb[82] 32,18,4	208.980 Bi[83] 32,18,5	209 Po[84] 32,18,6	210 At[85] 32,18,7	222 Rn[86] 32,18,8
7 / 2,8,18,32	(223) Fr[87] 18,8,1	(226) Ra[88] 18,8,2	† (89–103)	(265) Rf[104] 32,10,2	(268) Db[105] 32,11,2	(271) Sg[106] 32,12,2	(272) Bh[107] 32,13,2	(270) Hs[108] 32,14,2	(276) Mt[109] 32,15,2	(281) Ds[110] 32,17,1	(280) Rg[111] 32,18,1	(285) Cn[112] 32,18,2						

*lanthanide series

138.905 La[57] 18,9,2	140.116 Ce[58] 20,8,2	140.908 Pr[59] 21,8,2	144.242 Nd[60] 22,8,2	(145) Pm[61] 23,8,2	150.36 Sm[62] 24,8,2	151.964 Eu[63] 25,8,2	157.25 Gd[64] 25,9,2	158.925 Tb[65] 27,8,2	162.500 Dy[66] 28,8,2	164.930 Ho[67] 29,8,2	167.259 Er[68] 30,8,2	168.934 Tm[69] 31,8,2	173.054 Yb[70] 32,8,2	174.967 Lu[71] 32,9,2

†actinide series

(227) Ac[89] 18,9,2	232.038 Th[90] 18,10,2	231.036 Pa[91] 20,9,2	238.029 U[92] 21,9,2	(237) Np[93] 23,8,2	(244) Pu[94] 24,8,2	(243) Am[95] 25,8,2	(247) Cm[96] 25,9,2	(247) Bk[97] 26,9,2	(251) Cf[98] 28,8,2	(252) Es[99] 29,8,2	(257) Fm[100] 30,8,2	(258) Md[101] 31,8,2	(259) No[102] 32,8,2	(262) Lr[103] 32,9,2

APPENDIX 22.C
Water Chemistry $CaCO_3$ Equivalents

cations	formula	ionic weight	equivalent weight	substance to $CaCO_3$ factor
aluminum	Al^{+3}	27.0	9.0	5.56
ammonium	NH_4^+	18.0	18.0	2.78
calcium	Ca^{+2}	40.1	20.0	2.50
cupric copper	Cu^{+2}	63.6	31.8	1.57
cuprous copper	Cu^{+3}	63.6	21.2	2.36
ferric iron	Fe^{+3}	55.8	18.6	2.69
ferrous iron	Fe^{+2}	55.8	27.9	1.79
hydrogen	H^+	1.0	1.0	50.00
manganese	Mn^{+2}	54.9	27.5	1.82
magnesium	Mg^{+2}	24.3	12.2	4.10
potassium	K^+	39.1	39.1	1.28
sodium	Na^+	23.0	23.0	2.18

anions	formula	ionic weight	equivalent weight	substance to $CaCO_3$ factor
bicarbonate	HCO_3^-	61.0	61.0	0.82
carbonate	CO_3^{-2}	60.0	30.0	1.67
chloride	Cl^-	35.5	35.5	1.41
fluoride	F^-	19.0	19.0	2.66
hydroxide	OH^-	17.0	17.0	2.94
nitrate	NO_3^-	62.0	62.0	0.81
phosphate (tribasic)	PO_4^{-3}	95.0	31.7	1.58
phosphate (dibasic)	HPO_4^{-2}	96.0	48.0	1.04
phosphate (monobasic)	$H_2PO_4^-$	97.0	97.0	0.52
sulfate	SO_4^{-2}	96.1	48.0	1.04
sulfite	SO_3^{-2}	80.1	40.0	1.25

compounds	formula	molecular weight	equivalent weight	substance to $CaCO_3$ factor
aluminum hydroxide	$Al(OH)_3$	78.0	26.0	1.92
aluminum sulfate	$Al_2(SO_4)_3$	342.1	57.0	0.88
aluminum sulfate	$Al_2(SO_4)_3 \cdot 18H_2O$	666.1	111.0	0.45
alumina	Al_2O_3	102.0	17.0	2.94
sodium aluminate	$Na_2Al_2O_4$	164.0	27.3	1.83
calcium bicarbonate	$Ca(HCO_3)_2$	162.1	81.1	0.62
calcium carbonate	$CaCO_3$	100.1	50.1	1.00
calcium chloride	$CaCl_2$	111.0	55.5	0.90
calcium hydroxide (pure)	$Ca(OH)_2$	74.1	37.1	1.35
calcium hydroxide (90%)	$Ca(OH)_2$	–	41.1	1.22
calcium oxide (lime)	CaO	56.1	28.0	1.79
calcium sulfate (anhydrous)	$CaSO_4$	136.2	68.1	0.74
calcium sulfate (gypsum)	$CaSO_4 \cdot 2H_2O$	172.2	86.1	0.58

APPENDIX 22.C *(continued)*
Water Chemistry $CaCO_3$ Equivalents

compounds	formula	molecular weight	equivalent weight	substance to $CaCO_3$ factor
calcium phosphate	$Ca_3(PO_4)_2$	310.3	51.7	0.97
disodium phosphate	$Na_2HPO_4 \cdot 12H_2O$	358.2	119.4	0.42
disodium phosphate (anhydrous)	Na_2HPO_4	142.0	47.3	1.06
ferric oxide	Fe_2O_3	159.6	26.6	1.88
iron oxide (magnetic)	Fe_3O_4	321.4	–	–
ferrous sulfate (copperas)	$FeSO_4 \cdot 7H_2O$	278.0	139.0	0.36
magnesium oxide	MgO	40.3	20.2	2.48
magnesium bicarbonate	$Mg(HCO_3)_2$	146.3	73.2	0.68
magnesium carbonate	$MgCO_3$	84.3	42.2	1.19
magnesium chloride	$MgCl_2$	95.2	47.6	1.05
magnesium hydroxide	$Mg(OH)_2$	58.3	29.2	1.71
magnesium phosphate	$Mg_3(PO_4)_2$	263.0	43.8	1.14
magnesium sulfate	$MgSO_4$	120.4	60.2	0.83
monosodium phosphate	$NaH_2PO_4 \cdot H_2O$	138.1	46.0	1.09
monosodium phosphate (anhydrous)	NaH_2PO_4	120.1	40.0	1.25
metaphosphate	$NaPO_3$	102.0	34.0	1.47
silica	SiO_2	60.1	30.0	1.67
sodium bicarbonate	$NaHCO_3$	84.0	84.0	0.60
sodium carbonate	Na_2CO_3	106.0	53.0	0.94
sodium chloride	$NaCl$	58.5	58.5	0.85
sodium hydroxide	$NaOH$	40.0	40.0	1.25
sodium nitrate	$NaNO_3$	85.0	85.0	0.59
sodium sulfate	Na_2SO_4	142.0	71.0	0.70
sodium sulfite	Na_2SO_3	126.1	63.0	0.79
tetrasodium EDTA	$(CH_2)_2N_2(CH_2COONa)_4$	380.2	95.1	0.53
trisodium phosphate	$Na_3PO_4 \cdot 12H_2O$	380.2	126.7	0.40
trisodium phosphate (anhydrous)	Na_3PO_4	164.0	54.7	0.91
trisodium NTA	$(CH_2)_3N(COONa)_3$	257.1	85.7	0.58

gases	formula	molecular weight	equivalent weight	substance to $CaCO_3$ factor
ammonia	NH_3	17	17	2.94
carbon dioxide	CO_2	44	22	2.27
hydrogen	H_2	2	1	50.00
hydrogen sulfide	H_2S	34	17	2.94
oxygen	O_2	32	8	6.25

APPENDIX 22.C *(continued)*
Water Chemistry $CaCO_3$ Equivalents

acids	formula	molecular weight	equivalent weight	substance to $CaCO_3$ factor
carbonic	H_2CO_3	62.0	31.0	1.61
hydrochloric	HCl	36.5	36.5	1.37
phosphoric	H_3PO_4	98.0	32.7	1.53
sulfuric	H_2SO_4	98.1	49.1	1.02

(Multiply the concentration (in mg/L) of the substance by the corresponding factors to obtain the equivalent concentration in mg/L as $CaCO_3$. For example, 70 mg/L of Mg^{++} would be (70 mg/L) (4.1) = 287 mg/L as $CaCO_3$.)

APPENDIX 22.D
Saturation Concentrations of Dissolved Oxygen in Water[a]

temperature (°C)	chloride concentration in water (mg/L)			difference per 100 mg chloride	vapor pressure (mm Hg)
	0[b]	5000	10,000		
	dissolved oxygen (mg/L)				
0	14.60	13.79	12.97	0.0163	4.58
1	14.19	13.40	12.61	0.0158	4.93
2	13.81	13.05	12.28	0.0153	5.29
3	13.44	12.71	11.98	0.0146	5.69
4	13.09	12.39	11.69	0.0140	6.10
5	12.75	12.07	11.39	0.0136	6.54
6	12.43	11.78	11.12	0.0131	7.02
7	12.12	11.49	10.85	0.0127	7.52
8	11.83	11.22	10.61	0.0122	8.05
9	11.55	10.96	10.36	0.0119	8.61
10	11.27	10.70	10.13	0.0114	9.21
11	11.01	10.47	9.92	0.0109	9.85
12	10.76	10.24	9.72	0.0104	10.52
13	10.52	10.02	9.52	0.0100	11.24
14	10.29	9.81	9.32	0.0097	11.99
15	10.07	9.61	9.14	0.0093	12.79
16	9.85	9.41	8.96	0.0089	13.64
17	9.65	9.22	8.78	0.0087	14.54
18	9.45	9.04	8.62	0.0083	15.49
19	9.26	8.86	8.45	0.0081	16.49
20	9.07	8.69	8.30	0.0077	17.54
21	8.90	8.52	8.14	0.0076	18.66
22	8.72	8.36	7.99	0.0073	19.84
23	8.56	8.21	7.85	0.0071	21.08
24	8.40	8.06	7.71	0.0069	22.34
25	8.24	7.90	7.56	0.0068	23.77
26	8.09	7.76	7.42	0.0067	25.22
27	7.95	7.62	7.28	0.0067	26.75
28	7.81	7.48	7.14	0.0067	28.36
29	7.67	7.34	7.00	0.0067	30.05
30	7.54	7.20	6.86	0.0068	31.83

[a]For saturation at barometric pressures other than 760 mm Hg (29.92 in Hg), C_s' is related to the corresponding tabulated value, C_s, by the equation: $C_s' = C_s\left(\dfrac{P-p}{760-p}\right)$

C_s' = solubility at barometric pressure P and given temperature, mg/L

C_s = saturation solubility at given temperature from appendix, mg/L

P = barometric pressure, mm Hg

p = pressure of saturated water vapor at temperature of the water selected from appendix, mm Hg

[b]Zero-chloride values from *Volunteer Stream Monitoring: A Methods Manual* (EPA 841-B-97-003), Environmental Protection Agency, Office of Water, Sec. 5.2 "Dissolved Oxygen and Biochemical Oxygen Demand;" 1997.

APPENDIX 22.E
Names and Formulas of Important Chemicals

common name	chemical name	chemical formula
acetone	acetone	$(CH_3)_2CO$
acetylene	acetylene	C_2H_2
ammonia	ammonia	NH_3
ammonium	ammonium hydroxide	NH_4OH
aniline	aniline	$C_6H_5NH_2$
bauxite	hydrated aluminum oxide	$Al_2O_3 \cdot 2H_2O$
bleach	calcium hypochlorite	$Ca(ClO)_2$
borax	sodium tetraborate	$Na_2B_4O_7 \cdot 10H_2O$
carbide	calcium carbide	CaC_2
carbolic acid	phenol	C_6H_5OH
carbon dioxide	carbon dioxide	CO_2
carborundum	silicon carbide	SiC
caustic potash	potassium hydroxide	KOH
caustic soda/lye	sodium hydroxide	$NaOH$
chalk	calcium carbonate	$CaCO_3$
cinnabar	mercuric sulfide	HgS
ether	diethyl ether	$(C_2H_5)_2O$
formic acid	methanoic acid	$HCOOH$
Glauber's salt	decahydrated sodium sulfate	$Na_2SO_4 \cdot 10H_2O$
glycerine	glycerine	$C_3H_5(OH)_3$
grain alcohol	ethanol	C_2H_5OH
graphite	crystalline carbon	C
gypsum	calcium sulfate	$CaSO_4 \cdot 2H_2O$
halite	sodium chloride	$NaCl$
iron chloride	ferrous chloride	$FeCl_2 \cdot 4H_2O$
laughing gas	nitrous oxide	N_2O
limestone	calcium carbonate	$CaCO_3$
magnesia	magnesium oxide	MgO
marsh gas	methane	CH_4
muriate of potash	potassium chloride	KCl
muriatic acid	hydrochloric acid	HCl
niter	sodium nitrate	$NaNO_3$
niter cake	sodium bisulfate	$NaHSO_4$
oleum	fuming sulfuric acid	SO_3 in H_2SO_4
potash	potassium carbonate	K_2CO_3
prussic acid	hydrogen cyanide	HCN
pyrites	ferrous sulfide	FeS
pyrolusite	manganese dioxide	MnO_2
quicklime	calcium oxide	CaO
sal soda	decahydrated sodium carbonate	$NaCO_3 \cdot 10H_2O$
salammoniac	ammonium chloride	NH_4Cl
sand or silica	silicon dioxide	SiO_2
salt cake	sodium sulfate (crude)	Na_2SO_4
slaked lime	calcium hydroxide	$Ca(OH)_2$

APPENDIX 22.E *(continued)*
Names and Formulas of Important Chemicals

common name	chemical name	chemical formula
soda ash	sodium carbonate	Na_2CO_3
soot	amorphous carbon	C
stannous chloride	stannous chloride	$SnCl_2 \cdot 2H_2O$
superphosphate	monohydrated primary calcium phosphate	$Ca(H_2PO_4)_2 \cdot H_2O$
table salt	sodium chloride	NaCl
table sugar	sucrose	$C_{12}H_{22}O_{11}$
trilene	trichloroethylene	C_2HCl_3
urea	urea	$CO(NH_2)_2$
vinegar (acetic acid)	ethanoic acid	CH_3COOH
washing soda	decahydrated sodium carbonate	$Na_2CO_3 \cdot 10H_2O$
wood alcohol	methanol	CH_3OH
zinc blende	zinc sulfide	ZnS

APPENDIX 22.F
Approximate Solubility Product Constants at 25°C

substance	formula	K_{sp}
aluminum hydroxide	$Al(OH)_3$	1.3×10^{-33}
aluminum phosphate	$AlPO_4$	6.3×10^{-19}
barium carbonate	$BaCO_3$	5.1×10^{-9}
barium chromate	$BaCrO_4$	1.2×10^{-10}
barium fluoride	BaF_2	1.0×10^{-6}
barium hydroxide	$Ba(OH)_2$	5×10^{-3}
barium sulfate	$BaSO_4$	1.1×10^{-10}
barium sulfite	$BaSO_3$	8×10^{-7}
barium thiosulfate	BaS_2O_3	1.6×10^{-6}
bismuthyl chloride	$BiOCl$	1.8×10^{-31}
bismuthyl hydroxide	$BiOOH$	4×10^{-10}
cadmium carbonate	$CdCO_3$	5.2×10^{-12}
cadmium hydroxide	$Cd(OH)_2$	2.5×10^{-14}
cadmium oxalate	CdC_2O_4	1.5×10^{-8}
cadmium sulfide[a]	CdS	8×10^{-28}
calcium carbonate[b]	$CaCO_3$	2.8×10^{-9}
calcium chromate	$CaCrO_4$	7.1×10^{-4}
calcium fluoride	CaF_2	5.3×10^{-9}
calcium hydrogen phosphate	$CaHPO_4$	1×10^{-7}
calcium hydroxide	$Ca(OH)_2$	5.5×10^{-6}
calcium oxalate	CaC_2O_4	2.7×10^{-9}
calcium phosphate	$Ca_3(PO_4)_2$	2.0×10^{-29}
calcium sulfate	$CaSO_4$	9.1×10^{-6}
calcium sulfite	$CaSO_3$	6.8×10^{-8}
chromium (II) hydroxide	$Cr(OH)_2$	2×10^{-16}
chromium (III) hydroxide	$Cr(OH)_3$	6.3×10^{-31}
cobalt (II) carbonate	$CoCO_3$	1.4×10^{-13}
cobalt (II) hydroxide	$Co(OH)_2$	1.6×10^{-15}
cobalt (III) hydroxide	$Co(OH)_3$	1.6×10^{-44}
cobalt (II) sulfide[a]	CoS	4×10^{-21}
copper (I) chloride	$CuCl$	1.2×10^{-6}
copper (I) cyanide	$CuCN$	3.2×10^{-20}
copper (I) iodide	CuI	1.1×10^{-12}
copper (II) arsenate	$Cu_3(AsO_4)_2$	7.6×10^{-36}
copper (II) carbonate	$CuCO_3$	1.4×10^{-10}
copper (II) chromate	$CuCrO_4$	3.6×10^{-6}
copper (II) ferrocyanide	$Cu[Fe(CN)_6]$	1.3×10^{-16}
copper (II) hydroxide	$Cu(OH)_2$	2.2×10^{-20}
copper (II) sulfide[a]	CuS	6×10^{-37}
iron (II) carbonate	$FeCO_3$	3.2×10^{-11}
iron (II) hydroxide	$Fe(OH)_2$	8.0×10^{-16}
iron (II) sulfide[a]	FeS	6×10^{-19}
iron (III) arsenate	$FeAsO_4$	5.7×10^{-21}
iron (III) ferrocyanide	$Fe_4[Fe(CN)_6]_3$	3.3×10^{-41}
iron (III) hydroxide	$Fe(OH)_3$	4×10^{-38}
iron (III) phosphate	$FePO_4$	1.3×10^{-22}

APPENDIX 22.F *(continued)*
Approximate Solubility Product Constants at 25°C

substance	formula	K_{sp}
lead (II) arsenate	$Pb_3(AsO_4)_2$	4×10^{-36}
lead (II) azide	$Pb(N_3)_2$	2.5×10^{-9}
lead (II) bromide	$PbBr_2$	4.0×10^{-5}
lead (II) carbonate	$PbCO_3$	7.4×10^{-14}
lead (II) chloride	$PbCl_2$	1.6×10^{-5}
lead (II) chromate	$PbCrO_4$	2.8×10^{-13}
lead (II) fluoride	PbF_2	2.7×10^{-8}
lead (II) hydroxide	$Pb(OH)_2$	1.2×10^{-15}
lead (II) iodide	PbI_2	7.1×10^{-9}
lead (II) sulfate	$PbSO_4$	1.6×10^{-8}
lead (II) sulfide[a]	PbS	3×10^{-28}
lithium carbonate	Li_2CO_3	2.5×10^{-2}
lithium fluoride	LiF	3.8×10^{-3}
lithium phosphate	Li_3PO_4	3.2×10^{-9}
magnesium ammonium phosphate	$MgNH_4PO_4$	2.5×10^{-13}
magnesium arsenate	$Mg_3(AsO_4)_2$	2×10^{-20}
magnesium carbonate	$MgCO_3$	3.5×10^{-8}
magnesium fluoride	MgF_2	3.7×10^{-8}
magnesium hydroxide	$Mg(OH)_2$	1.8×10^{-11}
magnesium oxalate	MgC_2O_4	8.5×10^{-5}
magnesium phosphate	$Mg_3(PO_4)_2$	1×10^{-25}
manganese (II) carbonate	$MnCO_3$	1.8×10^{-11}
manganese (II) hydroxide	$Mn(OH)_2$	1.9×10^{-13}
manganese (II) sulfide[a]	MnS	3×10^{-14}
mercury (I) bromide	Hg_2Br_2	5.6×10^{-23}
mercury (I) chloride	Hg_2Cl_2	1.3×10^{-18}
mercury (I) iodide	Hg_2I_2	4.5×10^{-29}
mercury (II) sulfide[a]	HgS	2×10^{-53}
nickel (II) carbonate	$NiCO_3$	6.6×10^{-9}
nickel (II) hydroxide	$Ni(OH)_2$	2.0×10^{-15}
nickel (II) sulfide[a]	NiS	3×10^{-19}
scandium fluoride	ScF_3	4.2×10^{-18}
scandium hydroxide	$Sc(OH)_3$	8.0×10^{-31}
silver acetate	$AgC_2H_3O_2$	2.0×10^{-3}
silver arsenate	Ag_3AsO_4	1.0×10^{-22}
silver azide	AgN_3	2.8×10^{-9}
silver bromide	$AgBr$	5.0×10^{-13}
silver chloride	$AgCl$	1.8×10^{-10}
silver chromate	Ag_2CrO_4	1.1×10^{-12}
silver cyanide	$AgCN$	1.2×10^{-16}
silver iodate	$AgIO_3$	3.0×10^{-8}
silver iodide	AgI	8.5×10^{-17}
silver nitrite	$AgNO_2$	6.0×10^{-4}
silver sulfate	Ag_2SO_4	1.4×10^{-5}
silver sulfide[a]	Ag_2S	6×10^{-51}

APPENDIX 22.F *(continued)*
Approximate Solubility Product Constants at 25°C

substance	formula	K_{sp}
silver sulfite	Ag_2SO_3	1.5×10^{-14}
silver thiocyanate	$AgSCN$	1.0×10^{-12}
strontium carbonate	$SrCO_3$	1.1×10^{-10}
strontium chromate	$SrCrO_4$	2.2×10^{-5}
strontium fluoride	SrF_2	2.5×10^{-9}
strontium sulfate	$SrSO_4$	3.2×10^{-7}
thallium (I) bromide	$TlBr$	3.4×10^{-6}
thallium (I) chloride	$TlCl$	1.7×10^{-4}
thallium (I) iodide	TlI	6.5×10^{-8}
thallium (III) hydroxide	$Tl(OH)_3$	6.3×10^{-46}
tin (II) hydroxide	$Sn(OH)_2$	1.4×10^{-28}
tin (II) sulfide[a]	SnS	1×10^{-26}
zinc carbonate	$ZnCO_3$	1.4×10^{-11}
zinc hydroxide	$Zn(OH)_2$	1.2×10^{-17}
zinc oxalate	ZnC_2O_4	2.7×10^{-8}
zinc phosphate	$Zn_3(PO_4)_2$	9.0×10^{-33}
zincsulfide[a]	ZnS	2×10^{-25}

[a]Sulfide equilibrium of the type:

$$MS(s) + H_2O(l) \rightleftharpoons M^{2+}(aq) + HS^-(aq) + OH^-(aq)$$

[b]Solubility product depends on mineral form.

APPENDIX 22.G
Dissociation Constants of Acids at 25°C

acid		K_a
acetic	K_1	1.8×10^{-5}
arsenic	K_1	5.6×10^{-3}
	K_2	1.2×10^{-7}
	K_3	3.2×10^{-12}
arsenious	K_1	1.4×10^{-9}
benzoic	K_1	6.3×10^{-5}
boric	K_1	5.9×10^{-10}
carbonic	K_1^*	4.5×10^{-7}
	K_2	5.6×10^{-11}
chloroacetic	K_1	1.4×10^{-3}
chromic	K_2	3.2×10^{-7}
citric	K_1	7.4×10^{-4}
	K_2	1.7×10^{-5}
	K_3	3.9×10^{-7}
ethylenedinitrilotetracetic	K_1	1.0×10^{-2}
	K_2	2.1×10^{-3}
	K_3	6.9×10^{-7}
	K_4	7.4×10^{-11}
formic	K_1	1.8×10^{-4}
hydrocyanic	K_1	4.9×10^{-10}
hydrofluoric	K_1	6.8×10^{-4}
hydrogen sulfide	K_1	1.0×10^{-8}
	K_2	1.2×10^{-14}
hypochlorous	K_1	2.8×10^{-8}
iodic	K_1	1.8×10^{-1}
nitrous	K_1	4.5×10^{-4}
oxalic	K_1	5.4×10^{-2}
	K_2	5.1×10^{-5}
phenol	K_1	1.1×10^{-10}
phosphoric (ortho)	K_1	7.1×10^{-3}
	K_2	6.3×10^{-8}
	K_3	4.4×10^{-13}
o-phthalic	K_1	1.1×10^{-3}
	K_2	3.9×10^{-6}
salicylic	K_1	1.0×10^{-3}
	K_2	4.0×10^{-14}
sulfamic	K_1	1.0×10^{-1}
sulfuric	K_1	1.1×10^{-2}
sulfurous	K_1	1.7×10^{-2}
	K_2	6.3×10^{-8}
tartaric	K_1	9.2×10^{-4}
	K_2	4.3×10^{-5}
thiocyanic	K_1	1.4×10^{-1}

*apparent constant based on $C_{H_2CO_3} = [CO_2] + [H_2CO_3]$

APPENDIX 22.H
Dissociation Constants of Bases at 25°C

base		K_b
2-amino-2-(hydroxymethyl)-1,3-propanediol	K_1	1.2×10^{-6}
ammonia	K_1	1.8×10^{-5}
aniline	K_1	4.2×10^{-10}
diethylamine	K_1	1.3×10^{-3}
hexamethylenetetramine	K_1	1.0×10^{-9}
hydrazine	K_1	9.8×10^{-7}
hydroxylamine	K_1	9.6×10^{-9}
lead hydroxide	K_1	1.2×10^{-4}
piperidine	K_1	1.3×10^{-3}
pyridine	K_1	1.5×10^{-9}
silver hydroxide	K_1	6.0×10^{-5}

APPENDIX 24.A
Heats of Combustion for Common Compounds[a]

				heat of combustion			
				Btu/ft^3		Btu/lbm	
substance	formula	molecular weight	specific volume (ft^3/lbm)	gross (high)	net (low)	gross (high)	net (low)
carbon	C	12.01				14,093	14,093
carbon dioxide	CO_2	44.01	8.548				
carbon monoxide	CO	28.01	13.506	322	322	4347	4347
hydrogen	H_2	2.016	187.723	325	275	60,958	51,623
nitrogen	N_2	28.016	13.443				
oxygen	O_2	32.000	11.819				
paraffin series (alkanes)							
methane	CH_4	16.041	23.565	1013	913	23,879	21,520
ethane	C_2H_6	30.067	12.455	1792	1641	22,320	20,432
propane	C_3H_8	44.092	8.365	2590	2385	21,661	19,944
n-butane	C_4H_{10}	58.118	6.321	3370	3113	21,308	19,680
isobutane	C_4H_{10}	58.118	6.321	3363	3105	21,257	19,629
n-pentane	C_5H_{12}	72.144	5.252	4016	3709	21,091	19,517
isopentane	C_5H_{12}	72.144	5.252	4008	3716	21,052	19,478
neopentane	C_5H_{12}	72.144	5.252	3993	3693	20,970	19,396
n-hexane	C_6H_{14}	86.169	4.398	4762	4412	20,940	19,403
olefin series (alkenes and alkynes)							
ethylene	C_2H_4	28.051	13.412	1614	1513	21,644	20,295
propylene	C_3H_6	42.077	9.007	2336	2186	21,041	19,691
n-butene	C_4H_8	56.102	6.756	3084	2885	20,840	19,496
isobutene	C_4H_8	56.102	6.756	3068	2869	20,730	19,382
n-pentene	C_5H_{10}	70.128	5.400	3836	3586	20,712	19,363
aromatic series							
benzene	C_6H_6	78.107	4.852	3751	3601	18,210	17,480
toluene	C_7H_8	92.132	4.113	4484	4284	18,440	17,620
xylene	C_8H_{10}	106.158	3.567	5230	4980	18,650	17,760
miscellaneous fuels							
acetylene	C_2H_2	26.036	14.344	1499	1448	21,500	20,776
air		28.967	13.063				
ammonia	NH_3	17.031	21.914	441	365	9668	8001
digester gas[b]	–	25.8	18.3	658	593	15,521	13,988
ethyl alcohol	C_2H_5OH	46.067	8.221	1600	1451	13,161	11,929
hydrogen sulfide	H_2S	34.076	10.979	647	596	7100	6545
iso-octane	C_8H_{18}	114.2	0.0232[c]	106	98.9	20,590	19,160
methyl alcohol	CH_3OH	32.041	11.820	868	768	10,259	9078
naphthalene	$C_{10}H_8$	128.162	2.955	5854	5654	17,298	16,708

(Multiply Btu/lbm by 2.326 to obtain kJ/kg.)
(Multiply Btu/ft^3 by 37.25 to obtain kJ/m^3.)
[a]Gas volumes listed are at 60°F (16°C) and 1 atm.
[b]Digester gas from wastewater treatment plants is approximately 65% methane and 35% carbon dioxide by volume. Use composite properties of these two gases.
[c]liquid form; stoichiometric mixture

APPENDIX 24.A *(continued)*
Heats of Combustion for Common Compounds[a]

| substance | formula | molecular weight | specific volume (ft³/lbm) | heat of combustion | | | |
| | | | | Btu/ft³ | | Btu/lbm | |
				gross (high)	net (low)	gross (high)	net (low)
sulfur	S	32.06				3983	3983
sulfur dioxide	SO_2	64.06	5.770				
water vapor	H_2O	18.016	21.017				

(Multiply Btu/lbm by 2.326 to obtain kJ/kg.)
(Multiply Btu/ft³ by 37.25 to obtain kJ/m³.)
[a]Gas volumes listed are at 60°F (16°C) and 1 atm.
[b]Digester gas from wastewater treatment plants is approximately 65% methane and 35% carbon dioxide by volume. Use composite properties of these two gases.
[c]liquid form; stoichiometric mixture

APPENDIX 24.B
Approximate Properties of Selected Gases

				customary U.S. units			SI units			
				R	c_p	c_v	c_p	c_v	k	
		temp		ft-lbf	Btu	Btu	J	J	J	
gas	symbol	°F	MW	lbm-°R	lbm-°R	lbm-°R	kg·K	kg·K	kg·K	
acetylene	C_2H_2	68	26.038	59.35	0.350	0.274	319.32	1465	1146	1.279
air		100	28.967	53.35	0.240	0.171	287.03	1005	718	1.400
ammonia	NH_3	68	17.032	90.73	0.523	0.406	488.16	2190	1702	1.287
argon	Ar	68	39.944	38.69	0.124	0.074	208.15	519	311	1.669
n-butane	C_4H_{10}	68	58.124	26.59	0.395	0.361	143.04	1654	1511	1.095
carbon dioxide	CO_2	100	44.011	35.11	0.207	0.162	188.92	867	678	1.279
carbon monoxide	CO	100	28.011	55.17	0.249	0.178	296.82	1043	746	1.398
chlorine	Cl_2	100	70.910	21.79	0.115	0.087	117.25	481	364	1.322
ethane	C_2H_6	68	30.070	51.39	0.386	0.320	276.50	1616	1340	1.206
ethylene	C_2H_4	68	28.054	55.08	0.400	0.329	296.37	1675	1378	1.215
Freon (R-12)[*]	CCl_2F_2	200	120.925	12.78	0.159	0.143	68.76	666	597	1.115
helium	He	100	4.003	386.04	1.240	0.744	2077.03	5192	3115	1.667
hydrogen	H_2	100	2.016	766.53	3.420	2.435	4124.18	14 319	10 195	1.405
hydrogen sulfide	H_2S	68	34.082	45.34	0.243	0.185	243.95	1017	773	1.315
krypton	Kr		83.800	18.44	0.059	0.035	99.22	247	148	1.671
methane	CH_4	68	16.043	96.32	0.593	0.469	518.25	2483	1965	1.264
neon	Ne	68	20.183	76.57	0.248	0.150	411.94	1038	626	1.658
nitrogen	N_2	100	28.016	55.16	0.249	0.178	296.77	1043	746	1.398
nitric oxide	NO	68	30.008	51.50	0.231	0.165	277.07	967	690	1.402
nitrous oxide	NO_2	68	44.01	35.11	0.221	0.176	188.92	925	736	1.257
octane vapor	C_8H_{18}		114.232	13.53	0.407	0.390	72.78	1704	1631	1.045
oxygen	O_2	100	32.000	48.29	0.220	0.158	259.82	921	661	1.393
propane	C_3H_8	68	44.097	35.04	0.393	0.348	188.55	1645	1457	1.129
sulfur dioxide	SO_2	100	64.066	24.12	0.149	0.118	129.78	624	494	1.263
water vapor[*]	H_2	212	18.016	85.78	0.445	0.335	461.50	1863	1402	1.329
xenon	Xe		131.300	11.77	0.038	0.023	63.32	159	96	1.661

(Multiply Btu/lbm-°F by 4186.8 to obtain J/kg·K.)
(Multiply ft-lbf/lbm-°R by 5.3803 to obtain J/kg·K.)
[*]Values for steam and Freon are approximate and should be used only for low pressures and high temperatures.

APPENDIX 25.A
National Primary Drinking Water Regulations
Code of Federal Regulations, Title 40, Ch. I, Part 141, Subpart G (CFR)

microorganisms	MCLG[a] (mg/L)[b]	MCL or TT[a] (mg/L)[b]	potential health effects from ingestion of water	sources of contaminant in drinking water
Cryptosporidium	0	TT[c]	gastrointestinal illness (e.g., diarrhea, vomiting, cramps)	human and animal fecal waste
Giardia lamblia	0	TT[c]	gastrointestinal illness (e.g., diarrhea, vomiting, cramps)	human and animal fecal waste
heterotrophic plate count	n/a	TT[c]	HPC has no health effects; it is an analytic method used to measure the variety of bacteria that are common in water. The lower the concentration of bacteria in drinking water, the better maintained the water is.	HPC measures a range of bacteria that are naturally present in the environment.
Legionella	0	TT[c]	Legionnaire's disease, a type of pneumonia	found naturally in water; multiplies in heating systems
total coliforms (including fecal coliform and *E. coli*)	0	5.0%[d]	Not a health threat in itself; it is used to indicate whether other potentially harmful bacteria may be present.[e]	Coliforms are naturally present in the environment as well as in feces; fecal coliforms and *E. coli* only come from human and animal fecal waste.
turbidity	n/a	TT[c]	Turbidity is a measure of the cloudiness of water. It is used to indicate water quality and filtration effectiveness (e.g., whether disease causing organisms are present). Higher turbidity levels are often associated with higher levels of disease causing microorganisms such as viruses, parasites, and some bacteria. These organisms can cause symptoms such as nausea, cramps, diarrhea, and associated headaches.	soil runoff
viruses (enteric)	0	TT[c]	gastrointestinal illness (e.g., diarrhea, vomiting, cramps)	human and animal fecal waste

disinfection products	MCLG[a] (mg/L)[b]	MCL or TT[a] (mg/L)[b]	potential health effects from ingestion of water	sources of contaminant in drinking water
bromate	0	0.010	increased risk of cancer	by-product of drinking-water disinfection
chlorite	0.8	1.0	anemia in infants and young children; nervous system effects	by-product of drinking-water disinfection
haloacetic acids (HAA5)	n/a[f]	0.060	increased risk of cancer	by-product of drinking-water disinfection
total trihalomethanes (TTHMs)	n/a[f]	0.080	liver, kidney, or central nervous system problems; increased risk of cancer	by-product of drinking-water disinfection

APPENDIX 25.A *(continued)*
National Primary Drinking Water Regulations
Code of Federal Regulations, Title 40, Ch. I, Part 141, Subpart G (CFR)

disinfectants	MCLG[a] $(mg/L)^b$	MCL or TT[a] $(mg/L)^b$	potential health effects from ingestion of water	sources of contaminant in drinking water
chloramines (as Cl_2)	4.0^a	4.0^a	eye/nose irritation, stomach discomfort, anemia	water additive used to control microbes
chlorine (as Cl_2)	4.0^a	4.0^a	eye/nose irritation, stomach discomfort	water additive used to control microbes
chlorine dioxide (as ClO_2)	0.8^a	4.0^a	anemia in infants and young children, nervous system effects	water additive used to control microbes

inorganic chemicals	MCLG[a] $(mg/L)^b$	MCL or TT[a] $(mg/L)^b$	potential health effects from ingestion of water	sources of contaminant in drinking water
antimony	0.006	0.006	increase in blood cholesterol; decrease in blood sugar	discharge from petroleum refineries; fire retardants; ceramics; electronics; solder
arsenic	0^g	0.010 as of January 23, 2006	skin damage or problems with circulatory systems; may increase cancer risk	erosion of natural deposits; runoff from orchards; runoff from glass and electronics production wastes
asbestos (fiber > 10 micrometers)	7 million fibers perliter	7 MFL	increased risk of developing benign intestinal polyps	decay of asbestos cement in water mains; erosion of natural deposits
barium	2	2	increase in blood pressure	discharge of drilling wastes; discharge from metal refineries; erosion of natural deposits
beryllium	0.004	0.004	intestinal lesions	discharge from metal refineries and coal-burning factories; discharge from electrical, aerospace, and defense industries
cadmium	0.005	0.005	kidney damage	corrosion of galvanized pipes; erosion of natural deposits; discharge from metal refineries; runoff from waste batteries and paints
chromium (total)	0.1	0.1	allergic dermatitis	discharge from steel and pulp mills; erosion of natural deposits
copper	1.3	TT[h], action level = 1.3	short-term exposure: gastrointestinal distress long-term exposure: liver or kidney damage People with Wilson's disease should consult their personal doctor if the amount of copper in their water exceeds the action level.	corrosion of household plumbing systems; erosion of natural deposits
cyanide (as free cyanide)	0.2	0.2	nerve damage or thyroid problems	discharge from steel/metal factories; discharge from plastic and fertilizer factories

APPENDIX 25.A *(continued)*
National Primary Drinking Water Regulations
Code of Federal Regulations, Title 40, Ch. I, Part 141, Subpart G (CFR)

inorganic chemicals	MCLG[a] $(mg/L)^b$	MCL or TT[a] $(mg/L)^b$	potential health effects from ingestion of water	sources of contaminant in drinking water
fluoride	4.0	4.0	bone disease (pain and tenderness of the bones); children may get mottled teeth	water additive that promotes strong teeth; erosion of natural deposits; discharge from fertilizer and aluminum factories
lead	0	TT[h], action level = 0.015	infants and children: delays in physical or mental development; children could show slight deficits in attention span and learning disabilities adults: kidney problems, high blood pressure	corrosion of household plumbing systems; erosion of natural deposits
mercury (inorganic)	0.002	0.002	kidney damage	erosion of natural deposits; discharge from refineries and factories; runoff from landfills and croplands
nitrate (measured as nitrogen)	10	10	Infants below the age of six months who drink water containing nitrate in excess of the MCL could become seriously ill and, if untreated, may die. Symptoms include shortness of breath and blue baby syndrome.	runoff from fertilizer use; leaching from septic tanks/sewage; erosion of natural deposits
nitrite (measured as nitrogen)	1	1	Infants below the age of six months who drink water containing nitrite in excess of the MCL could become seriously ill and, if untreated, may die. Symptoms include shortness of breath and blue baby syndrome.	runoff from fertilizer use; leaching from septic tanks/sewage; erosion of natural deposits
selenium	0.05	0.05	hair and fingernail loss; numbness in fingers or toes; circulatory problems	discharge from petroleum refineries; erosion of natural deposits; discharge from mines
thalium	0.0005	0.002	hair loss; changes in blood; kidney, intestine, or liver problems	leaching from ore-processing sites; discharge from electronics, glass, and drug factories

organic chemicals	MCLG[a] $(mg/L)^b$	MCL or TT[a] $(mg/L)^b$	potential health effects from ingestion of water	sources of contaminant in drinking water
acrylamide	0	TT[i]	nervous system or blood problems; increased risk of cancer	added to water during sewage/wastewater treatment
alachlor	0	0.002	eye, liver, kidney, or spleen problems; anemia; increased risk of cancer	runoff from herbicide used on row crops
atrazine	0.003	0.003	cardiovascular system or reproductive problems	runoff from herbicide used on row crops
benzene	0	0.005	anemia; decrease in blood platelets; increased risk of cancer	discharge from factories; leaching from gas storage tanks and landfills
benzo(a)pyrene (PAHs)	0	0.0002	reproductive difficulties; increased risk of cancer	leaching from linings of water storage tanks and distribution lines

APPENDIX 25.A *(continued)*
National Primary Drinking Water Regulations
Code of Federal Regulations, Title 40, Ch. I, Part 141, Subpart G (CFR)

organic chemicals	MCLG[a] (mg/L)[b]	MCL or TT[a] (mg/L)[b]	potential health effects from ingestion of water	sources of contaminant in drinking water
carbofuran	0.04	0.04	problems with blood, nervous system, or reproductive system	leaching of soil fumigant used on rice and alfalfa
carbon tetrachloride	0	0.005	liver problems; increased risk of cancer	discharge from chemical plants and other industrial activities
chlordane	0	0.002	liver or nervous system problems; increased risk of cancer	residue of banned termiticide
chlorobenzene	0.1	0.1	liver or kidney problems	discharge from chemical and agricultural chemical factories
2,4-D	0.07	0.07	kidney, liver, or adrenal gland problems	runoff from herbicide used on row crops
dalapon	0.2	0.2	minor kidney changes	runoff from herbicide used on rights of way
1,2-dibromo-3-chloropropane (DBCP)	0	0.0002	reproductive difficulties; increased risk of cancer	runoff/leaching from soil fumigant used on soybeans, cotton, pineapples, and orchards
o-dichlorobenzene	0.6	0.6	liver, kidney, or circulatory system problems	discharge from industrial chemical factories
p-dichlorobenzene	0.007	0.075	anemia; liver, kidney, or spleen damage; changes in blood	discharge from industrial chemical factories
1,2-dichloroethane	0	0.005	increased risk of cancer	discharge from industrial chemical factories
1,1-dichloroethylene	0.007	0.007	liver problems	discharge from industrial chemical factories
cis-1,2-dichloroethylene	0.07	0.07	liver problems	discharge from industrial chemical factories
trans-1,2-dichloroethylene	0.1	0.1	liver problems	discharge from industrial chemical factories
dichloromethane	0	0.005	liver problems; increased risk of cancer	discharge from industrial chemical factories
1,2-dichloropropane	0	0.005	increased risk of cancer	discharge from industrial chemical factories
di(2-ethylhexyl) adipate	0.4	0.04	general toxic effects or reproductive difficulties	discharge from industrial chemical factories
di(2-ethylhexyl) phthalate	0	0.006	reproductive difficulties; liver problems; increased risk of cancer	discharge from industrial chemical factories
dinoseb	0.007	0.007	reproductive difficulties	runoff from herbicide used on soybeans and vegetables
dioxin (2,3,7,8-TCDD)	0	0.00000003	reproductive difficulties; increased risk of cancer	emissions from waste incineration and other combustion; discharge from chemical factories
diquat	0.02	0.02	cataracts	runoff from herbicide use
endothall	0.1	0.1	stomach and intestinal problems	runoff from herbicide use
endrin	0.002	0.002	liver problems	residue of banned insecticide
epichlorohydrin	0	TT[i]	increased cancer risk; over a long period of time, stomach problems	discharge from industrial chemical factories; an impurity of some water treatment chemicals

APPENDIX 25.A *(continued)*
National Primary Drinking Water Regulations
Code of Federal Regulations, Title 40, Ch. I, Part 141, Subpart G (CFR)

organic chemicals	MCLG[a] (mg/L)[b]	MCL or TT[a] (mg/L)[b]	potential health effects from ingestion of water	sources of contaminant in drinking water
ethylbenzene	0.7	0.7	liver or kidney problems	discharge from petroleum refineries
ethylene dibromide	0	0.00005	problems with liver, stomach, reproductive system, or kidneys; increased risk of cancer	discharge from petroleum refineries
glyphosphate	0.7	0.7	kidney problems; reproductive difficulties	runoff from herbicide use
heptachlor	0	0.0004	liver damage; increased risk of cancer	residue of banned termiticide
heptachlor epoxide	0	0.0002	liver damage; increased risk of cancer	breakdown of heptachlor
hexachlorobenzene	0	0.001	liver or kidney problems; reproductive difficulties; increased risk of cancer	discharge from metal refineries and agricultural chemical factories
hexachlorocyclo-pentadiene	0.05	0.05	kidney or stomach problems	discharge from chemical factories
lindane	0.0002	0.0002	liver or kidney problems	runoff/leaching from insecticide used on cattle, lumber, and gardens
methoxychlor	0.04	0.04	reproductive difficulties	runoff/leaching from insecticide used on fruits, vegetables, alfalfa, and livestock
oxamyl (vydate)	0.2	0.2	slight nervous system effects	runoff/leaching from insecticide used on apples, potatoes, and tomatoes
polychlorinated biphenyls (PCBs)	0	0.0005	skin changes; thymus gland problems; immune deficiencies; reproductive or nervous system difficulties; increased risk of cancer	runoff from landfills; discharge of waste chemicals
pentachlorophenol	0	0.001	liver or kidney problems; increased cancer risk	discharge from wood preserving factories
picloram	0.5	0.5	liver problems	herbicide runoff
simazine	0.004	0.004	problems with blood	herbicide runoff
styrene	0.1	0.1	liver, kidney, or circulatory system problems	discharge from rubber and plastic factories; leaching from landfills
tetrachloroethylene	0	0.005	liver problems; increased risk of cancer	discharge from factories and dry cleaners
toluene	1	1	nervous system, kidney, or liver problems	discharge from petroleum factories
toxaphene	0	0.003	kidney, liver, or thyroid problems; increased risk of cancer	runoff/leaching from insecticide used on cotton and cattle
2,4,5-TP (silvex)	0.05	0.05	liver problems	residue of banned herbicide
1,2,4-trichlorobenzene	0.07	0.07	changes in adrenal glands	discharge from textile finishing factories
1,1,1-trichloroethane	0.2	0.2	liver, nervous system, or circulatory problems	discharge from metal degreasing sites and other factories

APPENDIX 25.A *(continued)*
National Primary Drinking Water Regulations
Code of Federal Regulations, Title 40, Ch. I, Part 141, Subpart G (CFR)

organic chemicals	MCLG[a] (mg/L)[b]	MCL or TT[a] (mg/L)[b]	potential health effects from ingestion of water	sources of contaminant in drinking water
1,1,2-trichloroethane	0.003	0.005	liver, kidney, or immune system problems	discharge from industrial chemical factories
trichloroethylene	0	0.005	liver problems; increased risk of cancer	discharge from metal degreasing sites and other factories
vinyl chloride	0	0.002	increased risk of cancer	leaching from PVC pipes; discharge from plastic factories
xylenes (total)	10	10	nervous system damage	discharge from petroleum factories; discharge from chemical factories

radionuclides	MCLG[a] (mg/L)[b]	MCL or TT[a] (mg/L)[b]	potential health effects from ingestion of water	sources of contaminant in drinking water
alpha particles	none[g]	15 pCi/L	increased risk of cancer	erosion of natural deposits of certain minerals that are radioactive and may emit a form of radiation known as alpha radiation
beta particles and photon emitters	none[g]	4 mrem/yr	increased risk of cancer	decay of natural and artificial deposits of certain minerals that are radioactive and may emit forms of radiation known as photons and beta radiation
radium 226 and radium 228 (combined)	none[g]	5 pCi/L	increased risk of cancer	erosion of natural deposits
uranium	0	30 μg/L as of December 8, 2003	increased risk of cancer; kidney toxicity	erosion of natural deposits

[a]Definitions:

Maximum Contaminant Level (MCL): The highest level of a contaminant that is allowed in drinking water. MCLs are set as close to MCLGs as feasible using the best available treatment technology and taking cost into consideration. MCLs are enforceable standards.

Maximum Contaminant Level Goal (MCLG): The level of a contaminant in drinking water below which there is no known or expected risk to health. MCLGs allow for a margin of safety and are non-enforceable public health goals.

Maximum Residual Disinfectant Level (MRDL): The highest level of a disinfectant allowed in drinking water. There is convincing evidence that addition of a disinfectant is necessary for control of microbial contaminants.

Maximum Residual Disinfectant Level Goal (MRDLG): The level of a drinking water disinfectant below which there is no known or expected risk to health. MRDLGs do not reflect the benefits of the use of disinfectants to control microbial contaminants.

Treatment Technique: A required process intended to reduce the level of a contaminant in drinking water.

[b]Units are in milligrams per liter (mg/L) unless otherwise noted. Milligrams per liter are equivalent to parts per million.

[c]The EPA's surface water treatment rules require systems using surface water or ground water under the direct influence of surface water to (1) disinfect their water, and (2) filter their water or meet criteria for avoiding filtration so that the following contaminants are controlled at the following levels.

- Cryptosporidium (as of January 1, 2002, for systems serving >10,000 and January 14, 2005, for systems serving < 10,000): 99% removal
- *Giardia lamblia*: 99.9% removal/inactivation
- *Legionella*: No limit, but the EPA believes that if *Giardia* and viruses are removed/inactivated, *Legionella* will also be controlled.
- Turbidity: At no time can turbidity (cloudiness of water) go above 5 nephelolometric turbidity units (NTU); systems that filter must ensurethat the turbidity go no higher than 1 NTU (0.5 NTU for conventional or direct filtration) in at least 95% of the daily samples in any month. As of January 1, 2002, turbidity may never exceed 1 NTU, and must not exceed 0.3 NTU in 95% of daily samples in any month.
- Heterotrophic plate count (HPC): No more than 500 bacterial colonies per milliliter.

APPENDIX 25.A *(continued)*
National Primary Drinking Water Regulations
Code of Federal Regulations, Title 40, Ch. I, Part 141, Subpart G (CFR)

- Long Term 1 Enhanced Surface Water Treatment (as of January 14, 2005): Surface water systems or ground water under direct influence (GWUDI) systems serving fewer than 10,000 people must comply with the applicable Long Term 1 Enhanced Surface Water Treatment Rule provisions (e.g., turbidity standards, individual filter monitoring, cryptosporidium removal requirements, updated watershed control requirements for unfiltered systems).
- Filter Backwash Recycling: The Filter Backwash Recycling Rule requires systems that recycle to return specific recycle flows through all processes of the systems' existing conventional or direct filtration system or at an alternate location approved by the state.

[d]More than 5.0% of samples are total coliform-positive in a month. (For water systems that collect fewer than 40 routine samples per month, no more than one sample can be total coliform-positive per month.) Every sample that has total coliform must be analyzed for either fecal coliforms or *E. coli*: If two consecutive samples are TC-positive, and one is also positive for *E. coli* or fecal coliforms, the system has an acute MCL violation.

[e]Fecal coliform and *E. coli* are bacteria whose presence indicates that the water may be contaminated with human or animal wastes. Disease-causing microbes (pathogens) in these wastes can cause diarrhea, cramps, nausea, headaches, or other symptoms. These pathogens may pose a special health risk for infants, young children, and people with severely compromised immune systems.

[f]Although there is no collective MCLG for this contaminant group, there are individual MCLGs for some of the individual contaminants.
- Haloacetic acids: dichloroacetic acid (0); trichloroacetic acid (0.3 mg/L). Monochloroacetic acid, bromoacetic acid, and dibromoacetic acid are regulated with this group but have no MCLGs.
- Trihalomethanes: bromodichloromethane (0); bromoform (0); dibromochloromethane (0.06 mg/L). Chloroform is regulated with this group but has no MCLG.

[g]MCLGs were not established before the 1986 Amendments to the Safe Drinking Water Act. Therefore, there is no MCLG for this contaminant.

[h]Lead and copper are regulated by a treatment technique that requires systems to control the corrosiveness of their water. If more than 10% of tap water samples exceed the action level, water systems must take additional steps. For copper, the action level is 1.3 mg/L, and for lead it is 0.015 mg/L.

[i]Each water system agency must certify, in writing, to the state (using third party or manufacturers' certification) that, when acrylamide and epichlorohydrin are used in drinking water systems, the combination (or product) of dose and monomer level does not exceed the levels specified, as follows.
- acrylamide = 0.05% dosed at 1 mg/L (or equivalent)
- epichlorohydrin = 0.01% dosed at 20 mg/L (or equivalent)

Appendices

APPENDIX 26.A
Properties of Chemicals Used in Water Treatment

chemical name	formula	use	molecular weight	equivalent weight
activated carbon	C	taste and odor control	12.0	–
aluminum hydroxide	$Al(OH)_3$	–	78.0	26.0
aluminum sulfate (filter alum)	$Al_2(SO_4)_3 \cdot 14.3H_2O$	coagulation	600	100
ammonia	NH_3	chloramine disinfection	17.0	–
ammonium fluosilicate	$(NH_4)_2SiF_6$	fluoridation	178	89.0
ammonium sulfate	$(NH_4)_2SO_4$	coagulation	132	66.1
calcium bicarbonate*	$Ca(HCO_3)_2$	–	162	81.0
calcium carbonate	$CaCO_3$	corrosion control	100	50.0
calcium fluoride	CaF_2	fluoridation	78.1	39.0
calcium hydroxide	$Ca(OH)_2$	softening	74.1	37.0
calcium hypochlorite	$Ca(ClO)_2 \cdot 2H_2O$	disinfection	179	–
calcium oxide (lime)	CaO	softening	56.1	28.0
carbon dioxide	CO_2	recarbonation	44.0	22.0
chlorine	Cl_2	disinfection	71.0	–
chlorine dioxide	ClO_2	taste and odor control	67.0	–
copper sulfate	$CuSO_4$	algae control	160	79.8
ferric chloride	$FeCl_3$	coagulation	162	54.1
ferric hydroxide	$Fe(OH)_3$	arsenic removal	107	35.6
ferric sulfate	$Fe_2(SO_4)_3$	coagulation	400	66.7
ferrous sulfate (copperas)	$FeSO_4 \cdot 7H_2O$	coagulation	278	139
fluosilicic acid	H_2SiF_6	fluoridation	144	72.0
hydrochloric acid	HCl	pH adjustment	36.5	36.5
magnesium hydroxide	$Mg(OH)_2$	defluoridation	58.3	29.2
oxygen	O_2	aeration	32.0	16.0
ozone	O_3	disinfection	48.0	16.0
potassium permanganate	$KMnO_4$	oxidation	158	158
sodium aluminate	$NaAlO_2$	coagulation	82.0	82.0
sodium bicarbonate (baking soda)	$NaHCO_3$	alkalinity adjustment	84.0	84.0
sodium carbonate (soda ash)	Na_2CO_3	softening	106	53.0
sodium chloride (common salt)	$NaCl$	ion-exchange regeneration	58.4	58.4
sodium diphosphate	Na_2HPO_4	corrosion control	142	71.0
sodium fluoride	NaF	fluoridation	42.0	42.0
sodium fluosilicate	Na_2SiF_6	fluoridation	188	99.0
sodium hexametaphosphate	$(NaPO_3)_n$	corrosion control	–	–
sodium hydroxide	$NaOH$	pH adjustment	40.0	40.0
sodium hypochlorite	$NaClO$	disinfection	74.0	–
sodium phosphate	NaH_2PO_4	corrosion control	120	120
sodium silicate	Na_2OSiO_2	corrosion control	184	92.0

Appendices

APPENDIX 26.A *(continued)*
Properties of Chemicals Used in Water Treatment

chemical name	formula	use	weight	weight
sodium thiosulfate	$Na_2S_2O_3$	dechlorination	158	79.0
sodium tripolyphosphate	$Na_5P_3O_{10}$	corrosion control	368	–
sulfur dioxide	SO_2	dechlorination	64.1	–
sulfuric acid	H_2SO_4	pH adjustment	98.1	49.0
trisodium phosphate	Na_3PO_4	corrosion control	118	–
water	H_2O	–	18.0	–

*Exists only in aqueous form as a mixture of ions.

APPENDIX 29.A
Selected *Ten States' Standards*[*]

[11.243] **Hydraulic Load:** Use 100 gpcd (0.38 m³/d) for new systems in undeveloped areas unless other information is available.

[42.3] **Pumps:** At least two pumps are required. Both pumps should have the same capacity if only two pumps are used. This capacity must exceed the total design flow. If three or more pumps are used, the capacities may vary, but capacity (peak hourly flow) pumping must be possible with one pump out of service.

[42.7] **Pump Well Ventilation:** Provide 30 complete air changes per hour for both wet and dry wells using intermittent ventilation. For continuous ventilation, the requirement is reduced to 12 (for wet wells) and 6 (for dry wells) air changes per hour. In general, ventilation air should be forced in, as opposed to air being extracted and replaced by infiltration.

[61.12] **Racks and Bar Screens:** All racks and screens shall have openings less than 1.75 in (45 mm) wide. The smallest opening for manually cleaned screens is 1 in (2.54 cm). The smallest opening for mechanically cleaned screens may be smaller. Flow velocity must be 1.25–3.0 ft/sec (38–91 cm/s).

[62.2–62.3] **Grinders and Shredders:** Comminutors are required if there is no screening. Gravel traps or grit-removal equipment should precede comminutors.

[63.3–63.4] **Grit Chambers:** Grit chambers are required when combined storm and sanitary sewers are used. A minimum of two grit chambers in parallel should be used, with a provision for bypassing. For channel-type grit chambers, the optimum velocity is 1 ft/sec (30 cm/s)throughout. The detention time for channel-type grit chambers is dependent on the particle sizes to be removed.

[71–72] **Settling Tanks:** Multiple units are desirable, and multiple units must be provided if the average flow exceeds 100,000 gal/day (379 m³/d). For primary settling, the sidewater depth should be 10 ft (3.0 m) or greater. For tanks not receiving activated sludge, the design average overflow rate is 1000 gal/day-ft² (41 m³/d·m²); the maximum peak overflow rate is 1500–2000 gal/day-ft² (61–81 m³/d·m²). The maximum weir loading for flows greater than 1 MGD (3785 m³/d) is 30,000 gal/day-ft(375 m³/d·m). The basin size shall also be calculated based on the average design flow rate and a maximum settling rate of 1000 gal/day-ft² (41 m³/d·m²). The larger of the two sizes shall be used. If the flow rate is less than 1 MGD (3785 m³/d), the maximum weir loading is reduced to 20,000 gal/day-ft (250 m³/d·m).

For settling tanks following trickling filters and rotating biological contactors, the peak settling rate is 1500 gal/day-ft²(61 m³/d·m²).

For settling tanks following activated sludge processes, the maximum hydraulic loadings are: 1200 gal/day-ft² (49 m³/d·m²) for conventional, step, complete mix, and contact units; 1000 gal/day-ft² (41 m³/d·m²) for extended aeration units; and 800 gal/day-ft²(33 m³/d·m²) for separate two-stage nitrification units.

[84] **Anaerobic Digesters:** Multiple units are required. Minimum sidewater depth is 20 ft (6.1 m). For completely mixed digesters, maximum loading of volatile solids is 80 lbm/day-1000 ft³ (1.3 kg/d·m³) of digester volume. For moderately mixed digesters, the limit is 40 lbm/day-1000 ft³ (0.65 kg/d·m³).

[88.22] **Sludge Drying Beds:** For design purposes, the maximum sludge depth is 8 in (200 mm).

[91.3] **Trickling Filters:** All media should have a minimum depth of 6 ft (1.8 m). Rock media depth should not exceed 10 ft (3 m). Manufactured media depth should not exceed the manufacturer's recommendations. The rock media should be 1–4.5 in (2.5–11.4 cm) in size. Freeboard of 4 ft (1.2 ft) or more is required for manufactured media. The drain should slope at 1% or more, and the average drain velocity should be 2 ft/sec (0.6 m/s).

[92.3] **Activated Sludge Processes:** The maximum BOD loading shall be 40 lbm/day-1000 ft³ (0.64 kg/d·m³) for conventional, step, and complete-mix units; 50 lbm/day-1000 ft³ (0.8 kg/d·m³) for contact stabilization units; and 15 lbm/day-1000 ft³(0.24 kg/d·m³) for extended aeration units and oxidation ditch designs.

Aeration tank depths should be 10–30 ft (3–9 m). Freeboard should be 18 in (460 mm) or more. At least two aeration tanks shall be used. The dissolved oxygen content should not be allowed to drop below 2 mg/L at any time. The aeration rate should be 1500 ft³ of oxygen per lbm of BOD_5 (94 m³/kg). For extended aeration, the rate should be 2050 ft³ of oxygen per lbm of BOD_5 (128 m³/kg).

[93] **Wastewater Treatment Ponds:** (Applicable to controlled-discharge, flow-through (facultative), and aerated pond systems.) Pond bottoms must be at least 4 ft (1.2 m) above the highest water table elevation. Pond primary cells are designed based onaverage BOD_5 loads of 15–35 lbm/ac-day (17–40 kg/ha·d) at the mean operating depth. Detention time for controlled-discharge ponds shall be at least 180 days between a depth of 2 ft (0.61 m) and the maximum operating depth. Detention time for flow-through (facultative) ponds should be 90–120 days, modified for cold weather. Detention time for aerated ponds may be estimated from the percentage, E, of BOD_5 to be removed: $t_{days} = E/2.3k_1(100\% - E)$, where k_1 is the base-10 reaction coefficient, assumed to be 0.12/d at 68°F (20°C) and 0.06/d at 34°F (1°C). Although two cells can be used in very small systems, a minimum of three cells is normally required. Maximum cell size is 40 ac (16 ha). Ponds may be round, square, or rectangular (with length no greater than three times the width). Islands and sharp corners are not permitted. Dikes shall be at least 8 ft (2.4 m) wide at the top. Inner and outer dike slopes cannot be steeper than 1:3 (*V:H*). Inner slopes cannot be flatter than 1:4 (*V:H*). At least 3 ft (0.91 m) freeboard is required unless the pond system is very small, in which case 2 ft (0.6 m) is acceptable. Pond depth shall never be less than 2 ft (0.6 m). Depth shall be: (a) for controlled discharged and flow-through (facultative) ponds, a maximum of

APPENDIX 29.A *(continued)*
Selected *Ten States' Standards*[*]

6 ft (1.8 m) for primary cells; greater for subsequent cells with mixing and aeration as required; and (b) for aerated ponds, a design depth of 10–15 ft (3–4.5 m).

[102.4] **Chlorination:** Minimum contact time is 15 min at peak flow.

---

[*]Numbers in square brackets refer to sections in *Recommended Standards for Sewage Works*, 2004 ed., Great Lakes-Upper Mississippi River Board of State Sanitary Engineers, published by Health Education Service, NY, from which these guidelines were extracted. Refer to the original document for complete standards.

APPENDIX 35.A
USCS Soil Boring, Well, and Geotextile Symbols

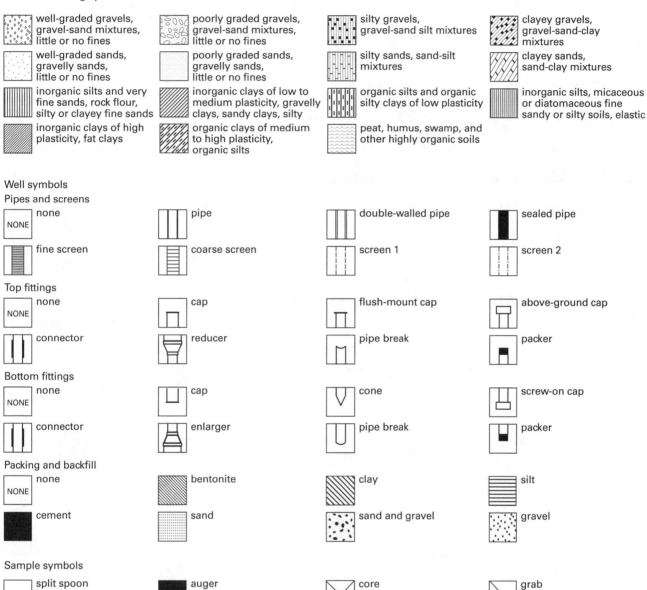

APPENDIX 35.A *(continued)*
USCS Soil Boring, Well, and Geotextile Symbols

Geotextile symbols

GT	– – – – – – – – – – – – – – – – – –	geotextile
GM	————————————	geomembrane
GG	—•—•—•—•—•—•—	geogrid
GCD	∿∿∿∿∿∿∿∿∿∿∿∿	geocomposite drain – with geotextile cover
GN	××××××××××××××××××××	geonet
GCL	ℒℒℒℒℒℒℒℒℒℒℒℒℒℒℒℒℒℒℒ	geocomposite clay liner
GEC	###########################	surficial geosynthetic erosion control
GL	IIIIIIIIIIIIIIIIIIIIIII	geocell
GA	∿∿∿∿∿∿∿∿∿∿∿	geomat
EKG	ƶƶƶƶƶƶƶƶƶƶƶƶƶƶƶƶƶƶƶƶƶƶƶƶƶƶ	electrokinetic geosynthetic

The following function symbols may be used where a description of the role of the geosynthetic material is needed.

S: separation geotextile

Ⓢ
↘
– – – – – – – – – – –

R: reinforcement geotextile

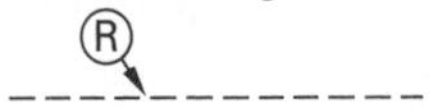

Ⓡ
↘
– – – – – – – – – – –

B: barrier (fluid)

D: drainage (fluid)

E: surficial erosion control

F: filtration

P: protection

R: reinforcement

S: separation

APPENDIX 37.A
Active Components for Retaining Walls (straight slope backfill)
(walls not over 20 ft high) (lbf/ft² per linear foot)

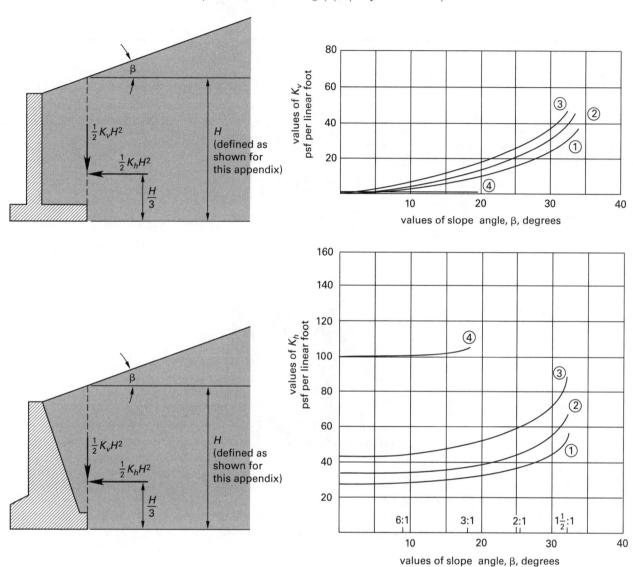

Circled numbers indicate the following soil types.

1. Clean sand and gravel: GW, GP, SW, SP.
2. Dirty sand and gravel of restricted permeability: GM, GM-GP, SM, SM-SP.
3. Stiff residual silts and clays, silty fine sands, and clayey sands and gravels: CL, ML, CH, MH, SM, SC, GC.
4. Soft or very soft clay, organic silt, or silty clay.

Source: *Foundations and Earth Structures*, NAVFAC DM-7.2 (1986), p. 7.2-86, Fig. 16

APPENDIX 37.B
Active Components for Retaining Walls (broken slope backfill)
(lbf/ft² per linear foot)

soil type 1   soil type 2   soil type 3

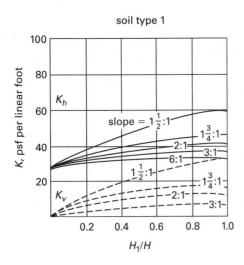
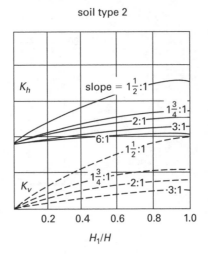
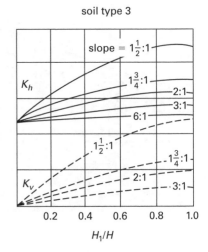

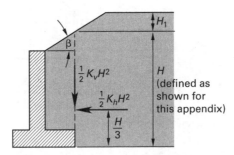

Soil types:

1. Clean sand and gravel: GW, GP, SW, SP.
2. Dirty sand and gravel of restricted permeability: GM, GM-GP, SM, SM-SP.
3. Stiff residual silts and clays, silty fine sands, and clayey sands and gravels: CL, ML, CH, MH, SM, SC, GC.

Source: *Foundations and Earth Structures*, NAVFAC DM-7.2 (1986), p. 7.2-87, Fig. 17

APPENDIX 37.C
Curves for Determining Active and Passive Earth Pressure Coefficients, k_a and k_p
(with inclined wall face, θ, wall friction, δ, and horizontal backfill) (after Terzaghi)

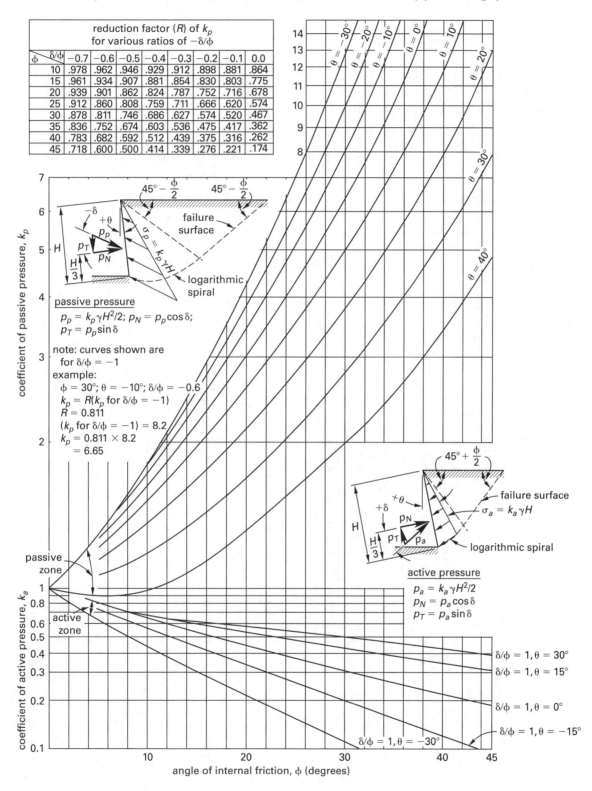

reduction factor (R) of k_p for various ratios of $-\delta/\phi$								
ϕ \ δ/ϕ	-0.7	-0.6	-0.5	-0.4	-0.3	-0.2	-0.1	0.0
10	.978	.962	.946	.929	.912	.898	.881	.864
15	.961	.934	.907	.881	.854	.830	.803	.775
20	.939	.901	.862	.824	.787	.752	.716	.678
25	.912	.860	.808	.759	.711	.666	.620	.574
30	.878	.811	.746	.686	.627	.574	.520	.467
35	.836	.752	.674	.603	.536	.475	.417	.362
40	.783	.682	.592	.512	.439	.375	.316	.262
45	.718	.600	.500	.414	.339	.276	.221	.174

passive pressure
$p_p = k_p \gamma H^2/2$; $p_N = p_p \cos \delta$;
$p_T = p_p \sin \delta$

note: curves shown are
for $\delta/\phi = -1$
example:
$\phi = 30°$; $\theta = -10°$; $\delta/\phi = -0.6$
$k_p = R(k_p \text{ for } \delta/\phi = -1)$
$R = 0.811$
$(k_p \text{ for } \delta/\phi = -1) = 8.2$
$k_p = 0.811 \times 8.2$
$\quad = 6.65$

active pressure
$p_a = k_a \gamma H^2/2$
$p_N = p_a \cos \delta$
$p_T = p_a \sin \delta$

$\delta/\phi = 1, \theta = 30°$
$\delta/\phi = 1, \theta = 15°$
$\delta/\phi = 1, \theta = 0°$
$\delta/\phi = 1, \theta = -15°$
$\delta/\phi = 1, \theta = -30°$

angle of internal friction, ϕ (degrees)

Source: *Foundations and Earth Structures*, NAVFAC DM-7.2 (1986), p. 7.2-66, Fig. 5

APPENDIX 37.D
Curves for Determining Active and Passive Earth Pressure Coefficients, k_a and k_p
(with vertical face, wall friction, δ, and sloping backfill, β) (after Terzaghi)

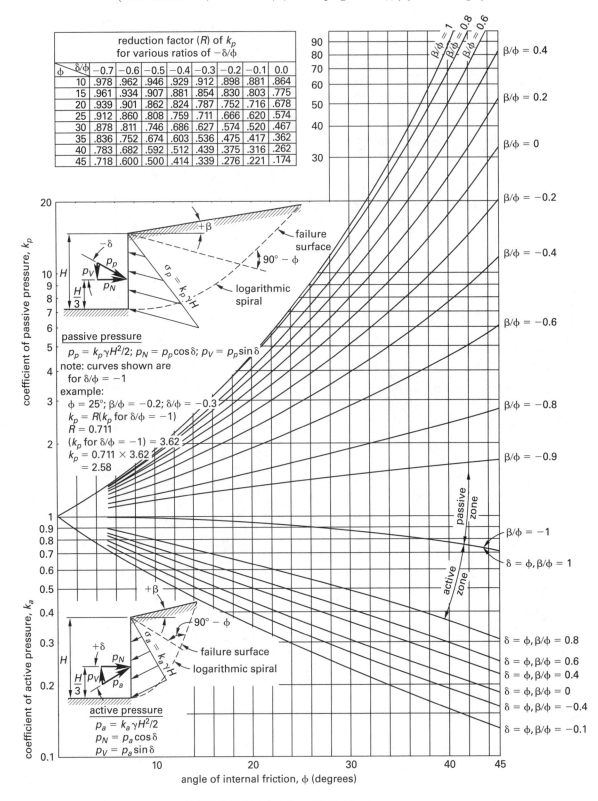

Source: *Foundations and Earth Structures*, NAVFAC DM-7.2 (1986), p. 7.2-67, Fig. 6

APPENDIX 40.A
Boussinesq Stress Contour Chart
(infinitely long and square footings)

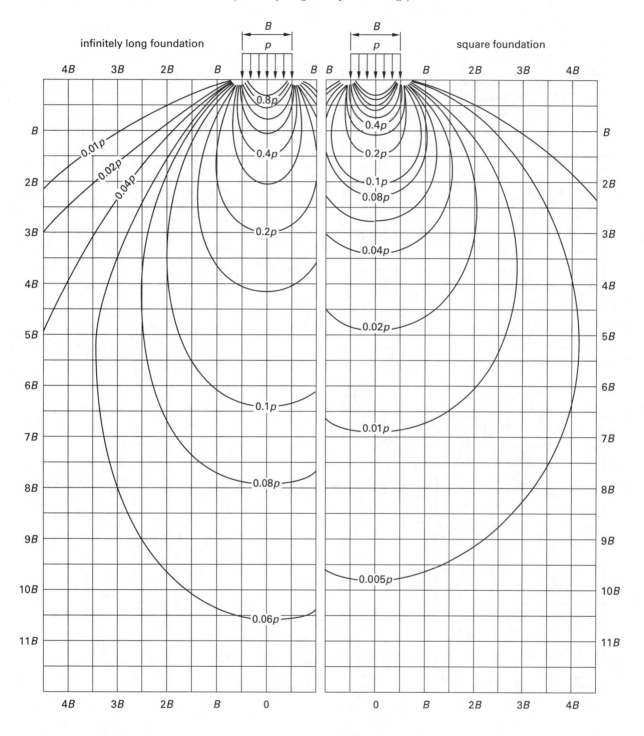

Appendices

APPENDIX 40.B
Boussinesq Stress Contour Chart
(uniformly loaded circular footings)

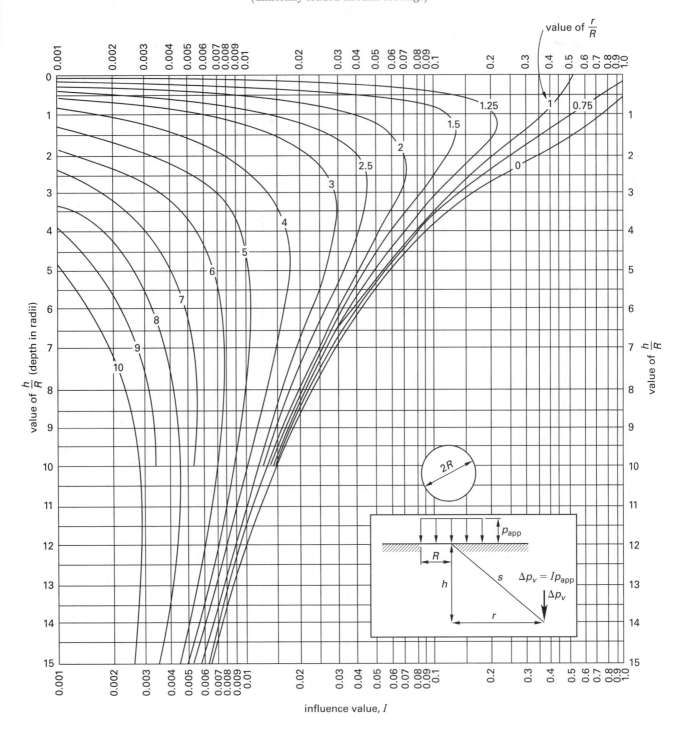

influence value, I

APPENDIX 42.A
Centroids and Area Moments of Inertia for Basic Shapes

shape		centroidal location		area, A	area moment of inertia (rectangular and polar), I, J	radius of gyration, r
		x_c	y_c			
rectangle		$\dfrac{b}{2}$	$\dfrac{h}{2}$	bh	$I_x = \dfrac{bh^3}{3}$ $I_{cx} = \dfrac{bh^3}{12}$ $J_c = \left(\dfrac{1}{12}\right)bh(b^2 + h^2)*$ (see note below)	$r_x = \dfrac{h}{\sqrt{3}}$ $r_{cx} = \dfrac{h}{2\sqrt{3}}$
triangular area		$\dfrac{2b}{3}$	$\dfrac{h}{3}$	$\dfrac{bh}{2}$	$I_x = \dfrac{bh^3}{12}$ $I_{cx} = \dfrac{bh^3}{36}$	$r_x = \dfrac{h}{\sqrt{6}}$ $r_{cx} = \dfrac{h}{3\sqrt{2}}$
trapezoid		$\dfrac{2tz + t^2 + zb + tb + b^2}{3b + 3t}$	$\left(\dfrac{h}{3}\right)\left(\dfrac{b + 2t}{b + t}\right)$	$\dfrac{(b + t)h}{2}$	$I_x = \dfrac{(b + 3t)h^3}{12}$ $I_{cx} = \dfrac{(b^2 + 4bt + t^2)h^3}{36(b + t)}$	$r_x = \left(\dfrac{h}{\sqrt{6}}\right)\sqrt{\dfrac{b + 3t}{b + t}}$ $r_{cx} = \dfrac{h\sqrt{2(b^2 + 4bt + t^2)}}{6(b + t)}$
circle		0	0	πr^2	$I_x = I_y = \dfrac{\pi r^4}{4}$ $J_c = \dfrac{\pi r^4}{2}$	$r_x = \dfrac{r}{2}$
quarter-circular area		$\dfrac{4r}{3\pi}$	$\dfrac{4r}{3\pi}$	$\dfrac{\pi r^2}{4}$	$I_x = I_y = \dfrac{\pi r^4}{16}$ $J_o = \dfrac{\pi r^4}{8}$	
semicircular area		0	$\dfrac{4r}{3\pi}$	$\dfrac{\pi r^2}{2}$	$I_x = I_y = \dfrac{\pi r^4}{8}$ $I_{cx} = 0.1098r^4$ $J_o = \dfrac{\pi r^4}{4}$ $J_c = 0.5025r^4$	$r_x = \dfrac{r}{2}$ $r_{cx} = 0.264r$
quarter-elliptical area		$\dfrac{4a}{3\pi}$	$\dfrac{4b}{3\pi}$	$\dfrac{\pi ab}{4}$	$I_x = \dfrac{\pi ab^3}{8}$ $I_y = \dfrac{\pi a^3 b}{8}$ $J_o = \dfrac{\pi ab(a^2 + b^2)}{8}$	
semielliptical area		0	$\dfrac{4b}{3\pi}$	$\dfrac{\pi ab}{2}$		
semiparabolic area		$\dfrac{3a}{8}$	$\dfrac{3h}{5}$	$\dfrac{2ah}{3}$		
parabolic area		0	$\dfrac{3h}{5}$	$\dfrac{4ah}{3}$	$I_x = \dfrac{4ah^3}{7}$ $I_y = \dfrac{4ha^3}{15}$ $I_{cx} = \dfrac{16ah^3}{175}$	$r_x = h\sqrt{\dfrac{3}{7}}$ $r_y = \dfrac{a}{\sqrt{5}}$
parabolic spandrel		$\dfrac{3a}{4}$	$\dfrac{3h}{10}$	$\dfrac{ah}{3}$	$I_x = \dfrac{ah^3}{21}$ $I_y = \dfrac{3ha^3}{15}$	
general spandrel		$\left(\dfrac{n + 1}{n + 2}\right)a$	$\left(\dfrac{n + 1}{4n + 2}\right)h$	$\dfrac{ah}{n + 1}$	(note to accompany rectangular area above) *Theoretical definition based on $J = I_x + I_y$. However, in torsion, not all parts of the shape are effective. Effective values will be lower.	
circular sector [α in radians]		$\dfrac{2r \sin \alpha}{3\alpha}$	0	αr^2	$J = C\left(\dfrac{b^2 + h^2}{b^3 h^3}\right)$	

b/h	C
1	3.56
2	3.50
4	3.34
8	3.21

APPENDIX 43.A
Typical Properties of Structural Steel, Aluminum, and Magnesium
(all values in ksi)

structural steel

designation	application	S_u	S_y	approximate S_e
A36	shapes	58–80	36	29–40
	plates	58–80	36	29–40
A53	pipe	60	35	30
A242	shapes	70	50	35
	plates to $\frac{3}{4}$ in	70	50	35
A440	shapes	70	50	35
	plates to $\frac{3}{4}$ in	70	50	35
A441	shapes	70	50	35
	plates to $\frac{3}{4}$ in	70	50	35
A500	tubes	45	33	22
A501	tubes	58	36	29
A514	plates to $\frac{3}{4}$ in	115–135	100	55
A529	shapes	60–85	42	30–42
	plates to $\frac{1}{2}$ in	60–85	42	30–42
A570	sheet/strip	55	40	27
A572	shapes	65	50	30
	plates	60	42	30
A588	shapes	70	50	35
	plates to 4 in	70	50	35
A606	hot-rolled sheet	70	50	35
	cold-rolled sheet	65	45	32
A607	sheet	60	45	30
A618	shapes	70	50	35
	tubes	70	50	35
A913	shapes	65	50	30
A992	shapes	65	50	–

structural aluminum

designation	application	S_u	S_y	approximate S_e (at 10^8 cyc.)
2014-T6	shapes/bars	63	55	19
6061-T6	all	42	35	14.5

structural magnesium

designation	application	S_u	S_y	approximate S_e (at 10^7 cyc.)
AZ31	shapes	38	29	19
AZ61	shapes	45	33	19
AZ80	shapes	55	40	–

(Multiply ksi by 6.895 to obtain MPa.)

APPENDIX 43.B
Typical Mechanical Properties of Representative Metals
(room temperature)

The following mechanical properties are not guaranteed since they are averages for various sizes, product forms, and methods of manufacture. Thus, this data is not for design use, but is intended only as a basis for comparing alloys and tempers.

material designation, composition, typical use, and source if applicable	condition, heat treatment	S_{ut} (ksi)	S_{yt} (ksi)
IRON BASED			
Armco ingot iron, for fresh and saltwater piping	normalized	44	24
AISI 1020, plain carbon steel, for general machine parts and screws and carburized parts	hot rolled / cold worked	65 / 78	43 / 66
AISI 1030, plain carbon steel, for gears, shafts, levers, seamless tubing, and carburized parts	cold drawn	87	74
AISI 1040, plain carbon steel, for high-strength parts, shafts, gears, studs, connecting rods, axles, and crane hooks	hot rolled / cold worked / hardened	91 / 100 / 113	58 / 88 / 86
AISI 1095, plain carbon steel, for handtools, music wire springs, leaf springs, knives, saws, and agricultural tools such as plows and disks	annealed / hot rolled / hardened	100 / 142 / 180	53 / 84 / 118
AISI 1330, manganese steel, for axles and drive shafts	annealed / cold drawn / hardened	97 / 113 / 122	83 / 93 / 100
AISI 4130, chromium-molybdenum steel, for high-strength aircraft structures	annealed / hardened	81 / 161	52 / 137
AISI 4340, nickel-chromium-molybdenum steel, for large-scale, heavy-duty, high-strength structures	annealed / as rolled / hardened	119 / 192 / 220	99 / 147 / 200
AISI 2315, nickel steel, for carburized parts	as rolled / cold drawn	85 / 95	56 / 75
AISI 2330, nickel steel	as rolled / cold drawn / annealed / normalized	98 / 110 / 80 / 95	65 / 90 / 50 / 61
AISI 3115, nickel-chromium steel for carburized parts	cold drawn / as rolled / annealed	95 / 75 / 71	70 / 60 / 62
STAINLESS STEELS			
AISI 302, stainless steel, most widely used, same as 18-8	annealed / cold drawn	90 / 105	35 / 60
AISI 303, austenitic stainless steel, good machineability	annealed / cold worked	90 / 110	35 / 75
AISI 304, austenitic stainless steel, good machineability and weldability	annealed / cold worked	85 / 110	30 / 75
AISI 309, stainless steel, good weldability, high strength at high temperatures, used in furnaces and ovens	annealed / cold drawn	90 / 110	35 / 65
AISI 316, stainless steel, excellent corrosion resistance	annealed / cold drawn	85 / 105	35 / 60
AISI 410, magnetic, martensitic, can be quenched and tempered to give varying strength	annealed / cold drawn / oil quenched and drawn	60 / 180 / 110	32 / 150 / 91
AISI 430, magnetic, ferritic, used for auto and architectural trim and for equipment in food and chemical industries	annealed / cold drawn	60 / 100	35
AISI 502, magnetic, ferritic, low cost, widely used in oil refineries	annealed	60	25

APPENDIX 43.B *(continued)*
Typical Mechanical Properties of Representative Metals
(room temperature)

material designation, composition, typical use, and source if applicable	condition, heat treatment	S_{ut} (ksi)	S_{yt} (ksi)
ALUMINUM BASED			
2011, for screw machine parts, excellent machineability, but not weldable, and corrosion sensitive	T3	55	43
	T8	59	45
2014, for aircraft structures, weldable	T3	63	40
	T4, T451	61	37
	T6, T651	68	60
2017, for screw machine parts	T4, T451	62	40
2018, for engine cylinders, heads, and pistons	T61	61	46
2024, for truck wheels, screw machine parts, and aircraft structures	T3	65	45
	T4, T351	64	42
	T361	72	57
2025, for forgings	T6	58	37
2117, for rivets	T4	43	24
2219, high-temperature applications (up to 600°F), excellent weldability and machineabilty	T31, T351	52	36
	T37	57	46
	T42	52	27
3003, for pressure vessels and storage tanks, poor machine-ability but good weldability, excellent corrosion resistance	0	16	6
	H12	19	18
	H14	22	21
	H16	26	25
3004, same characteristics as 3003	0	26	10
	H32	31	25
	H34	35	29
	H36	38	33
4032, pistons	T6	55	46
5083, unfired pressure vessels, cryogenics, towers, and drilling rigs	0	42	21
	H116, H117, H321	46	33
5154, saltwater services, welded structures, and storage tanks	0	35	17
	H32	39	30
	H34	42	33
5454, same characteristics as 5154	0	36	17
	H32	40	30
	H34	44	35
5456, same characteristics as 5154	0	45	23
	H111	47	33
	H321, H116, H117	51	37
6061, corrosion resistant and good weldability, used in railroad cars	T4	33	19
	T6	42	37
7178, Alclad, corrosion-resistant	0	33	15
	T6	88	78

CAST IRON (note redefinition of columns)		S_{ut} (ksi)	S_{us} (ksi)	S_{uc} (ksi)
gray cast iron	class 20	30	32.5	30
	class 25	25	34	100
	class 30	30	41	110
	class 35	35	49	125
	class 40	40	52	135
	class 50	50	64	160
	class 60	60	60	150

Appendices

APPENDIX 43.B *(continued)*
Typical Mechanical Properties of Representative Metals
(room temperature)

material designation, composition, typical use, and source if applicable	condition, heat treatment	S_{ut} (ksi)	S_{yt} (ksi)
COPPER BASED			
copper, commercial purity	annealed (furnace cool from 400°C)	32	10
	cold drawn	45	40
cartridge brass: 70% Cu, 30% Zn	cold rolled (annealed 400°C, furnace cool)	76	63
copper-beryllium (1.9% Be, 0.25% Co)	annealed, wqf 1450°F	70	
	cold rolled	200	
	hardened after annealing	200	150
phosphor-bronze, for springs	wire, 0.025 in and under	145	
	0.025 in to 0.0625 in	135	
	0.125 in to 0.250 in	125	
monel metal	cold-drawn bars, annealed	70	30
red brass	sheet and strip half-hard	51	
	hard	63	
	spring	78	
yellow brass	sheet and strip half-hard	55	
	hard	68	
	spring	86	
NICKEL BASED			
pure nickel, magnetic, high corrosion resistance	annealed (ht 1400°F, acrt)	46	8.5
	annealed at 2050°F	125	75
Inconel X, type 550, excellent high temperature properties	annealed and age hardened	175	110
	annealed (wqf 1600°F)	100	45
K-monel, excellent high temperature properties and corrosion resistance	age hardened spring stock	185	160
Invar, 36% Ni, 64% Fe, low coefficient of expansion (1.2×10^{-6} 1/°C, 0-200°C)	annealed (wqf 800°C)	71	40
REFRACTORY METALS (properties at room temperature)			
molybdenum	as rolled	100	75
tantalum	annealed at 1050°C in vacuum	60	45
	as rolled	110	100
titanium, commercial purity	annealed at 1200°F	95	80
titanium, 6% Al, 4% V	annealed at 1400°F, acrt	135	130
	heat treated (wqf 1750°F, ht 1000°F, acrt)	170	150
titanium, 4% Al, 4% Mn OR 5% Al, 2.75% Cr, 1.25% Fe OR 5% Al, 1.5% Fe, 1.4% Cr, 1.2% Mo	wqf 1450°F, ht 900°F, acrt	185	170
tungsten, commercial purity	hard wire	600	540

MAGNESIUM		S_{ut} (ksi)	S_{yt} (ksi)	S_{us} (ksi)
AZ92, for sand and permanent-mold casting	as cast	24	14	
	solution treated	39	14	
	aged	39	21	
AZ91, for die casting	as cast	33	21	
AZ31X (sheet)	annealed	35	20	
	hard	40	31	
AZ80X, for structural shapes	extruded	48	32	
	extruded and aged	52	37	
ZK60A, for structural shapes	extruded	49	38	
	extruded and aged	51	42	
AZ31B (sheet and plate), for structural shapes in use below 300°F	temper 0	32	15	17
	temper 1124	34	18	18
	temper 1126	35	21	18
	temper F	32	16	17

Abbreviations:
   wqf: water-quench from
   acrt: air-cooled to room temperature
   ht: heated to

(Multiply ksi by 6.895 to obtain MPa.)

APPENDIX 43.C
Typical Mechanical Properties of Thermoplastic Resins and Composites
(room temperature, after post-mold annealing)

base resin and glass content (% by wt)	specific gravity	tensile yield strength (lbf/in²)	flexural modulus (kips/in²)	flexural strength (lbf/in²)	impact strength, Izod notched/ unnotched (ft-lbf/in)	deflection temperature at 264 lbf/in² (°F)	coefficient of thermal expansion (10⁻⁵ in/in-°F)
ASTM test →	D792	D638	D790	D790	D256	D648	D696
polyimide (30)	1.48	14,000	950	22,000	1.8/9	530	2.3
ethylene tetrafluoro-ethylene (ETFE) (20)	1.82	11,000	750	15,000	7/13	440	2.0
fluorinated ethylenepropylene (FEP) (20)	2.21	5000	800	10,500	8/17	350	2.4
polyphenylene sulfide (40)	1.56	23,000	1800	32,000	1.5/11	505	1.1
polyethersulfone (40)	1.68	22,000	1600	30,000	1.6/12	420	1.6
nylon 6/6 (50)	1.57	32,000	2200	46,500	3.3/20	500	1.0
polyester (40)	1.62	22,000	1600	32,000	2/12	475	1.05
polysulfone (40)	1.55	20,000	1600	27,000	2/16	370	1.2
polyarylsulfone (0)	1.36	13,000	395	17,000	5/40	525	2.6
poly-p-oxybenzoate (0)	1.40	14,000	700	17,000	1/3	560	1.6
polyamide-imide (0)	1.40	27,400	665	30,500	2.5/14	525	1.8

(Multiply kips/in² by 6.895 to obtain MPa.)

(Multiply lbf/in² by 0.006895 to obtain MPa.)

(Multiply ft-lbf/in by 53.38 to obtain N·m/m.)

(Multiply in/in-°F by 1.8 to obtain m/m·K.)

APPENDIX 44.A
Elastic Beam Deflection Equations
(w is the load per unit length.) (y is positive downward.)

Case 1: Cantilever with End Load

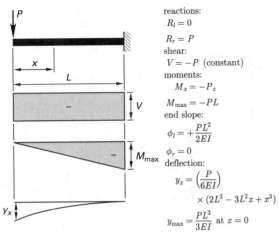

reactions:
$$R_l = 0$$
$$R_r = P$$
shear:
$$V = -P \text{ (constant)}$$
moments:
$$M_x = -P_x$$
$$M_{max} = -PL$$
end slope:
$$\phi_l = +\frac{PL^2}{2EI}$$
$$\phi_r = 0$$
deflection:
$$y_x = \left(\frac{P}{6EI}\right)$$
$$\times (2L^3 - 3L^2x + x^3)$$
$$y_{max} = \frac{PL^3}{3EI} \text{ at } x = 0$$

Case 2: Cantilever with Uniform Load

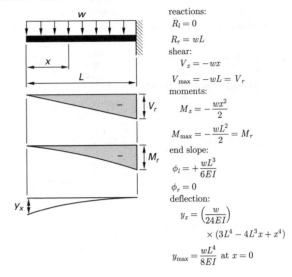

reactions:
$$R_l = 0$$
$$R_r = wL$$
shear:
$$V_x = -wx$$
$$V_{max} = -wL = V_r$$
moments:
$$M_x = -\frac{wx^2}{2}$$
$$M_{max} = -\frac{wL^2}{2} = M_r$$
end slope:
$$\phi_l = +\frac{wL^3}{6EI}$$
$$\phi_r = 0$$
deflection:
$$y_x = \left(\frac{w}{24EI}\right)$$
$$\times (3L^4 - 4L^3x + x^4)$$
$$y_{max} = \frac{wL^4}{8EI} \text{ at } x = 0$$

Case 3: Cantilever with Triangular Load

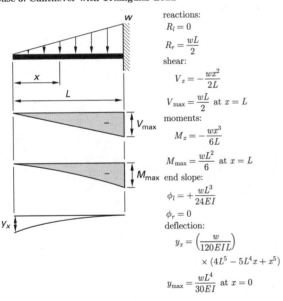

reactions:
$$R_l = 0$$
$$R_r = \frac{wL}{2}$$
shear:
$$V_x = -\frac{wx^2}{2L}$$
$$V_{max} = \frac{wL}{2} \text{ at } x = L$$
moments:
$$M_x = -\frac{wx^3}{6L}$$
$$M_{max} = \frac{wL^2}{6} \text{ at } x = L$$
end slope:
$$\phi_l = +\frac{wL^3}{24EI}$$
$$\phi_r = 0$$
deflection:
$$y_x = \left(\frac{w}{120EIL}\right)$$
$$\times (4L^5 - 5L^4x + x^5)$$
$$y_{max} = \frac{wL^4}{30EI} \text{ at } x = 0$$

Case 4: Propped Cantilever with Uniform Load

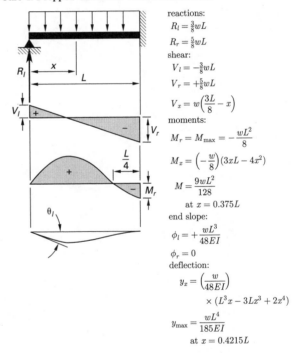

reactions:
$$R_l = \tfrac{3}{8}wL$$
$$R_r = \tfrac{5}{8}wL$$
shear:
$$V_l = -\tfrac{3}{8}wL$$
$$V_r = +\tfrac{5}{8}wL$$
$$V_x = w\left(\frac{3L}{8} - x\right)$$
moments:
$$M_r = M_{max} = -\frac{wL^2}{8}$$
$$M_x = \left(-\frac{w}{8}\right)(3xL - 4x^2)$$
$$M = \frac{9wL^2}{128}$$
$$\text{at } x = 0.375L$$
end slope:
$$\phi_l = +\frac{wL^3}{48EI}$$
$$\phi_r = 0$$
deflection:
$$y_x = \left(\frac{w}{48EI}\right)$$
$$\times (L^3x - 3Lx^3 + 2x^4)$$
$$y_{max} = \frac{wL^4}{185EI}$$
$$\text{at } x = 0.4215L$$

APPENDIX 44.A *(continued)*
Elastic Beam Deflection Equations
(w is the load per unit length.) (y is positive downward.)

Case 5: Cantilever with End Moment

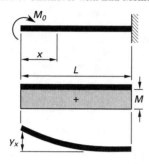

reactions:
$$R_l = 0$$
$$R_r = 0$$
shear:
$$V = 0$$
moments:
$$M = M_0 = M_{max}$$
end slope:
$$\phi_l = -\frac{M_0 L}{EI}$$
$$\phi_r = 0$$
deflection:
$$y_x = \left(-\frac{M_0}{2EI}\right)$$
$$\times (L^2 - 2xL + x^2)$$
$$y_{max} = -\frac{M_0 L^2}{2EI}$$
$$\text{at } x = 0$$

Case 6: Simple Beam with Center Load

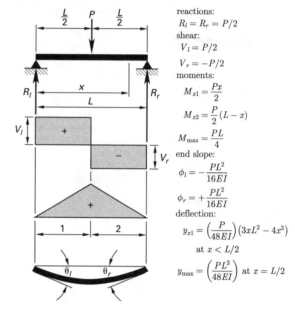

reactions:
$$R_l = R_r = P/2$$
shear:
$$V_l = P/2$$
$$V_r = -P/2$$
moments:
$$M_{x1} = \frac{Px}{2}$$
$$M_{x2} = \frac{P}{2}(L - x)$$
$$M_{max} = \frac{PL}{4}$$
end slope:
$$\phi_l = -\frac{PL^2}{16EI}$$
$$\phi_r = +\frac{PL^2}{16EI}$$
deflection:
$$y_{x1} = \left(\frac{P}{48EI}\right)\left(3xL^2 - 4x^3\right)$$
$$\text{at } x < L/2$$
$$y_{max} = \left(\frac{PL^3}{48EI}\right) \text{ at } x = L/2$$

APPENDIX 44.A *(continued)*
Elastic Beam Deflection Equations
(*w* is the load per unit length.) (*y* is positive downward.)

Case 7: Simple Beam with Intermediate Load

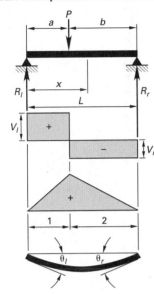

reactions:

$$R_l = \frac{Pb}{L}$$

$$R_r = \frac{Pa}{L}$$

shear:

$$V_l = +\frac{Pb}{L}$$

$$V_r = -\frac{Pa}{L}$$

moments:

$$M_{x1} = \frac{Pbx}{L}$$

$$M_{x2} = \frac{Pa(L-x)}{L}$$

$$M_{max} = \frac{Pab}{L} \text{ at } x = a$$

end slope:

$$\phi_l = -\frac{Pab\left(1+\frac{b}{L}\right)}{6EI}$$

$$\phi_r = \frac{Pab\left(1+\frac{a}{L}\right)}{6EI}$$

deflection:

$$y_{x1} = \left(\frac{Pb}{6EIL}\right)\left(L^2 x - b^2 x - x^3\right)$$

$$\text{at } x < a$$

$$y_{x2} = \left(\frac{Pb}{6EIL}\right)\left(\begin{array}{c}\left(\frac{L}{b}\right)(x-a)^3 \\ +(L^2-b^2)x - x^3\end{array}\right)$$

$$\text{at } x > a$$

$$y = \frac{Pa^2 b^2}{3EIL} \text{ at } x = a$$

$$y_{max} = \left(\frac{0.06415Pb}{EIL}\right)\left(L^2 - b^2\right)^{3/2}$$

$$\text{at } x = \sqrt{\frac{a(L+b)}{3}}$$

Case 8: Simple Beam with Two Loads

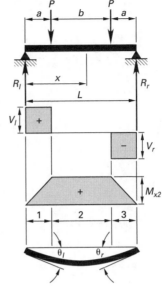

reactions:

$$R_l = R_r = P$$

shear:

$$V_l = +P$$

$$V_r = -P$$

moments:

$$M_{x1} = Px$$

$$M_{x2} = Pa$$

$$M_{x3} = P(L-x)$$

end slope:

$$\phi_l = -\frac{Pa(a+b)}{2EI}$$

$$\phi_r = +\frac{Pa(a+b)}{2EI}$$

deflection:

$$y_{x1} = \left(\frac{P}{6EI}\right)\left(3Lax - 3a^2 x - x^3\right)$$

$$\text{at } x < a$$

$$y_{x2} = \left(\frac{P}{6EI}\right)\left(3Lax - 3ax^2 - a^3\right)$$

$$\text{at } a < x < a+b$$

$$y_{max} = \left(\frac{P}{24EI}\right)\left(3L^2 a - 4a^3\right)$$

$$\text{at } x = L/2$$

APPENDIX 44.A *(continued)*
Elastic Beam Deflection Equations
(w is the load per unit length.) (y is positive downward.)

Case 9: Simple Beam with Uniform Load

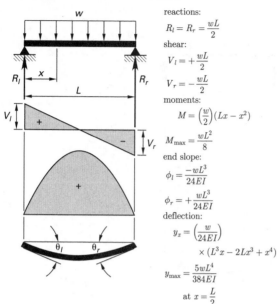

reactions:
$$R_l = R_r = \frac{wL}{2}$$
shear:
$$V_l = +\frac{wL}{2}$$
$$V_r = -\frac{wL}{2}$$
moments:
$$M = \left(\frac{w}{2}\right)(Lx - x^2)$$
$$M_{max} = \frac{wL^2}{8}$$
end slope:
$$\phi_l = \frac{-wL^3}{24EI}$$
$$\phi_r = +\frac{wL^3}{24EI}$$
deflection:
$$y_x = \left(\frac{w}{24EI}\right)$$
$$\times (L^3x - 2Lx^3 + x^4)$$
$$y_{max} = \frac{5wL^4}{384EI}$$
$$\text{at } x = \frac{L}{2}$$

Case 10: Simple Beam with Triangular Load
(w is the maximum loading per unit length at the right end, not the total load, $W = \frac{1}{2}Lw$.)

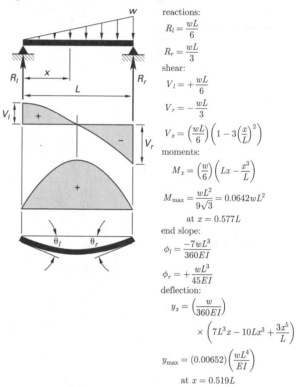

reactions:
$$R_l = \frac{wL}{6}$$
$$R_r = \frac{wL}{3}$$
shear:
$$V_l = +\frac{wL}{6}$$
$$V_r = -\frac{wL}{3}$$
$$V_x = \left(\frac{wL}{6}\right)\left(1 - 3\left(\frac{x}{L}\right)^2\right)$$
moments:
$$M_x = \left(\frac{w}{6}\right)\left(Lx - \frac{x^3}{L}\right)$$
$$M_{max} = \frac{wL^2}{9\sqrt{3}} = 0.0642wL^2$$
$$\text{at } x = 0.577L$$
end slope:
$$\phi_l = \frac{-7wL^3}{360EI}$$
$$\phi_r = +\frac{wL^3}{45EI}$$
deflection:
$$y_x = \left(\frac{w}{360EI}\right)$$
$$\times \left(7L^3x - 10Lx^3 + \frac{3x^5}{L}\right)$$
$$y_{max} = (0.00652)\left(\frac{wL^4}{EI}\right)$$
$$\text{at } x = 0.519L$$

APPENDIX 44.A *(continued)*
Elastic Beam Deflection Equations
(w is the load per unit length.) (y is positive downward.)

Case 11: Simple Beam with Overhung Load

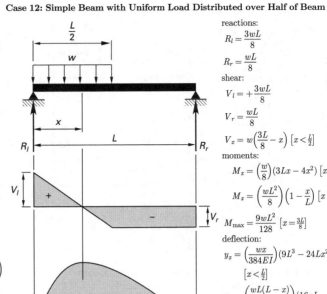

reactions:
$$R_l = \left(\frac{P}{b}\right)(b+a)$$
$$R_r = \frac{-Pa}{b}$$
shear:
$$V_l = -P$$
$$V_r = \frac{Pa}{b}$$
moments:
$$M_a = Px_a$$
$$M_b = \left(\frac{Pa}{b}\right)(b - x_b)$$
$$M_{\max} = Pa \text{ at } x_a = a$$
deflection:
$$y_a = \left(\frac{P}{3EI}\right)$$
$$\times \left(\begin{array}{c} (a^2 + ab)(a - x_a) \\ + \left(\frac{x_a}{2}\right)(x_a^2 - a^2) \end{array}\right)$$
$$y_b = \left(\frac{Pax_b}{6EI}\right)\left(3x_b - \left(\frac{x_b^2}{b}\right) - 2b\right)$$
$$y_{\text{tip}} = \left(\frac{Pa^2}{3EI}\right)(a+b) \text{ [max down]}$$
$$y_{\max} = (0.06415)\left(\frac{Pab^2}{EI}\right) \text{ at } x_b$$
$$= 0.4226b \text{ [max up]}$$

Case 12: Simple Beam with Uniform Load Distributed over Half of Beam

reactions:
$$R_l = \frac{3wL}{8}$$
$$R_r = \frac{wL}{8}$$
shear:
$$V_l = +\frac{3wL}{8}$$
$$V_r = \frac{wL}{8}$$
$$V_x = w\left(\frac{3L}{8} - x\right) \left[x < \frac{L}{2}\right]$$
moments:
$$M_x = \left(\frac{w}{8}\right)(3Lx - 4x^2) \left[x < \frac{L}{2}\right]$$
$$M_x = \left(\frac{wL^2}{8}\right)\left(1 - \frac{x}{L}\right) \left[x > \frac{L}{2}\right]$$
$$M_{\max} = \frac{9wL^2}{128} \left[x = \frac{3L}{8}\right]$$
deflection:
$$y_x = \left(\frac{wx}{384EI}\right)(9L^3 - 24Lx^2 + 16x^3)$$
$$\left[x < \frac{L}{2}\right]$$
$$y_x = \left(\frac{wL(L-x)}{384EI}\right)(16xL - 8x^2 - L^2)$$
$$\left[x > \frac{L}{2}\right]$$

APPENDIX 44.B
Stress Concentration Factors

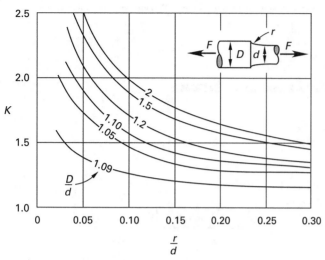

(a) stress concentration factor, *K*, for filleted shaft in tension (basis: smaller section)

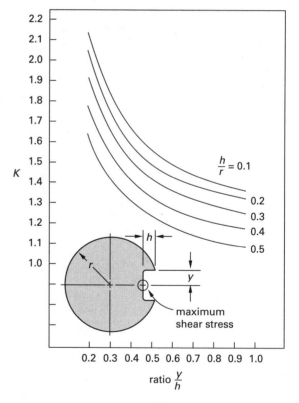

(b) stress concentration factor, *K*, for filleted shaft in torsion (basis: smaller section)

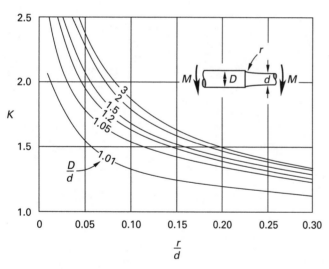

(c) stress concentration factor, *K*, for a shaft with shoulder fillet in bending (basis: smaller section)

(d) stress concentration factor, *K*, for slotted shaft in torsion (based on unslotted section)

APPENDIX 45.A
Properties of Weld Groups
(treated as lines)

weld configuration	centroid location	section modulus $S = I_{c,x}/\bar{y}$	polar moment of inertia $J = I_{c,x} + I_{c,y}$
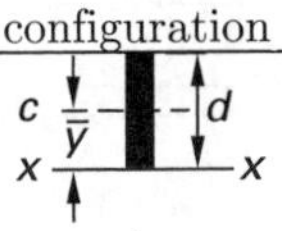	$\bar{y} = \dfrac{d}{2}$	$\dfrac{d^2}{6}$	$\dfrac{d^3}{12}$
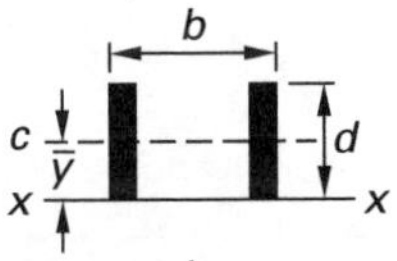	$\bar{y} = \dfrac{d}{2}$	$\dfrac{d^2}{3}$	$\dfrac{d(3b^2 + d^2)}{6}$
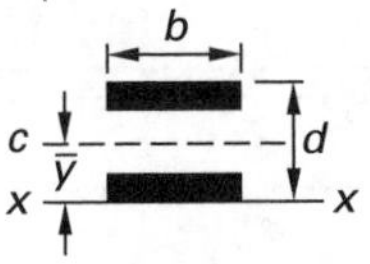	$\bar{y} = \dfrac{d}{2}$	bd	$\dfrac{b(3d^2 + b^2)}{6}$
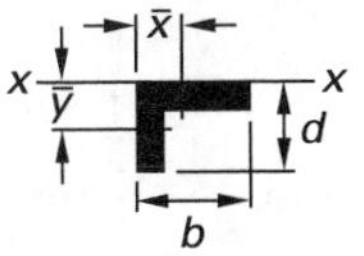	$\bar{y} = \dfrac{d^2}{2(b+d)}$ $\bar{x} = \dfrac{b^2}{2(b+d)}$	$\dfrac{4bd + d^2}{6}$	$\dfrac{(b+d)^4 - 6b^2 d^2}{12(b+d)}$
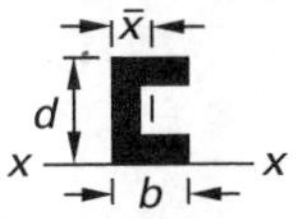	$\bar{x} = \dfrac{b^2}{2b+d}$	$bd + \dfrac{d^2}{6}$	$\dfrac{8b^3 + 6bd^2 + d^3}{12} - \dfrac{b^4}{2b+d}$
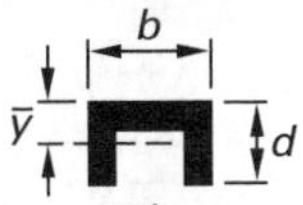	$\bar{y} = \dfrac{d^2}{b+2d}$	$\dfrac{2bd + d^2}{3}$	$\dfrac{b^3 + 6b^2 d + 8d^3}{12} - \dfrac{d^4}{2d+b}$
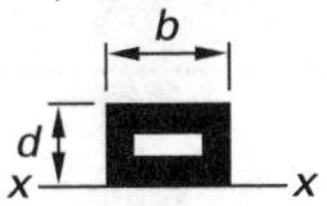	$\bar{y} = \dfrac{d}{2}$	$bd + \dfrac{d^2}{3}$	$\dfrac{(b+d)^3}{6}$
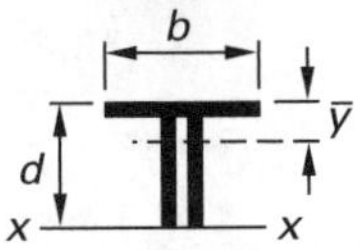	$\bar{y} = \dfrac{d^2}{b+2d}$	$\dfrac{2bd + d^2}{3}$	$\dfrac{b^3 + 8d^3}{12} - \dfrac{d^4}{b+2d}$
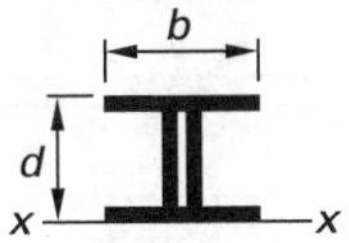	$\bar{y} = \dfrac{d}{2}$	$bd + \dfrac{d^2}{3}$	$\dfrac{b^3 + 3bd^2 + d^3}{6}$
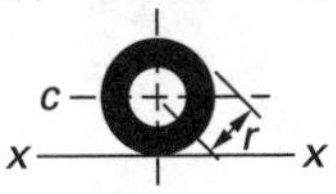	$\bar{y} = r$	πr^2	$2\pi r^3$

APPENDIX 47.A
Elastic Fixed-End Moments

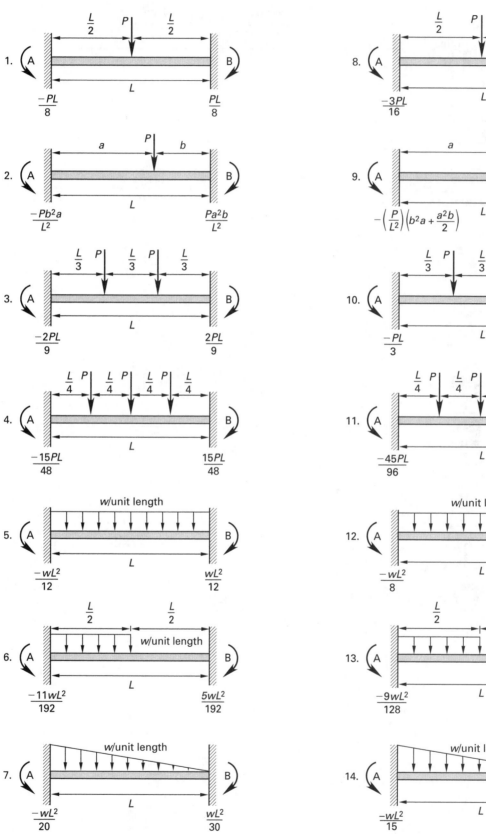

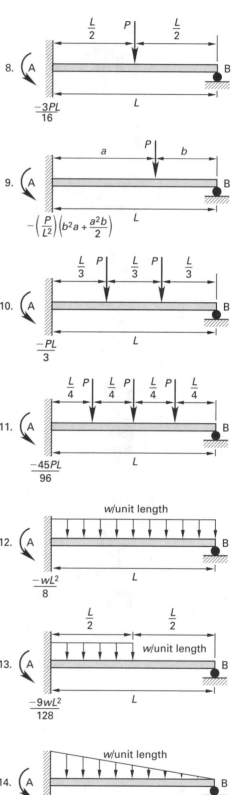

1. $\dfrac{-PL}{8}$; $\dfrac{PL}{8}$

2. $\dfrac{-Pb^2a}{L^2}$; $\dfrac{Pa^2b}{L^2}$

3. $\dfrac{-2PL}{9}$; $\dfrac{2PL}{9}$

4. $\dfrac{-15PL}{48}$; $\dfrac{15PL}{48}$

5. $\dfrac{-wL^2}{12}$; $\dfrac{wL^2}{12}$

6. $\dfrac{-11wL^2}{192}$; $\dfrac{5wL^2}{192}$

7. $\dfrac{-wL^2}{20}$; $\dfrac{wL^2}{30}$

8. $\dfrac{-3PL}{16}$

9. $-\left(\dfrac{P}{L^2}\right)\left(b^2a+\dfrac{a^2b}{2}\right)$

10. $\dfrac{-PL}{3}$

11. $\dfrac{-45PL}{96}$

12. $\dfrac{-wL^2}{8}$

13. $\dfrac{-9wL^2}{128}$

14. $\dfrac{-wL^2}{15}$

APPENDIX 47.A *(continued)*
Elastic Fixed-End Moments

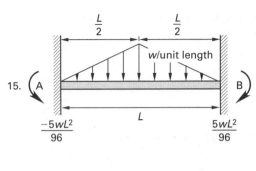

15.

$$\dfrac{-5wL^2}{96} \qquad \dfrac{5wL^2}{96}$$

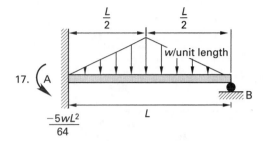

17.

$$\dfrac{-5wL^2}{64}$$

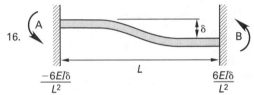

16.

$$\dfrac{-6EI\delta}{L^2} \qquad \dfrac{6EI\delta}{L^2}$$

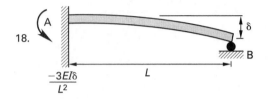

18.

$$\dfrac{-3EI\delta}{L^2}$$

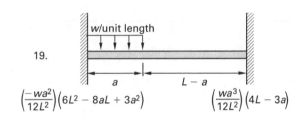

19.

$$\left(\dfrac{-wa^2}{12L^2}\right)\left(6L^2 - 8aL + 3a^2\right) \qquad \left(\dfrac{wa^3}{12L^2}\right)\left(4L - 3a\right)$$

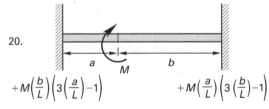

20.

$$+M\left(\dfrac{b}{L}\right)\left(3\left(\dfrac{a}{L}\right)-1\right) \qquad +M\left(\dfrac{a}{L}\right)\left(3\left(\dfrac{b}{L}\right)-1\right)$$

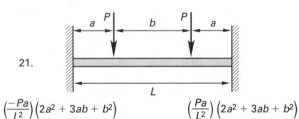

21.

$$\left(\dfrac{-Pa}{L^2}\right)\left(2a^2 + 3ab + b^2\right) \qquad \left(\dfrac{Pa}{L^2}\right)\left(2a^2 + 3ab + b^2\right)$$

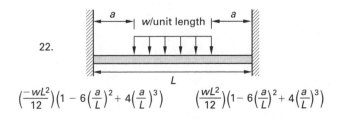

22.

$$\left(\dfrac{-wL^2}{12}\right)\left(1 - 6\left(\dfrac{a}{L}\right)^2 + 4\left(\dfrac{a}{L}\right)^3\right) \qquad \left(\dfrac{wL^2}{12}\right)\left(1 - 6\left(\dfrac{a}{L}\right)^2 + 4\left(\dfrac{a}{L}\right)^3\right)$$

APPENDIX 47.B
Indeterminate Beam Formulas
(see also type)

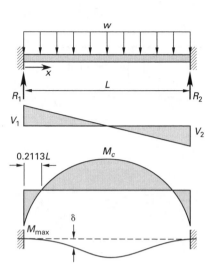

Uniformly distributed load: w in load/unit length
Total load: $W = wL$

Reactions: $R_1 = R_2 = \dfrac{W}{2}$

Shear forces: $V_1 = +\dfrac{W}{2}$

$V_2 = -\dfrac{W}{2}$

Maximum (negative) bending moment:

$M_{\max} = -\dfrac{wL^2}{12} = -\dfrac{WL}{12}$, at end

Maximum (positive) bending moment:

$M_{\max} = \dfrac{wL^2}{24} = \dfrac{WL}{24}$, at center

Maximum deflection: $\dfrac{wL^4}{384EI} = \dfrac{WL^3}{384EI}$, at center

$\delta = \left(\dfrac{wx^2}{24EI}\right)(L-x)^2,\ 0 \le x \le L$

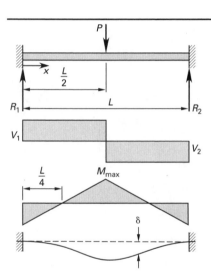

Concentrated load, P, at center

Reactions: $R_1 = R_2 = \dfrac{P}{2}$

Shear forces: $V_1 = +\dfrac{P}{2}$; $V_2 = -\dfrac{P}{2}$

Maximum bending moment:

$M_{\max} = \dfrac{PL}{8}$, at center

$M_{\max} = -\dfrac{PL}{8}$, at ends

Maximum deflection: $\dfrac{PL^3}{192EI}$, at center

$\delta = \left(\dfrac{Px^2}{48EI}\right)(3L - 4x),\ 0 \le x \le \dfrac{L}{2}$

APPENDIX 47.B *(continued)*
Indeterminate Beam Formulas
(see also type)

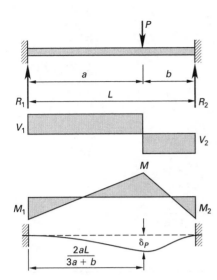

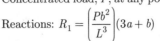

Concentrated load, P, at any point

Reactions: $R_1 = \left(\dfrac{Pb^2}{L^3}\right)(3a + b)$

$R_2 = \left(\dfrac{Pa^2}{L^3}\right)(3b + a)$

Shear forces: $V_1 = R_1$; $V_2 = -R_2$

Bending moments:

$M_1 = \dfrac{Pab^2}{L^2}$, maximum, when $a < b$

$M_2 = -\dfrac{Pa^2b}{L^2}$, maximum, when $a > b$

$M_P = +\dfrac{2Pa^2b^2}{L^3}$, at point of load

δ_P: $\dfrac{Pa^3b^3}{3EIL^2}$, at point of load

$\delta_{max} = \dfrac{2Pa^3b^2}{3EI(3a+b)^2}$, at $x = \dfrac{2aL}{3a+b}$, for $a > b$

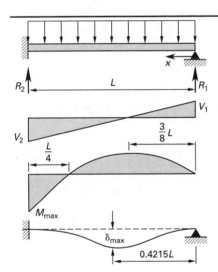

Uniformly distributed load: w in load/unit length
Total load: $W = wL$

Reactions: $R_1 = \dfrac{3wL}{8}$, $R_2 = \dfrac{5wL}{8}$

Shear forces: $V_1 = +R_1$; $V_2 = -R_2$

Bending moments:

Maximum negative moment: $-\dfrac{wL^2}{8}$, at left end

Maximum positive moment: $\dfrac{9}{128}wL^2$, $x = \dfrac{3}{8}L$

$M = \dfrac{3wLx}{8} - \dfrac{wx^2}{2}$, $0 \le x \le L$

Maximum deflection: $\dfrac{wL^4}{185EI}$, $x = 0.4215L$

$\delta = \left(\dfrac{wx}{48EI}\right)(L^3 - 3Lx^2 + 2x^3)$, $0 \le x \le L$

APPENDIX 47.B *(continued)*
Indeterminate Beam Formulas
(see also type)

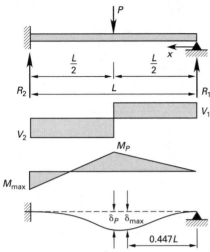

Concentrated load, P, at center

Reactions: $R_1 = \dfrac{5}{16}P$; $R_2 = \dfrac{11}{16}P$

Shear forces: $V_1 = R_1$; $V_2 = -R_2$

Bending moments:

Maximum negative moment: $-\dfrac{3PL}{16}$, at fixed end

Maximum positive moment: $\dfrac{5PL}{32}$, at center

Maximum deflection: $0.009317\left(\dfrac{PL^3}{EI}\right)$, at $x = 0.447L$

Deflection at center under load: $\dfrac{7PL^3}{768EI}$

Concentrated load, P, at any point

Reactions: $R_1 = \left(\dfrac{Pb^2}{2L^3}\right)(a+2L)$, $R_2 = \left(\dfrac{Pa}{2L^3}\right)(3L^2 - a^2)$

Shear forces: $V_1 = R_1$; $V_2 = -R_2$

Bending moments:

Maximum negative moment: $M_2 = \left(-\dfrac{Pab}{2L^2}\right)(a+L)$, at fixed end

Maximum positive moment: $M_1 = \left(\dfrac{Pab^2}{2L^3}\right)(a+2L)$, at load

Deflections: $\delta_P = \left(\dfrac{Pa^2b^3}{12EIL^3}\right)(3L+a)$, at load

$\delta_{max} = \dfrac{Pa(L^2-a^2)^3}{3EI(3L^2-a^2)^2}$, at $x = \dfrac{L^2+a^2}{3L^2-a^2}L$, when $a < 0.414L$

$\delta_{max} = \dfrac{Pab^2}{6EI}\sqrt{\dfrac{a}{2L+a}}$, at $x = L\sqrt{\dfrac{2}{2L+a}}$, when $a > 0.414L$

APPENDIX 47.B *(continued)*
Indeterminate Beam Formulas
(see also type)

Continuous beam of two equal spans—equal concentrated loads, P, at center of each span

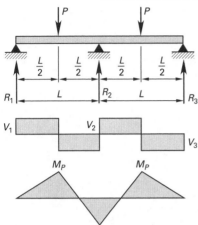

Reactions: $R_1 = R_3 = \dfrac{5}{16}P$

$R_2 = 1.375P$

Shear forces: $V_1 = -V_3 = \dfrac{5}{16}P$

$V_2 = \pm\dfrac{11}{16}P$

Bending moments:

$M_{max} = -\dfrac{6}{32}PL$, at R_2

$M_P = \dfrac{5}{32}PL$, at point of load

Continuous beam of two equal spans—concentrated loads, P, at third points of each span

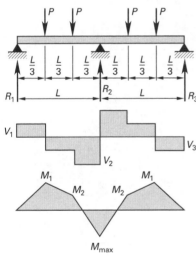

Reactions: $R_1 = R_3 = \dfrac{2}{3}P$

$R_2 = \dfrac{8}{3}P$

Shear forces: $V_1 = -V_3 = \dfrac{2}{3}P$

$V_2 = \pm\dfrac{4}{3}P$

Bending moments:

$M_{max} = -\dfrac{1}{3}PL$, at R_2

$M_1 = \dfrac{2}{9}PL$

$M_2 = \dfrac{1}{9}PL$

APPENDIX 47.B *(continued)*
Indeterminate Beam Formulas
(see also type)

Continuous beam of two equal spans—uniformly distributed load of w in load/unit length

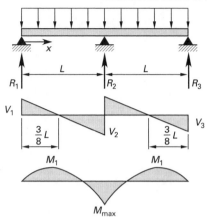

Reactions: $R_1 = R_3 = \dfrac{3}{8}wL$

$\qquad R_2 = 1.25wL$

Shear forces: $V_1 = -V_3 = \dfrac{3}{8}wL$

$\qquad V_2 = \pm\dfrac{5}{8}wL$

Bending moments:

$\qquad M_{\max} = -\dfrac{1}{8}wL^2$

$\qquad M_1 = \dfrac{9}{128}wL^2$

Maximum deflection: $0.00541\left(\dfrac{wL^4}{EI}\right)$, at $x = 0.4215L$

$\qquad \delta = \left(\dfrac{w}{48EI}\right)(L^3x - 3Lx^3 + 2x^4),\ 0 \le x \le L$

Continuous beam of two equal spans—uniformly distributed load of w in lbf/unit length on one span

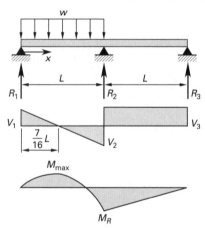

Reactions: $R_1 = \dfrac{7}{16}wL,\ R_2 = \dfrac{5}{8}wL,\ R_3 = -\dfrac{1}{16}wL$

Shear forces: $V_1 = \dfrac{7}{16}wL,\ V_2 = -\dfrac{9}{16}wL,\ V_3 = \dfrac{1}{16}wL$

Bending moments:

$\qquad M_{\max} = \dfrac{49}{512}wL^2$, at $x = \dfrac{7}{16}L$

$\qquad M_R = -\dfrac{1}{16}wL^2$, at R_2

$\qquad M = \left(\dfrac{wx}{16}\right)(7L - 8x),\ 0 \le x \le L$

APPENDIX 47.B *(continued)*
Indeterminate Beam Formulas
(see also type)

Continuous beam of two equal spans—concentrated load, P, at center of one span

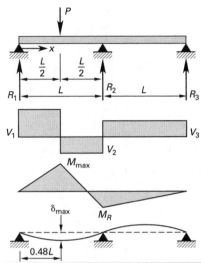

Reactions: $R_1 = \dfrac{13}{32}P$, $R_2 = \dfrac{11}{16}P$, $R_3 = -\dfrac{3}{32}P$

Shear forces: $V_1 = \dfrac{13}{32}P$, $V_2 = -\dfrac{19}{32}P$, $V_3 = \dfrac{3}{32}P$

Bending moments:

$$M_{max} = \frac{13}{64}PL, \text{ at point of load}$$

$$M_R = -\frac{3}{32}PL, \text{ at support of load}$$

Maximum deflection: $\dfrac{0.96PL^3}{64EI}$, at $x = 0.48L$

Continuous beam of two equal spans—concentrated load, P, at any point on one span

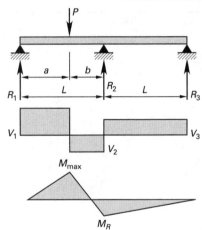

Reactions: $R_1 = \left(\dfrac{Pb}{4L^3}\right)\left(4L^2 - a(L+a)\right)$

$R_2 = \left(\dfrac{Pa}{2L^3}\right)\left(2L^2 + b(L+a)\right)$ $R_3 = \left(-\dfrac{Pab}{4L^3}\right)\left(L+a\right)$

Shear forces: $V_1 = \left(\dfrac{Pb}{4L^3}\right)\left(4L^2 - a(L+a)\right)$

$V_2 = \left(-\dfrac{Pa}{4L^3}\right)\left(4L^2 + b(L+a)\right)$

$V_3 = \left(\dfrac{Pab}{4L^3}\right)\left(L+a\right)$

Bending moments:

$$M_{max} = \left(\frac{Pab}{4L^3}\right)\left(4L^2 - a(L+a)\right)$$

$$M_R = \left(-\frac{Pab}{4L^2}\right)\left(L+a\right)$$

APPENDIX 47.B *(continued)*
Indeterminate Beam Formulas
(see also type)

Continuous beam of three equal spans—concentrated load, P, at center of each span

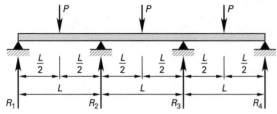
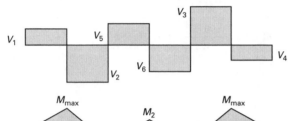
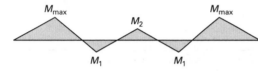

Reactions:

$$R_1 = R_4 = \frac{7}{20}P$$

$$R_2 = R_3 = \frac{23}{20}P$$

Shear forces:

$$V_1 = -V_4 = \frac{7}{20}P$$

$$V_3 = -V_2 = \frac{13}{20}P$$

$$V_5 = -V_6 = \frac{P}{2}$$

Bending moments:

$$M_{\max} = \frac{7}{40}PL$$

$$M_1 = -\frac{3}{20}PL$$

$$M_2 = \frac{1}{10}$$

Continuous beam of three equal spans—concentrated load, P, at third points of each span

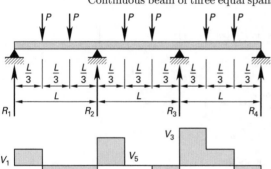
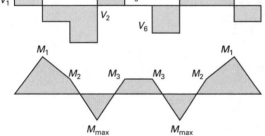

Reactions:

$$R_1 = R_4 = \frac{11}{15}P$$

$$R_2 = R_3 = \frac{34}{15}P$$

Shear forces:

$$V_1 = -V_4 = \frac{11}{15}P$$

$$V_3 = -V_2 = \frac{19}{15}P$$

$$V_5 = -V_6 = P$$

Bending moments:

$$M_{\max} = -\frac{12}{45}PL$$

$$M_1 = \frac{11}{45}PL$$

$$M_2 = \frac{7}{45}PL$$

$$M_3 = \frac{3}{45}PL$$

APPENDIX 47.C
Moment Distribution Worksheet

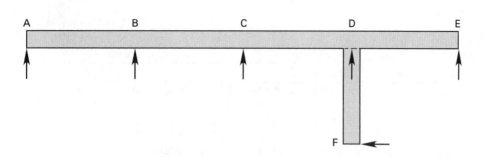

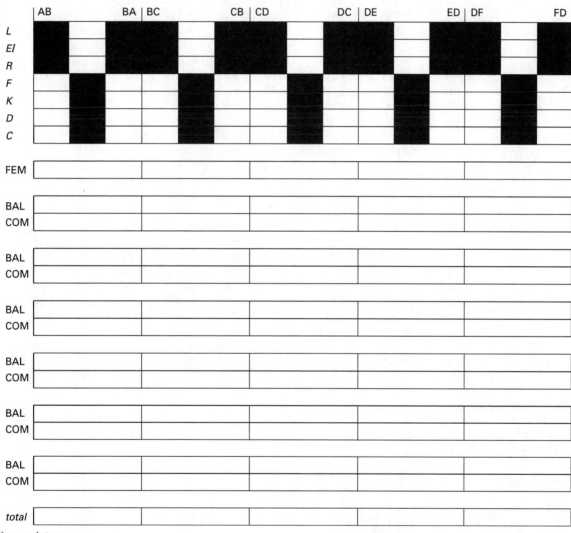

Nomenclature

BAL	balance amount	F	fixity factor
C	carryover factor	FEM	fixed-end moment
COM	carryover moment	K	stiffness factor
D	distribution factor	L	length
EI	product of modulus of elasticity and moment of inertia	R	rigidity factor

APPENDIX 48.A
ASTM Standards for Wire Reinforcement[*]

W&D size		nominal diameter	nominal area	nominal weight	area, in²/ft of width for various spacings						
					center-to-center spacing, in						
plain	deformed	(in)	(in²)	(lbm/ft)	2	3	4	6	8	10	12
W31	D31	0.628	0.310	1.054	1.86	1.24	0.93	0.62	0.46	0.37	0.31
W30	D30	0.618	0.300	1.020	1.80	1.20	0.90	0.60	0.45	0.36	0.30
W28	D28	0.597	0.280	0.952	1.68	1.12	0.84	0.56	0.42	0.33	0.28
W26	D26	0.575	0.260	0.934	1.56	1.04	0.78	0.52	0.39	0.31	0.26
W24	D24	0.553	0.240	0.816	1.44	0.96	0.72	0.48	0.36	0.28	0.24
W22	D22	0.529	0.220	0.748	1.32	0.88	0.66	0.44	0.33	0.26	0.22
W20	D20	0.504	0.200	0.680	1.20	0.80	0.60	0.40	0.30	0.24	0.20
W18	D18	0.478	0.180	0.612	1.08	0.72	0.54	0.36	0.27	0.21	0.18
W16	D16	0.451	0.160	0.544	0.96	0.64	0.48	0.32	0.24	0.19	0.16
W14	D14	0.422	0.140	0.476	0.84	0.56	0.42	0.28	0.21	0.16	0.14
W12	D12	0.390	0.120	0.408	0.72	0.48	0.36	0.24	0.18	0.14	0.12
W11	D11	0.374	0.110	0.374	0.66	0.44	0.33	0.22	0.16	0.13	0.11
W10.5		0.366	0.105	0.357	0.63	0.42	0.315	0.21	0.15	0.12	0.105
W10	D10	0.356	0.100	0.340	0.60	0.40	0.30	0.20	0.14	0.12	0.10
W9.5		0.348	0.095	0.323	0.57	0.38	0.285	0.19	0.14	0.11	0.095
W9	D9	0.338	0.090	0.306	0.54	0.36	0.27	0.18	0.13	0.10	0.09
W8.5		0.329	0.085	0.289	0.51	0.34	0.255	0.17	0.12	0.10	0.085
W8	D8	0.319	0.080	0.272	0.48	0.32	0.24	0.16	0.12	0.09	0.08
W7.5		0.309	0.075	0.255	0.45	0.30	0.225	0.15	0.11	0.09	0.075
W7	D7	0.298	0.070	0.238	0.42	0.28	0.21	0.14	0.10	0.08	0.07
W6.5		0.288	0.065	0.221	0.39	0.26	0.195	0.13	0.09	0.07	0.065
W6	D6	0.276	0.060	0.204	0.36	0.24	0.18	0.12	0.09	0.07	0.06
W5.5		0.264	0.055	0.187	0.33	0.22	0.165	0.11	0.08	0.06	0.055
W5	D5	0.252	0.050	0.170	0.30	0.20	0.15	0.10	0.07	0.06	0.05
W4.5		0.240	0.045	0.153	0.27	0.18	0.135	0.09	0.06	0.05	0.045
W4	D4	0.225	0.040	0.136	0.24	0.16	0.12	0.08	0.06	0.04	0.04
W3.5		0.211	0.035	0.119	0.21	0.14	0.105	0.07	0.05	0.04	0.035
W3		0.195	0.030	0.102	0.18	0.12	0.09	0.06	0.04	0.03	0.03
W2.9		0.192	0.029	0.098	0.174	0.116	0.087	0.058	0.04	0.03	0.029
W2.5		0.178	0.025	0.085	0.15	0.10	0.075	0.05	0.03	0.03	0.025
W2		0.159	0.020	0.068	0.12	0.08	0.06	0.04	0.03	0.02	0.02
W1.4		0.135	0.014	0.049	0.084	0.056	0.042	0.028	0.02	0.01	0.014

(Multiply in by 25.4 to obtain mm.)
(Multiply in² by 6.45 to obtain cm².)
(Multiply lbm/ft by 1.49 to obtain kg/m.)
(Multiply in²/ft by 21.2 to obtain cm²/m.)

[*]as adopted by ACI 318

Reprinted with permission from the Wire Reinforcement Institute, *Manual for Standard Practice—Structural Welded Wire Reinforcement*, 8th ed., © 2010, by the Wire Reinforcement Institute.

APPENDIX 52.A
Reinforced Concrete Interaction Diagram, $\gamma = 0.60$
(round, 4 ksi concrete, 60 ksi steel)

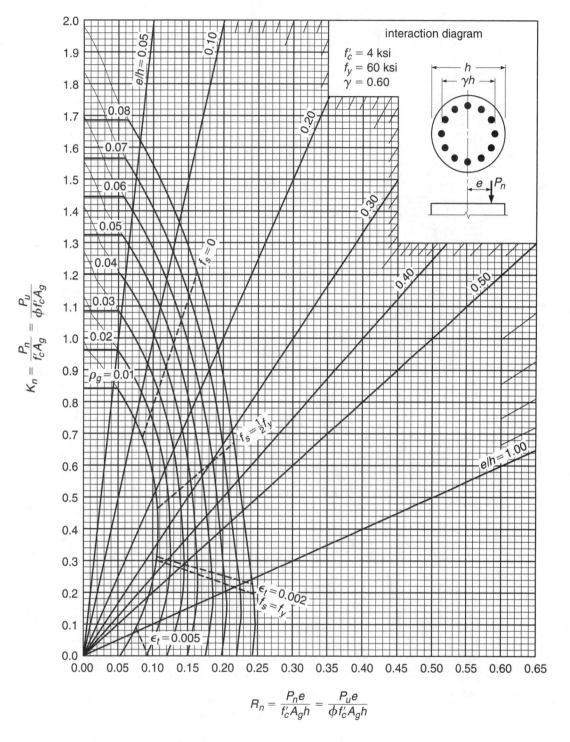

Reprinted with permission from Arthur H. Nilson, David Darwin, and Charles W. Dolan, *Design of Concrete Structures*, 13th ed., copyright © 2004, by the McGraw-Hill Companies.

APPENDIX 52.B
Reinforced Concrete Interaction Diagram, $\gamma = 0.70$
(round, 4 ksi concrete, 60 ksi steel)

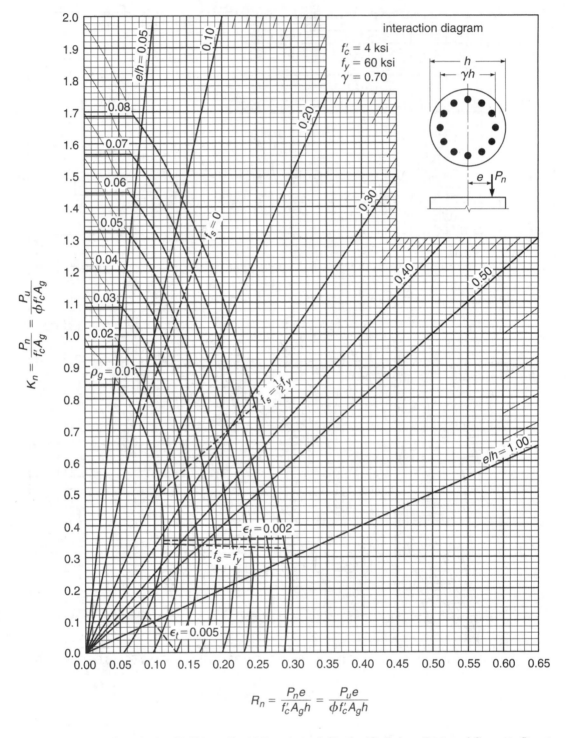

interaction diagram

$f'_c = 4$ ksi
$f_y = 60$ ksi
$\gamma = 0.70$

$$K_n = \frac{P_n}{f'_c A_g} = \frac{P_u}{\phi f'_c A_g}$$

$$R_n = \frac{P_n e}{f'_c A_g h} = \frac{P_u e}{\phi f'_c A_g h}$$

Reprinted with permission from Arthur H. Nilson, David Darwin, and Charles W. Dolan, *Design of Concrete Structures*, 13th ed., copyright © 2004, by the McGraw-Hill Companies.

Appendices

APPENDIX 52.C
Reinforced Concrete Interaction Diagram, $\gamma = 0.75$
(round, 4 ksi concrete, 60 ksi steel)

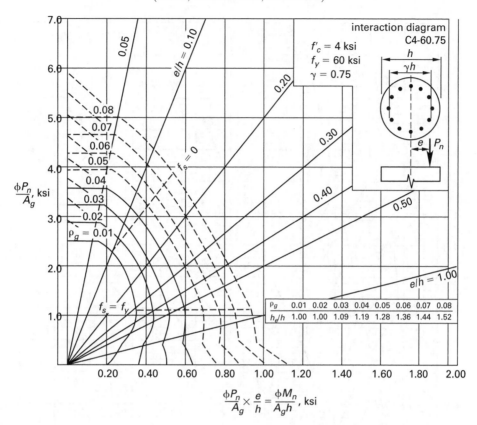

$$\frac{\phi P_n}{A_g} \times \frac{e}{h} = \frac{\phi M_n}{A_g h}, \text{ ksi}$$

Reprinted by permission of the American Concrete Institute from *ACI Publication SP-17A: Design Handbook*, Volume 2, Columns, copyright 1990.

APPENDIX 52.D
Reinforced Concrete Interaction Diagram, $\gamma = 0.80$
(round, 4 ksi concrete, 60 ksi steel)

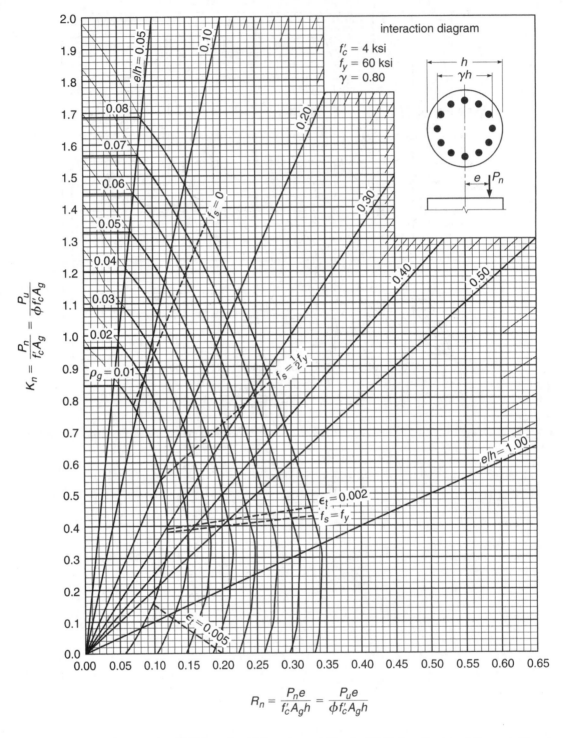

Reprinted with permission from Arthur H. Nilson, David Darwin, and Charles W. Dolan, *Design of Concrete Structures*, 13th ed., copyright © 2004, by the McGraw-Hill Companies.

APPENDIX 52.E
Reinforced Concrete Interaction Diagram, $\gamma = 0.90$
(round, 4 ksi concrete, 60 ksi steel)

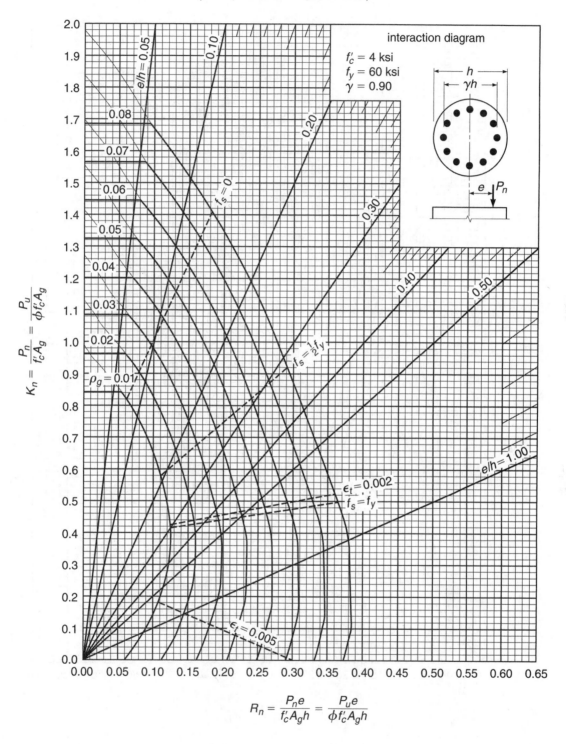

interaction diagram

$f'_c = 4$ ksi
$f_y = 60$ ksi
$\gamma = 0.90$

$$K_n = \frac{P_n}{f'_c A_g} = \frac{P_u}{\phi f'_c A_g}$$

$$R_n = \frac{P_n e}{f'_c A_g h} = \frac{P_u e}{\phi f'_c A_g h}$$

Reprinted with permission from Arthur H. Nilson, David Darwin, and Charles W. Dolan, *Design of Concrete Structures*, 13th ed., copyright © 2004, by the McGraw-Hill Companies.

Appendices

APPENDIX 52.F
Reinforced Concrete Interaction Diagram, $\gamma = 0.60$
(square, 4 ksi concrete, 60 ksi steel)

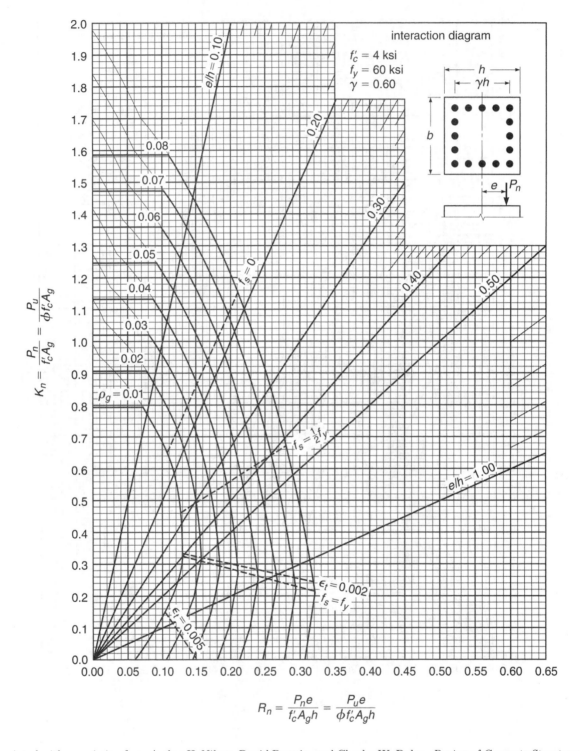

$$R_n = \frac{P_n e}{f'_c A_g h} = \frac{P_u e}{\phi f'_c A_g h}$$

Reprinted with permission from Arthur H. Nilson, David Darwin, and Charles W. Dolan, *Design of Concrete Structures*, 13th ed., copyright © 2004, by the McGraw-Hill Companies.

APPENDIX 52.G
Reinforced Concrete Interaction Diagram, $\gamma = 0.70$
(square, 4 ksi concrete, 60 ksi steel)

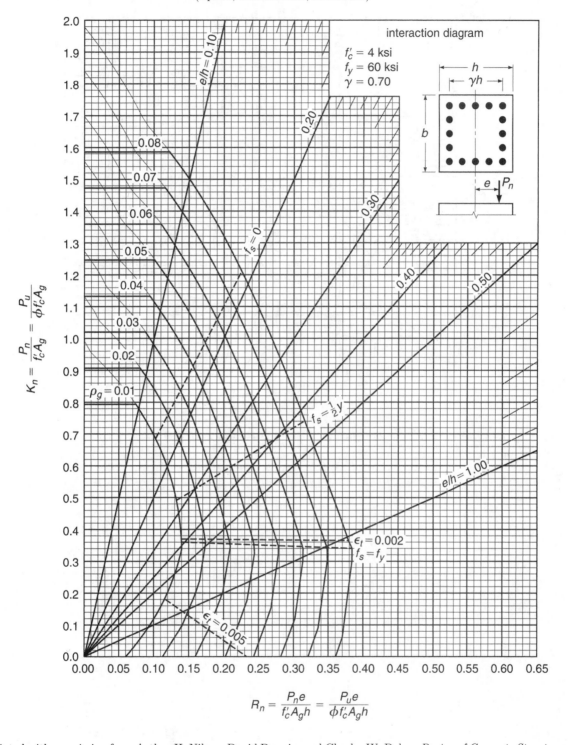

$$R_n = \frac{P_n e}{f'_c A_g h} = \frac{P_u e}{\phi f'_c A_g h}$$

Reprinted with permission from Arthur H. Nilson, David Darwin, and Charles W. Dolan, *Design of Concrete Structures*, 13th ed., copyright © 2004, by the McGraw-Hill Companies.

APPENDIX 52.H
Reinforced Concrete Interaction Diagram, $\gamma = 0.75$
(square, 4 ksi concrete, 60 ksi steel)

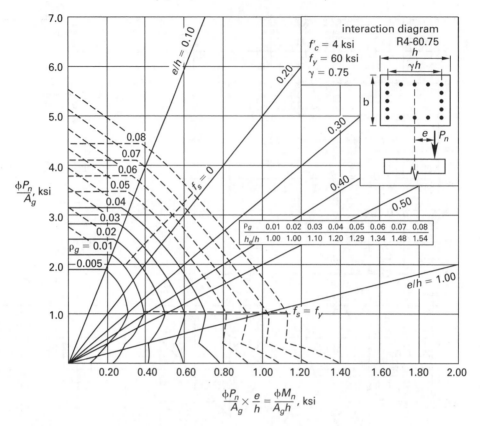

Reprinted by permission of the American Concrete Institute from *ACI Publication SP-17A: Design Handbook*, Volume 2, Columns, copyright 1990.

APPENDIX 52.I
Reinforced Concrete Interaction Diagram, $\gamma = 0.80$
(square, 4 ksi concrete, 60 ksi steel)

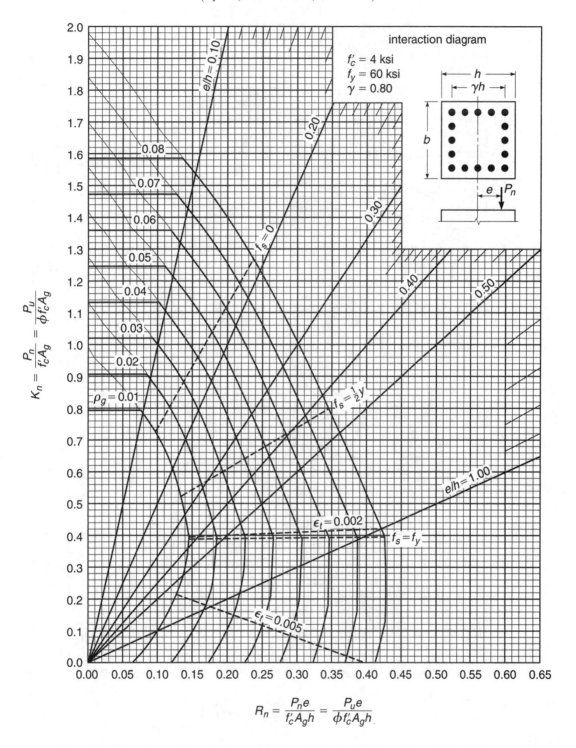

$$R_n = \frac{P_n e}{f'_c A_g h} = \frac{P_u e}{\phi f'_c A_g h}$$

Reprinted with permission from Arthur H. Nilson, David Darwin, and Charles W. Dolan, *Design of Concrete Structures*, 13th ed., copyright © 2004, by the McGraw-Hill Companies.

Appendices

APPENDIX 52.J
Reinforced Concrete Interaction Diagram, $\gamma = 0.90$
(square, 4 ksi concrete, 60 ksi steel)

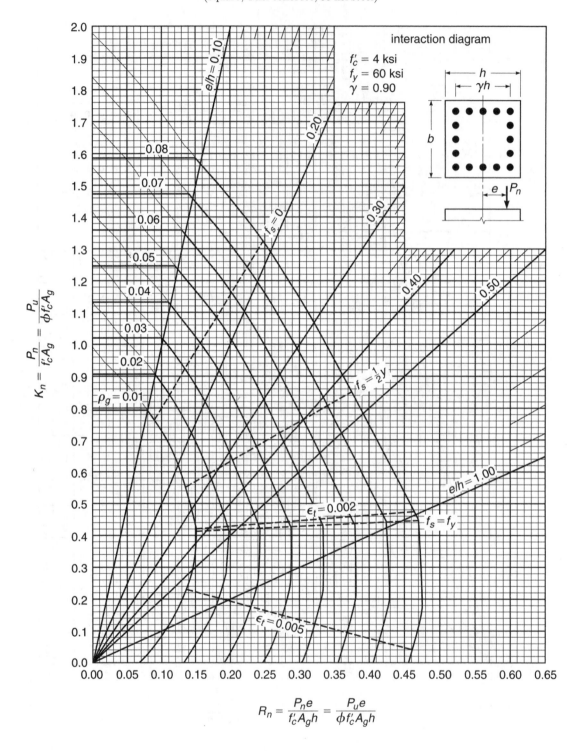

APPENDIX 52.K
Reinforced Concrete Interaction Diagram, $\gamma = 0.60$
(uniplane, 4 ksi concrete, 60 ksi steel)

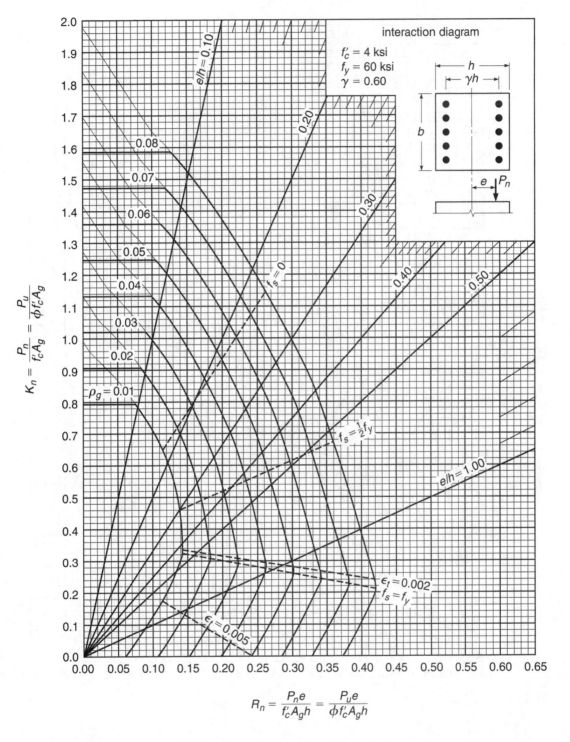

interaction diagram

$f'_c = 4$ ksi
$f_y = 60$ ksi
$\gamma = 0.60$

$$K_n = \frac{P_n}{f'_c A_g} = \frac{P_u}{\phi f'_c A_g}$$

$$R_n = \frac{P_n e}{f'_c A_g h} = \frac{P_u e}{\phi f'_c A_g h}$$

Reprinted with permission from Arthur H. Nilson, David Darwin, and Charles W. Dolan, *Design of Concrete Structures*, 13th ed., copyright © 2004, by the McGraw-Hill Companies.

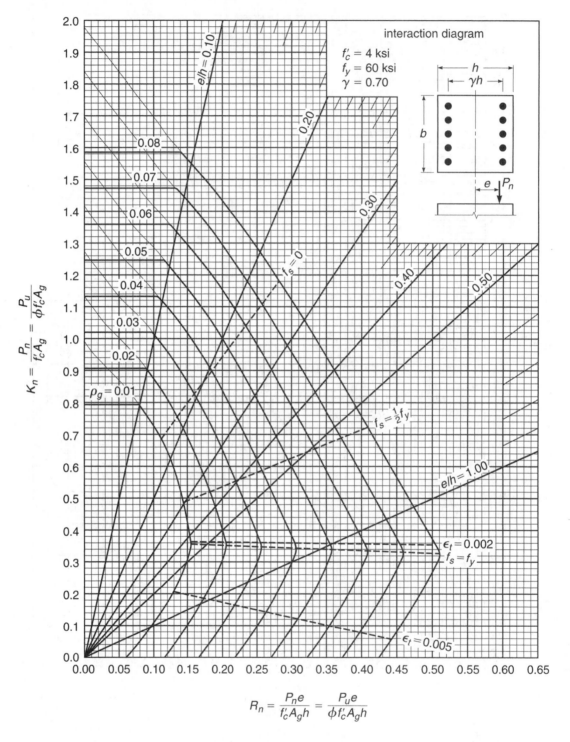

APPENDIX 52.L
Reinforced Concrete Interaction Diagram, $\gamma = 0.70$
(uniplane, 4 ksi concrete, 60 ksi steel)

Reprinted with permission from Arthur H. Nilson, David Darwin, and Charles W. Dolan, *Design of Concrete Structures*, 13th ed., copyright © 2004, by the McGraw-Hill Companies.

Appendices

APPENDIX 52.M
Reinforced Concrete Interaction Diagram, $\gamma = 0.75$
(uniplane, 4 ksi concrete, 60 ksi steel)

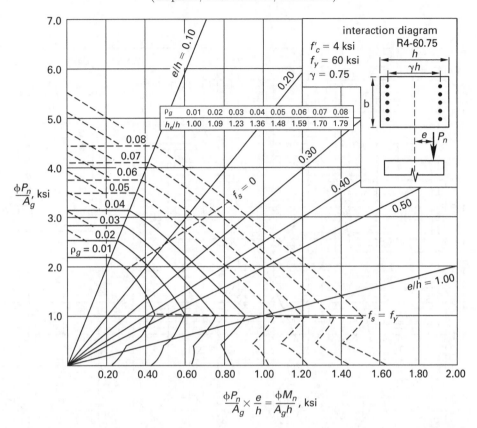

$$\frac{\phi P_n}{A_g} \times \frac{e}{h} = \frac{\phi M_n}{A_g h} \text{, ksi}$$

APPENDIX 52.N
Reinforced Concrete Interaction Diagram, $\gamma = 0.80$
(uniplane, 4 ksi concrete, 60 ksi steel)

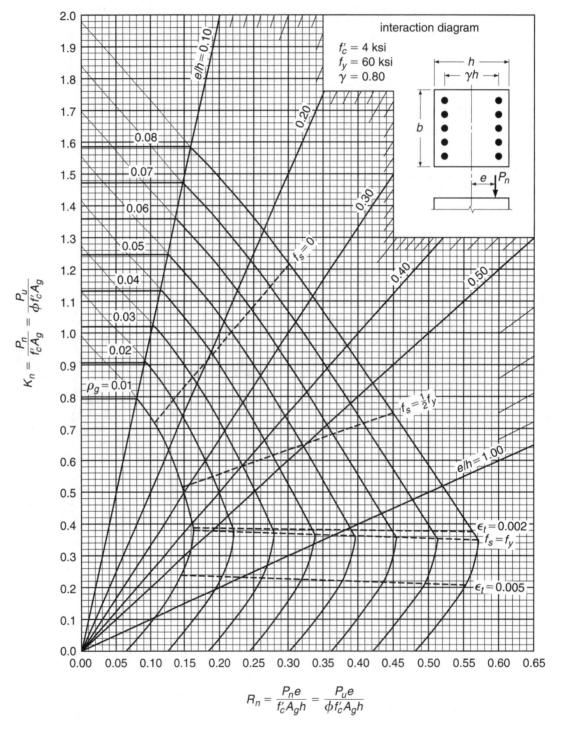

Reprinted with permission from Arthur H. Nilson, David Darwin, and Charles W. Dolan, *Design of Concrete Structures*, 13th ed., copyright © 2004, by the McGraw-Hill Companies.

APPENDIX 52.O
Reinforced Concrete Interaction Diagram, $\gamma = 0.90$
(uniplane, 4 ksi concrete, 60 ksi steel)

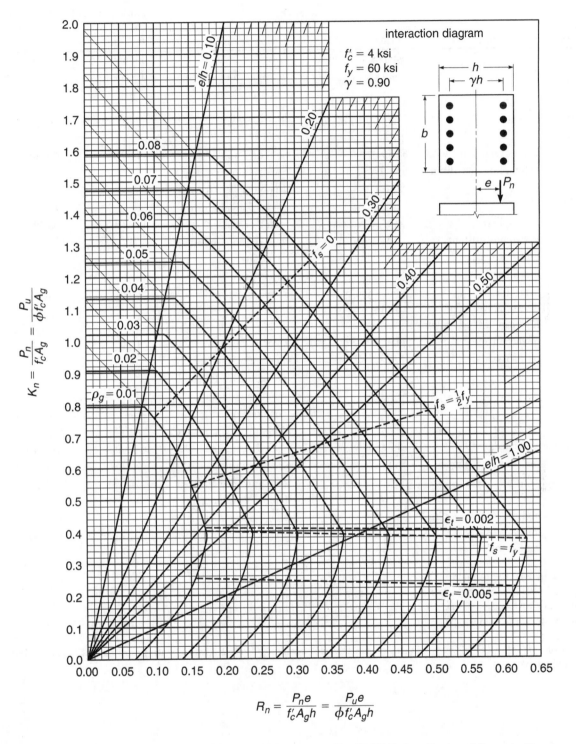

$$R_n = \frac{P_n e}{f'_c A_g h} = \frac{P_u e}{\phi f'_c A_g h}$$

Reprinted with permission from Arthur H. Nilson, David Darwin, and Charles W. Dolan, *Design of Concrete Structures*, 13th ed., copyright © 2004, by the McGraw-Hill Companies.

APPENDIX 58.A
Common Structural Steels

ASTM designation	type of steel	forms	recommended uses	minimum yield stress, F_y, (ksi)[a]	minimum tensile strength, F_u, (ksi)[b]
A36	carbon	shapes, bars, and plates	bolted or welded buildings and bridges and other structural uses	36, but 32 if thickness > 8 in	58–80
A529	carbon	shapes and plates to ½ in	similar to A36	42–50	60–100
A572	high-strength low-alloy columbium-vanadium	shapes, plates, and bars to 6 in	bolted or welded construction; not for welded bridges for F_y grades 55 and above	42–65	60–80
A242	atmospheric corrosion-resistant high-strength low-alloy	shapes, plates, and bars to 5 in	bolted or welded construction; welding technique very important	42–50	63–70
A588	atmospheric corrosion-resistant high-strength low-alloy	plates and bars to 4 in	bolted construction	42–50	63–70
A852	quenched and tempered alloy	plates only to 4 in	welded or bolted construction, primarily for welded bridges and buildings; welding technique of fundamental importance	70	90–110
A514	quenched and tempered low-alloy	plates only 2½ to 6 in	welded structures with great attention technique, discouraged if ductility important	90–100	100–130
A992	high-strength low-alloy manganese-silicon-vanadium[d]	shapes	wide flange shapes	42–65 (design 50)	60–80 (design 65)

(Multiply in by 25.4 to obtain mm.)
(Multiply ksi by 6.9 to obtain MPa.)
[a]F_y values vary with thickness and group.
[b]F_u values vary by grade and type.

Adapted from *Structural Steel Design*, fourth edition, by McCormac, © 1995. Reprinted with permission of Prentice-Hall, Inc., Upper Saddle River, NJ.

APPENDIX 58.B

Properties of Structural Steel at High Temperatures

ASTM designation	temperature (°F)	yield strength 0.2% offset (ksi)	ultimate tensile strength (ksi)
A36	80	36.0	64.0
	300	30.2	64.0
	500	27.8	63.8
	700	25.4	57.0
	900	21.5	44.0
	1100	16.3	25.2
	1300	7.7	9.0
A242	80	54.1	81.3
	200	50.8	76.2
	400	47.6	76.4
	600	41.1	81.3
	800	39.9	76.4
	1000	35.2	52.8
	1200	20.6	27.6
A588	80	58.6	78.5
	200	57.3	79.5
	400	50.4	74.8
	600	42.5	77.7
	800	37.6	70.7
	1000	32.6	46.4
	1200	17.9	23.3

(Multiply in by 25.4 to obtain mm.)
(Multiply ksi by 6.9 to obtain MPa.)

APPENDIX 59.A
Values of C_b for Simply Supported Beams
($\times$ designates a point of lateral bracing)

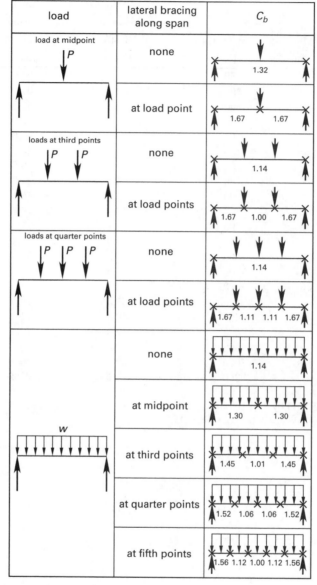

Note: Per *AISC Specification* Sec. F, lateral bracing must be provided at points of support.

APPENDIX 68.A
Section Properties of Masonry Horizontal Cross Sections
(loading parallel to wall face)

unit configuration	grouted cavity spacing[a]	mortar bedding	net cross-sectional properties			average cross-sectional properties			
			A_n (in²/ft)	I_n (in⁴/ft)	S_n (in³/ft)	A_{avg} (in²/ft)	I_{avg} (in⁴/ft)	S_{avg} (in³/ft)	r_{avg} (in)
4 in (102 mm) single wythe walls, ¾ in (19 mm) face shells (standard)[b]									
hollow	no grout	face shell	18.0	38.0	21.0	21.6	39.4	21.7	1.35
hollow	no grout	full	21.6	39.4	21.7	21.6	39.4	21.7	1.35
100% solid	solidly grouted	full	43.5	47.6	26.3	43.5	47.6	26.3	1.05
6 in (152 mm) single wythe walls, 1 in (25 mm) face shells (standard)									
hollow	no grout	face shell	24.0	130.3	46.3	32.2	139.3	49.5	2.08
hollow	no grout	full	32.2	139.3	49.5	32.2	139.3	49.5	2.08
100% solid	solidly grouted	full	67.5	178.0	63.3	67.5	178.0	63.3	1.62
hollow	16 in o.c.	face shell	46.6	155.1	55.1	49.3	158.1	56.2	1.79
hollow	24 in o.c.	face shell	39.1	146.8	52.2	43.6	151.8	54.0	1.87
hollow	32 in o.c.	face shell	35.3	142.7	50.7	40.7	148.7	52.9	1.91
hollow	40 in o.c.	face shell	33.0	140.2	49.9	39.0	146.8	52.2	1.94
hollow	48 in o.c.	face shell	31.5	138.6	49.3	37.9	145.5	51.7	1.96
hollow	72 in o.c.	face shell	29.0	135.8	48.3	36.0	143.5	51.0	2.00
hollow	96 in o.c.	face shell	27.8	134.5	47.8	35.0	142.4	50.6	2.02
hollow	120 in o.c.	face shell	27.0	133.6	47.5	34.4	141.8	50.4	2.03
8 in (203 mm) single wythe walls, 1¼ in (32 mm) face shells (standard)									
hollow	no grout	face shell	30.0	308.7	81.0	41.5	334.0	87.6	2.84
hollow	no grout	full	41.5	334.0	87.6	41.5	334.0	87.6	2.84
100% solid	solidly grouted	full	91.5	443.3	116.3	91.5	443.3	116.3	2.20
hollow	16 in o.c.	face shell	62.0	378.6	99.3	65.8	387.1	101.5	2.43
hollow	24 in o.c.	face shell	51.3	355.3	93.2	57.7	369.4	96.9	2.53
hollow	32 in o.c.	face shell	46.0	343.7	90.1	53.7	360.5	94.6	2.59
hollow	40 in o.c.	face shell	42.8	336.7	88.3	51.2	355.2	93.2	2.63
hollow	48 in o.c.	face shell	40.7	332.0	87.1	49.6	351.7	92.2	2.66
hollow	72 in o.c.	face shell	37.1	324.3	85.0	46.9	345.8	90.7	2.71
hollow	96 in o.c.	face shell	35.3	320.4	84.0	45.6	342.8	89.9	2.74
hollow	120 in o.c.	face shell	34.3	318.0	83.4	44.8	341.0	89.5	2.76
10 in (254 mm) single wythe walls, 1¼ in (32 mm) face shells (standard)									
hollow	no grout	face shell	30.0	530.0	110.1	48.0	606.3	126.0	3.55
hollow	no grout	full	48.0	606.3	126.0	48.0	606.3	126.0	3.55
100% solid/solid	grouted	full	115.5	891.7	185.3	115.5	891.7	185.3	2.78
hollow	16 in o.c.	face shell	74.8	719.3	149.5	80.8	744.7	154.7	3.04
hollow	24 in o.c.	face shell	59.8	656.2	136.3	69.9	698.6	145.2	3.16
hollow	32 in o.c.	face shell	52.4	624.6	129.8	64.4	675.5	140.4	3.24
hollow	40 in o.c.	face shell	47.9	605.7	125.9	61.1	661.6	137.5	3.29
hollow	48 in o.c.	face shell	44.9	593.1	123.2	58.9	652.4	135.6	3.33
hollow	72 in o.c.	face shell	39.9	572.0	118.8	55.3	637.0	132.4	3.39
hollow	96 in o.c.	face shell	37.5	561.5	116.7	53.5	629.3	130.8	3.43
hollow	120 in o.c.	face shell	36.0	555.2	115.4	52.4	624.7	129.8	3.45

APPENDIX 68.A *(continued)*
Section Properties of Masonry Horizontal Cross Sections
(loading parallel to wall face)

unit configuration	grouted cavity spacing[a]	mortar bedding	net cross-sectional properties			average cross-sectional properties			
			A_n (in²/ft)	I_n (in⁴/ft)	S_n (in³/ft)	A_{avg} (in²/ft)	I_{avg} (in⁴/ft)	S_{avg} (in³/ft)	r_{avg} (in)
12 in (305 mm) single wythe walls, ¼ in (32 mm) face shells (standard)									
hollow	no grout	face shell	30.0	811.2	139.6	53.1	971.5	167.1	4.28
hollow	no grout	full	53.1	971.5	167.1	53.1	971.5	167.1	4.28
100% solid	solidly grouted	full	139.5	1571.0	270.3	139.5	1571.0	270.3	3.36
hollow	16 in o.c.	face shell	87.3	1208.9	208.0	95.0	1262.3	217.2	3.64
hollow	24 in o.c.	face shell	68.2	1076.3	185.2	81.0	1165.4	200.5	3.79
hollow	32 in o.c.	face shell	58.7	1010.1	173.8	74.1	1116.9	192.2	3.88
hollow	40 in o.c.	face shell	52.9	970.3	166.9	69.9	1087.8	187.2	3.95
hollow	48 in o.c.	face shell	49.1	943.8	162.4	67.1	1068.4	183.8	3.99
hollow	72 in o.c.	face shell	42.7	899.6	154.8	62.4	1036.1	178.3	4.07
hollow	96 in o.c.	face shell	39.6	877.5	151.0	60.1	1020.0	175.5	4.12
hollow	120 in o.c.	face shell	37.6	864.2	148.7	58.7	1010.3	173.8	4.15

(Multiply in by 25.4 to obtain mm.)
(Multiply in²/ft by 2116.6 to obtain mm²/m.)
(Multiply in³/ft by 53,763 to obtain mm³/m.)
(Multiply in⁴/ft by 1.3655 × 10⁶ to obtain mm⁴/m.)

[a]o.c. = on center

[b]Values in these tables are based on minimum face shell and web thicknesses defined in ASTM C90. Manufactured units generally exceed these dimensions, making 4 in concrete masonry units difficult or impossible to grout.

Source: *NCMA TEK Section Properties of Concrete Masonry Walls*, TEK 14-1B, Structural, Tables 1, 2, 3, 4, 5, National Concrete Masonry Association, copyright © 2007.

APPENDIX 68.B
Section Properties of Masonry Vertical Cross Sections
(loading perpendicular to wall face; masonry spanning horizontally)

unit configuration	grouted cavity spacing[a]	mortar bedding	net cross-sectional properties			average cross-sectional properties			
			A_n (in²/ft)	I_n (in⁴/ft)	S_n (in³/ft)	A_{avg} (in²/ft)	I_{avg} (in⁴/ft)	S_{avg} (in³/ft)	r_{avg} (in)
4 in (102 mm) single wythe walls, ¾ in (19 mm) face shells (standard)[b]									
hollow	no grout	face shell	18.0	38.0	21.0	21.2	39.1	21.6	1.36
hollow	no grout	full	18.0	38.0	21.0	21.6	39.4	21.7	1.35
100% solid	solidly grouted	full	43.5	47.6	26.3	43.5	47.6	26.3	1.05
6 in (152 mm) single wythe walls, 1 in (25 mm) face shells (standard)									
hollow	no grout	face shell	24.0	130.3	46.3	31.4	137.7	49.0	2.09
hollow	no grout	full	24.0	130.3	46.3	32.2	139.3	49.5	2.08
100% solid	solidly grouted	full	67.5	178.0	63.3	67.5	178.0	63.3	1.62
hollow	16 in o.c.	face shell	45.8	154.2	54.8	53.1	161.5	57.4	1.74
hollow	24 in o.c.	face shell	38.5	146.2	52.0	45.9	153.6	54.6	1.83
hollow	32 in o.c.	face shell	34.9	142.3	50.6	42.3	149.6	53.2	1.88
hollow	40 in o.c.	face shell	32.7	139.9	49.7	40.1	147.2	52.4	1.92
hollow	48 in o.c.	face shell	31.3	138.3	49.2	38.6	145.7	51.8	1.94
hollow	96 in o.c.	face shell	27.6	134.3	47.8	35.0	141.7	50.4	2.01
hollow	120 in o.c.	face shell	26.9	133.5	47.5	34.3	140.9	50.1	2.03
8 in (203 mm) single wythe walls, 1¼ in (32 mm) face shells (standard)									
hollow	no grout	face shell	30.0	308.7	81.0	40.5	330.1	86.6	2.86
hollow	no grout	full	30.0	308.7	81.0	41.5	334.0	87.6	2.84
100% solid	solidly grouted	full	91.5	443.3	116.3	91.5	443.3	116.3	2.20
hollow	16 in o.c.	face shell	60.8	376.0	98.6	71.2	397.4	104.2	2.36
hollow	24 in o.c.	face shell	50.5	353.6	92.7	61.0	374.9	98.3	2.48
hollow	32 in o.c.	face shell	45.4	342.4	89.8	55.8	363.7	95.4	2.55
hollow	40 in o.c.	face shell	42.3	335.6	88.0	52.8	357.0	93.6	2.60
hollow	48 in o.c.	face shell	40.3	331.1	86.9	50.7	352.5	92.5	2.64
hollow	96 in o.c.	face shell	35.1	319.9	83.9	45.6	341.3	89.5	2.74
hollow	120 in o.c.	face shell	34.1	317.7	83.3	44.6	339.0	88.9	2.76
10 in (254 mm) single wythe walls, 1¼ in (32 mm) face shells (standard)									
hollow	no grout	face shell	30.0	530.0	110.1	46.3	597.4	124.1	3.59
hollow	no grout	full	30.0	530.0	110.1	48.0	606.3	126.0	3.55
100% solid	solidly grouted	full	115.5	891.7	185.3	115.5	891.7	185.3	2.78
hollow	16 in o.c.	face shell	72.8	710.8	147.7	89.1	778.3	161.7	2.96
hollow	24 in o.c.	face shell	58.5	650.5	135.2	74.8	718.0	149.2	3.10
hollow	32 in o.c.	face shell	51.4	620.4	128.9	67.7	687.9	142.9	3.19
hollow	40 in o.c.	face shell	47.1	602.3	125.2	63.4	669.8	139.2	3.25
hollow	48 in o.c.	face shell	44.3	590.2	122.6	60.6	657.7	136.7	3.29
hollow	96 in o.c.	face shell	37.1	560.1	116.4	53.5	627.6	130.4	3.43
hollow	120 in o.c.	face shell	35.7	554.1	115.1	52.0	621.6	129.2	3.46
12 in (305 mm) single wythe walls, ¼ in (32 mm) face shells (standard)									
hollow	no grout	face shell	36.0	811.2	139.9	55.8	1049.2	180.5	4.34
hollow	no grout	full	36.0	811.2	139.9	57.8	1064.7	183.2	4.29
100% solid	solidly grouted	full	139.5	1571.0	270.3	139.5	1571.0	270.3	3.36
hollow	16 in o.c.	face shell	84.8	1191.1	204.9	105.7	1335.8	229.8	3.56

APPENDIX 68.B *(continued)*
Section Properties of Masonry Vertical Cross Sections
(loading perpendicular to wall face; masonry spanning horizontally)

unit configuration	grouted cavity spacing[a]	mortar bedding	net cross-sectional properties			average cross-sectional properties			
			A_n (in²/ft)	I_n (in⁴/ft)	S_n (in³/ft)	A_{avg} (in²/ft)	I_{avg} (in⁴/ft)	S_{avg} (in³/ft)	r_{avg} (in)
hollow	24 in o.c.	face shell	66.5	1064.5	183.1	87.4	1209.2	208.0	3.72
hollow	32 in o.c.	face shell	57.4	1001.2	172.2	78.3	1145.9	197.1	3.83
hollow	40 in o.c.	face shell	51.9	963.2	165.7	72.8	1107.9	190.6	3.90
hollow	48 in o.c.	face shell	48.3	937.8	161.3	69.2	1082.6	186.3	3.96
hollow	96 in o.c.	face shell	39.1	874.5	150.5	60.1	1019.3	175.4	4.12
hollow	120 in o.c.	face shell	37.3	861.9	148.3	58.2	1006.6	173.2	4.16

(Multiply in by 25.4 to obtain mm.)
(Multiply in²/ft by 2116.6 to obtain mm²/m.)
(Multiply in³/ft by 53,763 to obtain mm³/m.)
(Multiply in⁴/ft by 1.3655×10^6 to obtain mm⁴/m.)

[a]o.c. = on center

[b]Values in these tables are based on minimum face shell and web thicknesses defined in ASTM C90. Manufactured units generally exceed these dimensions, making 4 in concrete masonry units difficult or impossible to grout.

Source: *NCMA TEK Section Properties of Concrete Masonry Walls*, TEK 14-1B, Structural, Tables 1, 2, 3, 4, and 5, National Concrete Masonry Association, copyright © 2007.

APPENDIX 68.C
Ungrouted Wall Section Properties
(loading parallel to wall face)

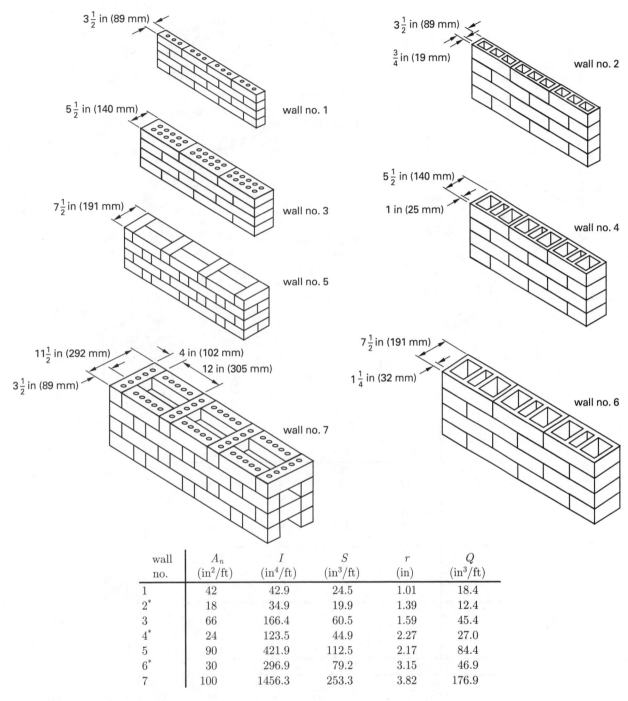

wall no.	A_n (in²/ft)	I (in⁴/ft)	S (in³/ft)	r (in)	Q (in³/ft)
1	42	42.9	24.5	1.01	18.4
2*	18	34.9	19.9	1.39	12.4
3	66	166.4	60.5	1.59	45.4
4*	24	123.5	44.9	2.27	27.0
5	90	421.9	112.5	2.17	84.4
6*	30	296.9	79.2	3.15	46.9
7	100	1456.3	253.3	3.82	176.9

(Multiply in by 25.4 to obtain mm.)
(Multiply in²/ft by 211.6 to obtain mm²/m.)
(Multiply in³/ft by 53,763 to obtain mm³/m.)
(Multiply in⁴/ft by 1.3655 × 10⁶ to obtain mm⁴/m.)
*Section properties are based on minimum solid face shell thickness and face shell bedding.

Derived from the Brick Industry Association from *Technical Notes on Brick Construction 3B*, Table 4, © 1993.

APPENDIX 68.D
Grouted Wall Section Properties[*]
(load parallel to wall face)

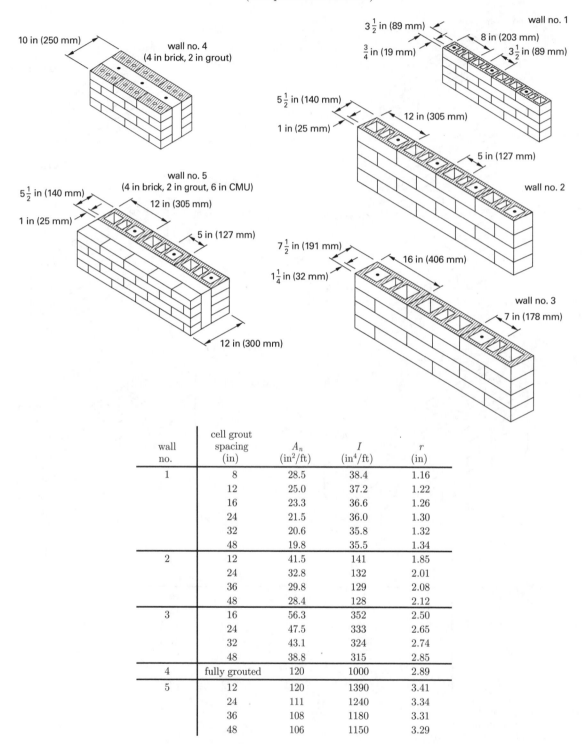

wall no.	cell grout spacing (in)	A_n (in^2/ft)	I (in^4/ft)	r (in)
1	8	28.5	38.4	1.16
	12	25.0	37.2	1.22
	16	23.3	36.6	1.26
	24	21.5	36.0	1.30
	32	20.6	35.8	1.32
	48	19.8	35.5	1.34
2	12	41.5	141	1.85
	24	32.8	132	2.01
	36	29.8	129	2.08
	48	28.4	128	2.12
3	16	56.3	352	2.50
	24	47.5	333	2.65
	32	43.1	324	2.74
	48	38.8	315	2.85
4	fully grouted	120	1000	2.89
5	12	120	1390	3.41
	24	111	1240	3.34
	36	108	1180	3.31
	48	106	1150	3.29

(Multiply in by 25.4 to obtain mm.) (Multiply in^2/ft by 211.6 to obtain mm^2/m.) (Multiply in^3/ft by 53,763 to obtain mm^3/m.) (Multiply in^4/ft by 1.3655 × 10^6 to obtain mm^4/m.)
[*]Section properties are based on minimum solid face shell thickness and face shell bedding of hollow unit masonry.

Derived from the Brick Industry Association from *Technical Notes on Brick Construction 3B*, Table 5, © 1993.

APPENDIX 69.A
Column Interaction Diagram
(compression controls, $g = 0.4$)

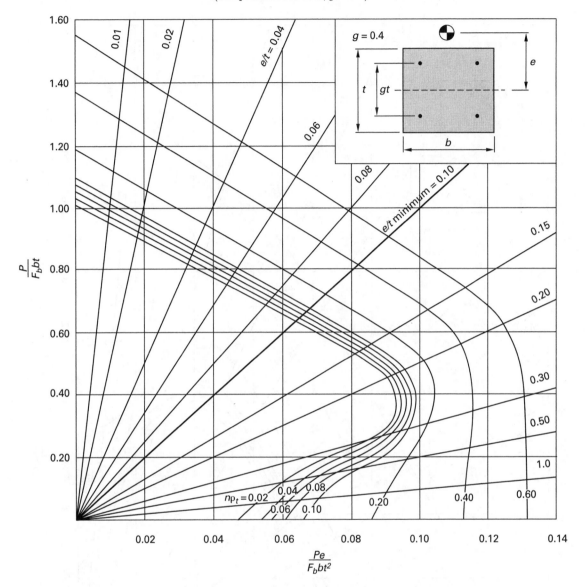

Reprinted with permission of the American Concrete Institute from *Masonry Designer's Guide*, copyright © 1993.

APPENDIX 69.B
Column Interaction Diagram
(compression controls, $g = 0.6$)

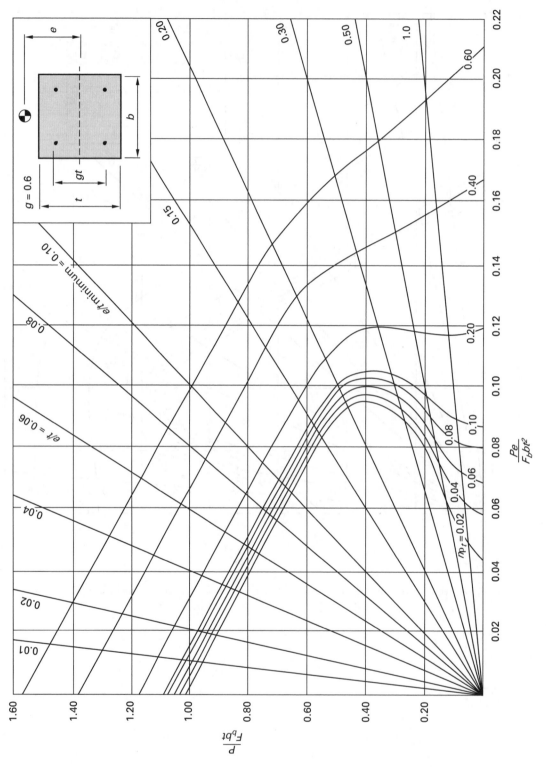

Reprinted with permission of the American Concrete Institute from *Masonry Designer's Guide*, copyright © 1993.

APPENDIX 69.C
Column Interaction Diagram
(compression controls, $g = 0.8$)

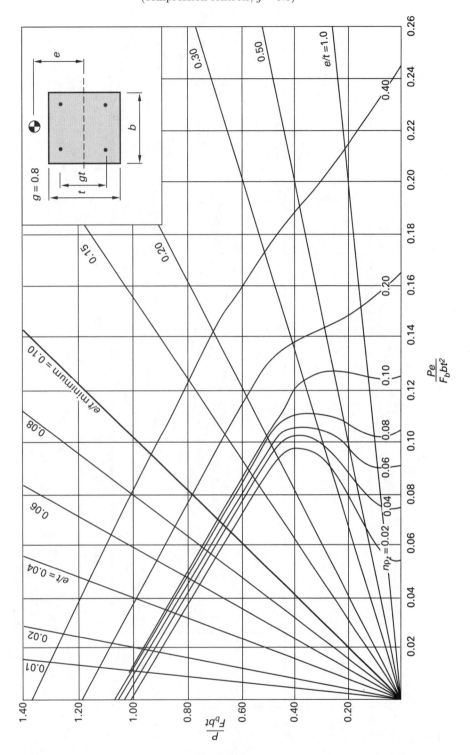

Reprinted with permission of the American Concrete Institute from *Masonry Designer's Guide*, copyright © 1993.

APPENDIX 69.D
Column Interaction Diagram
(tension controls, $g = 0.4$)

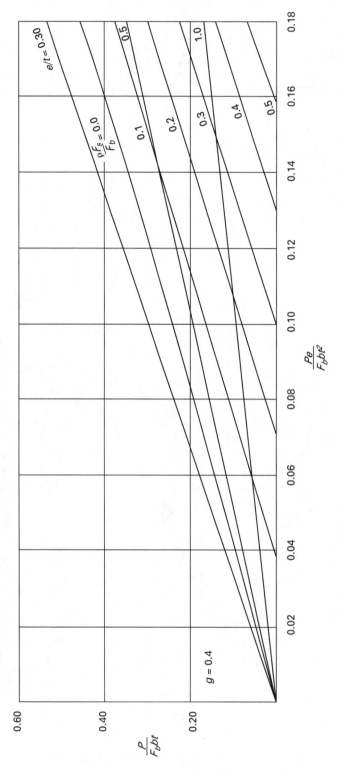

Reprinted with permission of the American Concrete Institute from *Masonry Designer's Guide*, copyright © 1993.

APPENDIX 69.E
Column Interaction Diagram
(tension controls, $g = 0.6$)

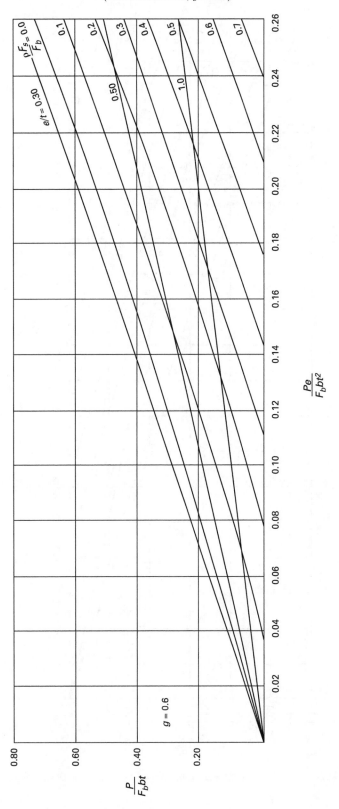

Reprinted with permission of the American Concrete Institute from *Masonry Designer's Guide*, copyright © 1993.

APPENDIX 69.F
Column Interaction Diagram
(tension controls, $g = 0.8$)

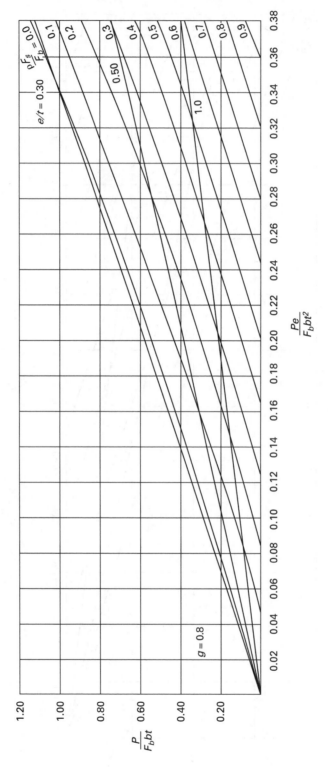

Reprinted with permission of the American Concrete Institute from *Masonry Designer's Guide*, copyright © 1993.

APPENDIX 70.A
Mass Moments of Inertia
(centroids at points labeled C)

slender rod		$I_y = I_z = \dfrac{mL^2}{12}$ $I_{y'} = I_{z'} = \dfrac{mL^2}{3}$
solid circular cylinder, radius r		$I_x = \dfrac{mr^2}{2}$ $I_y = I_z = \dfrac{m(3r^2 + L^2)}{12}$
hollow circular cylinder, inner radius r_i, outer radius r_o		$I_x = \dfrac{m\left(r_o^2 + r_i^2\right)}{2}$ $= \left(\dfrac{\pi\rho L}{2}\right)\left(r_o^4 - r_i^4\right)$ $I_y = I_z = \dfrac{\pi\rho L}{12}\left(3(r_2^4 - r_1^4)\right.$ $\left. + L^2(r_2^2 - r_1^2)\right)$
thin disk, radius r		$I_x = \dfrac{mr^2}{2}$ $I_y = I_z = \dfrac{mr^2}{4}$
solid circular cone, base radius r		$I_x = \dfrac{3mr^2}{10}$ $I_y = I_z = \left(\dfrac{3m}{5}\right)\left(\dfrac{r^2}{4} + h^2\right)$
thin rectangular plate		$I_x = \dfrac{m(b^2 + c^2)}{12}$ $I_y = \dfrac{mc^2}{12}$ $I_z = \dfrac{mb^2}{12}$
rectangular parallelepiped		$I_x = \dfrac{m(b^2 + c^2)}{12}$ $I_y = \dfrac{m(c^2 + a^2)}{12}$ $I_z = \dfrac{m(a^2 + b^2)}{12}$ $I_{x'} = \dfrac{m(4b^2 + c^2)}{12}$
sphere, radius r		$I_x = I_y = I_z = \dfrac{2mr^2}{5}$

Appendices

APPENDIX 76.A
Axle Load Equivalency Factors for Flexible Pavements
(single axles and p_t of 2.5)

axle load (kips)	pavement structural number (SN)					
	1	2	3	4	5	6
2	0.0004	0.0004	0.0003	0.0002	0.0002	0.0002
4	0.003	0.004	0.004	0.003	0.002	0.002
6	0.011	0.017	0.017	0.013	0.010	0.009
8	0.032	0.047	0.051	0.041	0.034	0.031
10	0.078	0.102	0.118	0.102	0.088	0.080
12	0.168	0.198	0.229	0.213	0.189	0.176
14	0.328	0.358	0.399	0.388	0.360	0.342
16	0.591	0.613	0.646	0.645	0.623	0.606
18	1.00	1.00	1.00	1.00	1.00	1.00
20	1.61	1.57	1.49	1.47	1.51	1.55
22	2.48	2.38	2.17	2.09	2.18	2.30
24	3.69	3.49	3.09	2.89	3.03	3.27
26	5.33	4.99	4.31	3.91	4.09	4.48
28	7.49	6.98	5.90	5.21	5.39	5.98
30	10.3	9.5	7.9	6.8	7.0	7.8
32	13.9	12.8	10.5	8.8	8.9	10.0
34	18.4	16.9	13.7	11.3	11.2	12.5
36	24.0	22.0	17.7	14.4	13.9	15.5
38	30.9	28.3	22.6	18.1	17.2	19.0
40	39.3	35.9	28.5	22.5	21.1	23.0
42	49.3	45.0	35.6	27.8	25.6	27.7
44	61.3	55.9	44.0	34.0	31.0	33.1
46	75.5	68.8	54.0	41.4	37.2	39.3
48	92.2	83.9	65.7	50.1	44.5	46.5
50	112	102	79	60	53	55

From *Guide for Design of Pavement and Structures*, Table D.4, copyright © 1993 by the American Association of State Highway and Transportation Officials, Washington, D.C. Used by permission.

APPENDIX 76.B
Axle Load Equivalency Factors for Flexible Pavements
(tandem axles and p_t of 2.5)

axle load (kips)	pavement structural number (SN)					
	1	2	3	4	5	6
2	0.0001	0.0001	0.0001	0.0000	0.0000	0.0000
4	0.0005	0.0005	0.0004	0.0003	0.0003	0.0002
6	0.002	0.002	0.002	0.001	0.001	0.001
8	0.004	0.006	0.005	0.004	0.003	0.003
10	0.008	0.013	0.011	0.009	0.007	0.006
12	0.015	0.024	0.023	0.018	0.014	0.013
14	0.026	0.041	0.042	0.033	0.027	0.024
16	0.044	0.065	0.070	0.057	0.047	0.043
18	0.070	0.097	0.109	0.092	0.077	0.070
20	0.107	0.141	0.162	0.141	0.121	0.110
22	0.160	0.198	0.229	0.207	0.180	0.166
24	0.231	0.273	0.315	0.292	0.260	0.242
26	0.327	0.370	0.420	0.401	0.364	0.342
28	0.451	0.493	0.548	0.534	0.495	0.470
30	0.611	0.648	0.703	0.695	0.658	0.633
32	0.813	0.843	0.889	0.887	0.857	0.834
34	1.06	1.08	1.11	1.11	1.09	1.08
36	1.38	1.38	1.38	1.38	1.38	1.38
38	1.75	1.73	1.69	1.68	1.70	1.73
40	2.21	2.16	2.06	2.03	2.08	2.14
42	2.76	2.67	2.49	2.43	2.51	2.61
44	3.41	3.27	2.99	2.88	3.00	3.16
46	4.18	3.98	3.58	3.40	3.55	3.79
48	5.08	4.80	4.25	3.98	4.17	4.49
50	6.12	5.76	5.03	4.64	4.86	5.28
52	7.33	6.87	5.93	5.38	5.63	6.17
54	8.72	8.14	6.95	6.22	6.47	7.15
56	10.3	9.6	8.1	7.2	7.4	8.2
58	12.1	11.3	9.4	8.2	8.4	9.4
60	14.2	13.1	10.9	9.4	9.6	10.7
62	16.5	15.3	12.6	10.7	10.8	12.1
64	19.1	17.6	14.5	12.2	12.2	13.7
66	22.1	20.3	16.6	13.8	13.7	15.4
68	25.3	23.3	18.9	15.6	15.4	17.2
70	29.0	26.6	21.5	17.6	17.2	19.2
72	33.0	30.3	24.4	19.8	19.2	21.3
74	37.5	34.4	27.6	22.2	21.3	23.6
76	42.5	38.9	31.1	24.8	23.7	26.1
78	48.0	43.9	35.0	27.8	26.2	28.8
80	54.0	49.4	39.2	30.9	29.0	31.7
82	60.6	55.4	43.9	34.4	32.0	34.8
84	67.8	61.9	49.0	38.2	35.3	38.1
86	75.7	69.1	54.5	42.3	38.8	41.7
88	84.3	76.9	60.6	46.8	42.6	45.6
90	93.7	85.4	67.1	51.7	46.8	49.7

From *Guide for Design of Pavement and Structures*, Table D.5, copyright © 1993 by the American Association of State Highway and Transportation Officials, Washington, D.C. Used by permission.

APPENDIX 76.C
Axle Load Equivalency Factors for Flexible Pavements
(triple axles and p_t of 2.5)

axle load (kips)	pavement structural number (SN)					
	1	2	3	4	5	6
2	0.0000	0.0000	0.0000	0.0000	0.0000	0.0000
4	0.0002	0.0002	0.0002	0.0001	0.0001	0.0001
6	0.0006	0.0007	0.0005	0.0004	0.0003	0.0003
8	0.001	0.002	0.001	0.001	0.001	0.001
10	0.003	0.004	0.003	0.002	0.002	0.002
12	0.005	0.007	0.006	0.004	0.003	0.003
14	0.008	0.012	0.010	0.008	0.006	0.006
16	0.012	0.019	0.018	0.013	0.011	0.010
18	0.018	0.029	0.028	0.021	0.017	0.016
20	0.027	0.042	0.042	0.032	0.027	0.024
22	0.038	0.058	0.060	0.048	0.040	0.036
24	0.053	0.078	0.084	0.068	0.057	0.051
26	0.072	0.103	0.114	0.095	0.080	0.072
28	0.098	0.133	0.151	0.128	0.109	0.099
30	0.129	0.169	0.195	0.170	0.145	0.133
32	0.169	0.213	0.247	0.220	0.191	0.175
34	0.219	0.266	0.308	0.281	0.246	0.228
36	0.279	0.329	0.379	0.352	0.313	0.292
38	0.352	0.403	0.461	0.436	0.393	0.368
40	0.439	0.491	0.554	0.533	0.487	0.459
42	0.543	0.594	0.661	0.644	0.597	0.567
44	0.666	0.714	0.781	0.769	0.723	0.692
46	0.811	0.854	0.918	0.911	0.868	0.838
48	0.979	1.015	1.072	1.069	1.033	1.005
50	1.17	1.20	1.24	1.25	1.22	1.20
52	1.40	1.41	1.44	1.44	1.43	1.41
54	1.66	1.66	1.66	1.66	1.66	1.66
56	1.95	1.93	1.90	1.90	1.91	1.93
58	2.29	2.25	2.17	2.16	2.20	2.24
60	2.67	2.60	2.48	2.44	2.51	2.58
62	3.09	3.00	2.82	2.76	2.85	2.95
64	3.57	3.44	3.19	3.10	3.22	3.36
66	4.11	3.94	3.61	3.47	3.62	3.81
68	4.71	4.49	4.06	3.88	4.05	4.30
70	5.38	5.11	4.57	4.32	4.52	4.84
72	6.12	5.79	5.13	4.80	5.03	5.41
74	6.93	6.54	5.74	5.32	5.57	6.04
76	7.84	7.37	6.41	5.88	6.15	6.71
78	8.83	8.28	7.14	6.49	6.78	7.43
80	9.92	9.28	7.95	7.15	7.45	8.21
82	11.1	10.4	8.8	7.9	8.2	9.0
84	12.4	11.6	9.8	8.6	8.9	9.9
86	13.8	12.9	10.8	9.5	9.8	10.9
88	15.4	14.3	11.9	10.4	10.6	11.9
90	17.1	15.8	13.2	11.3	11.6	12.9

APPENDIX 77.A
Axle Load Equivalency Factors for Rigid Pavements
(single axles and p_t of 2.5)

axle load (kips)	slab thickness, D (in)								
	6	7	8	9	10	11	12	13	14
2	0.0002	0.0002	0.0002	0.0002	0.0002	0.0002	0.0002	0.0002	0.0002
4	0.003	0.002	0.002	0.002	0.002	0.002	0.002	0.002	0.002
6	0.012	0.011	0.010	0.010	0.010	0.010	0.010	0.010	0.010
8	0.039	0.035	0.033	0.032	0.032	0.032	0.032	0.032	0.032
10	0.097	0.089	0.084	0.082	0.081	0.080	0.080	0.080	0.080
12	0.203	0.189	0.181	0.176	0.175	0.174	0.174	0.173	0.173
14	0.376	0.360	0.347	0.341	0.338	0.337	0.336	0.336	0.336
16	0.634	0.623	0.610	0.604	0.601	0.599	0.599	0.599	0.598
18	1.00	1.00	1.00	1.00	1.00	1.00	1.00	1.00	1.00
20	1.51	1.52	1.55	1.57	1.58	1.58	1.59	1.59	1.59
22	2.21	2.20	2.28	2.34	2.38	2.40	2.41	2.41	2.41
24	3.16	3.10	3.22	3.36	3.45	3.50	3.53	3.54	3.55
26	4.41	4.26	4.42	4.67	4.85	4.95	5.01	5.04	5.05
28	6.05	5.76	5.92	6.29	6.61	6.81	6.92	6.98	7.01
30	8.16	7.67	7.79	8.28	8.79	9.14	9.35	9.46	9.52
32	10.8	10.1	10.1	10.7	11.4	12.0	12.3	12.6	12.7
34	14.1	13.0	12.9	13.6	14.6	15.4	16.0	16.4	16.5
36	18.2	16.7	16.4	17.1	18.3	19.5	20.4	21.0	21.3
38	23.1	21.1	20.6	21.3	22.7	24.3	25.6	26.4	27.0
40	29.1	26.5	25.7	26.3	27.9	29.9	31.6	32.9	33.7
42	36.2	32.9	31.7	32.2	34.0	36.3	38.7	40.4	41.6
44	44.6	40.4	38.8	39.2	41.0	43.8	46.7	49.1	50.8
46	54.5	49.3	47.1	47.3	49.2	52.3	55.9	59.0	61.4
48	66.1	59.7	56.9	56.8	58.7	62.1	66.3	70.3	73.4
50	79.4	71.7	68.2	67.8	69.6	73.3	78.1	83.0	87.1

From *Guide for Design of Pavement and Structures*, Table D.13, copyright © 1993 by the American Association of State Highway and Transportation Officials, Washington, D.C. Used by permission.

APPENDIX 77.B
Axle Load Equivalency Factors for Rigid Pavements
(double axles and p_t of 2.5)

axle load (kips)	slab thickness, D (in)								
	6	7	8	9	10	11	12	13	14
2	0.0001	0.0001	0.0001	0.0001	0.0001	0.0001	0.0001	0.0001	0.0001
4	0.0006	0.0006	0.0005	0.0005	0.0005	0.0005	0.0005	0.0005	0.0005
6	0.002	0.002	0.002	0.002	0.002	0.002	0.002	0.002	0.002
8	0.007	0.006	0.006	0.005	0.005	0.005	0.005	0.005	0.005
10	0.015	0.014	0.013	0.013	0.012	0.012	0.012	0.012	0.012
12	0.031	0.028	0.026	0.026	0.025	0.025	0.025	0.025	0.025
14	0.057	0.052	0.049	0.048	0.047	0.047	0.047	0.047	0.047
16	0.097	0.089	0.084	0.082	0.081	0.081	0.080	0.080	0.080
18	0.155	0.143	0.136	0.133	0.132	0.131	0.131	0.131	0.131
20	0.234	0.220	0.211	0.206	0.204	0.203	0.203	0.203	0.203
22	0.340	0.325	0.313	0.308	0.305	0.304	0.303	0.303	0.303
24	0.475	0.462	0.450	0.444	0.441	0.440	0.439	0.439	0.439
26	0.644	0.637	0.627	0.622	0.620	0.619	0.618	0.618	0.618
28	0.855	0.854	0.852	0.850	0.850	0.850	0.849	0.849	0.849
30	1.11	1.12	1.13	1.14	1.14	1.14	1.14	1.14	1.14
32	1.43	1.44	1.47	1.49	1.50	1.51	1.51	1.51	1.51
34	1.82	1.82	1.87	1.92	1.95	1.96	1.97	1.97	1.97
36	2.29	2.27	2.35	2.43	2.48	2.51	2.52	2.52	2.53
38	2.85	2.80	2.91	3.03	3.12	3.16	3.18	3.20	3.20
40	3.52	3.42	3.55	3.74	3.87	3.94	3.98	4.00	4.01
42	4.32	4.16	4.30	4.55	4.74	4.86	4.91	4.95	4.96
44	5.26	5.01	5.16	5.48	5.75	5.92	6.01	6.06	6.09
46	6.36	6.01	6.14	6.53	6.90	7.14	7.28	7.36	7.40
48	7.64	7.16	7.27	7.73	8.21	8.55	8.75	8.86	8.92
50	9.11	8.50	8.55	9.07	9.68	10.14	10.42	10.58	10.66
52	10.8	10.0	10.0	10.6	11.3	11.9	12.3	12.5	12.7
54	12.8	11.8	11.7	12.3	13.2	13.9	14.5	14.8	14.9
56	15.0	13.8	13.6	14.2	15.2	16.2	16.8	17.3	17.5
58	17.5	16.0	15.7	16.3	17.5	18.6	19.5	20.1	20.4
60	20.3	18.5	18.1	18.7	20.0	21.4	22.5	23.2	23.6
62	23.5	21.4	20.8	21.4	22.8	24.4	25.7	26.7	27.3
64	27.0	24.6	23.8	24.4	25.8	27.7	29.3	30.5	31.3
66	31.0	28.1	27.1	27.6	29.2	31.3	33.2	34.7	35.7
68	35.4	32.1	30.9	31.3	32.9	35.2	37.5	39.3	40.5
70	40.3	36.5	35.0	35.3	37.0	39.5	42.1	44.3	45.9
72	45.7	41.4	39.6	39.8	41.5	44.2	47.2	49.8	51.7
74	51.7	46.7	44.6	44.7	46.4	49.3	52.7	55.7	58.0
76	58.3	52.6	50.2	50.1	51.8	54.9	58.6	62.1	64.8
78	65.5	59.1	56.3	56.1	57.7	60.9	65.0	69.0	72.3
80	73.4	66.2	62.9	62.5	64.2	67.5	71.9	76.4	80.2
82	82.0	73.9	70.2	69.6	71.2	74.7	79.4	84.4	88.8
84	91.4	82.4	78.1	77.3	78.9	82.4	87.4	93.0	98.1
86	102	92	87	86	87	91	96	102	108
88	113	102	96	95	96	100	105	112	119
90	125	112	106	105	106	110	115	123	130

From *Guide for Design of Pavement and Structures*, Table D.14, copyright © 1993 by the American Association of State Highway and Transportation Officials, Washington, D.C. Used by permission.

APPENDIX 77.C
Axle Load Equivalency Factors for Rigid Pavements
(triple axles and p_t of 2.5)

axle load (kips)	slab thickness, D (in)								
	6	7	8	9	10	11	12	13	14
2	0.0001	0.0001	0.0001	0.0001	0.0001	0.0001	0.0001	0.0001	0.0001
4	0.0003	0.0003	0.0003	0.0003	0.0003	0.0003	0.0003	0.0003	0.0003
6	0.001	0.001	0.001	0.001	0.001	0.001	0.001	0.001	0.001
8	0.003	0.002	0.002	0.002	0.002	0.002	0.002	0.002	0.002
10	0.006	0.005	0.005	0.005	0.005	0.005	0.005	0.005	0.005
12	0.011	0.010	0.010	0.009	0.009	0.009	0.009	0.009	0.009
14	0.020	0.018	0.017	0.017	0.016	0.016	0.016	0.016	0.016
16	0.033	0.030	0.029	0.028	0.027	0.027	0.027	0.027	0.027
18	0.053	0.048	0.045	0.044	0.044	0.043	0.043	0.043	0.043
20	0.080	0.073	0.069	0.067	0.066	0.066	0.066	0.066	0.066
22	0.116	0.107	0.101	0.099	0.098	0.097	0.097	0.097	0.097
24	0.163	0.151	0.144	0.141	0.139	0.139	0.138	0.138	0.138
26	0.222	0.209	0.200	0.195	0.194	0.193	0.192	0.192	0.192
28	0.295	0.281	0.271	0.265	0.263	0.262	0.262	0.262	0.262
30	0.384	0.371	0.359	0.354	0.351	0.350	0.349	0.349	0.349
32	0.490	0.480	0.468	0.463	0.460	0.459	0.458	0.458	0.458
34	0.616	0.609	0.601	0.596	0.594	0.593	0.592	0.592	0.592
36	0.765	0.762	0.759	0.757	0.756	0.755	0.755	0.755	0.755
38	0.939	0.941	0.946	0.948	0.950	0.951	0.951	0.951	0.951
40	1.14	1.15	1.16	1.17	1.18	1.18	1.18	1.18	1.18
42	1.38	1.38	1.41	1.44	1.45	1.46	1.46	1.46	1.46
44	1.65	1.65	1.70	1.74	1.77	1.78	1.78	1.78	1.79
46	1.97	1.96	2.03	2.09	2.13	2.15	2.16	2.16	2.16
48	2.34	2.31	2.40	2.49	2.55	2.58	2.59	2.60	2.60
50	2.76	2.71	2.81	2.94	3.02	3.07	3.09	3.10	3.11
52	3.24	3.15	3.27	3.44	3.56	3.62	3.66	3.68	3.68
54	3.79	3.66	3.79	4.00	4.16	4.26	4.30	4.33	4.34
56	4.41	4.23	4.37	4.63	4.84	4.97	5.03	5.07	5.09
58	5.12	4.87	5.00	5.32	5.59	5.76	5.85	5.90	5.93
60	5.91	5.59	5.71	6.08	6.42	6.64	6.77	6.84	6.87
62	6.80	6.39	6.50	6.91	7.33	7.62	7.79	7.88	7.93
64	7.79	7.29	7.37	7.82	8.33	8.70	8.92	9.04	9.11
66	8.90	8.28	8.33	8.83	9.42	9.88	10.17	10.33	10.42
68	10.1	9.4	9.4	9.9	10.6	11.2	11.5	11.7	11.9
70	11.5	10.6	10.6	11.1	11.9	12.6	13.0	13.3	13.5
72	13.0	12.0	11.8	12.4	13.3	14.1	14.7	15.0	15.2
74	14.6	13.5	13.2	13.8	14.8	15.8	16.5	16.9	17.1
76	16.5	15.1	14.8	15.4	16.5	17.6	18.4	18.9	19.2
78	18.5	16.9	16.5	17.1	18.2	19.5	20.5	21.1	21.5
80	20.6	18.8	18.3	18.9	20.2	21.6	22.7	23.5	24.0
82	23.0	21.0	20.3	20.9	22.2	23.8	25.2	26.1	26.7
84	25.6	23.3	22.5	23.1	24.5	26.2	27.8	28.9	29.6
86	28.4	25.8	24.9	25.4	26.9	28.8	30.5	31.9	32.8
88	31.5	28.6	27.5	27.9	29.4	31.5	33.5	35.1	36.1
90	34.8	31.5	30.3	30.7	32.2	34.4	36.7	38.5	39.8

From *Guide for Design of Pavement and Structures*, Table D.15, copyright © 1993 by the American Association of State Highway and Transportation Officials, Washington, D.C. Used by permission.

APPENDIX 78.A
Oblique Triangle Equations

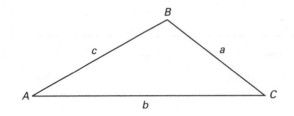

given	equation
A, B, a	$C = 180° - (A + B)$
	$b = \left(\dfrac{a}{\sin A}\right)\sin B$
	$c = \left(\dfrac{a}{\sin A}\right)\sin\left(A + B\right) = \left(\dfrac{a}{\sin A}\right)\sin C$
	$\text{area} = \dfrac{1}{2}ab\sin C = \dfrac{a^2\sin B\sin C}{2\sin A}$
A, a, b	$\sin B = \left(\dfrac{\sin A}{a}\right)b$
	$C = 180° - (A + B)$
	$c = \left(\dfrac{a}{\sin A}\right)\sin C$
	$\text{area} = \dfrac{1}{2}ab\sin C$
C, a, b	$c = \sqrt{a^2 + b^2 - 2ab\cos C}$
	$\dfrac{1}{2}\left(A + B\right) = 90° - \dfrac{1}{2}C$
	$\tan\dfrac{1}{2}\left(A - B\right) = \left(\dfrac{a - b}{a + b}\right)\tan\dfrac{1}{2}\left(A + B\right)$
	$A = \dfrac{1}{2}\left(A + B\right) + \dfrac{1}{2}\left(A - B\right)$
	$B = \dfrac{1}{2}\left(A + B\right) - \dfrac{1}{2}\left(A - B\right)$
	$c = (a + b)\left(\dfrac{\cos\dfrac{1}{2}\left(A + B\right)}{\cos\dfrac{1}{2}\left(A - B\right)}\right) = (a - b)\left(\dfrac{\sin\dfrac{1}{2}\left(A + B\right)}{\sin\dfrac{1}{2}\left(A - B\right)}\right)$
	$\text{area} = \dfrac{1}{2}ab\sin C$

APPENDIX 78.A *(continued)*
Oblique Triangle Equations

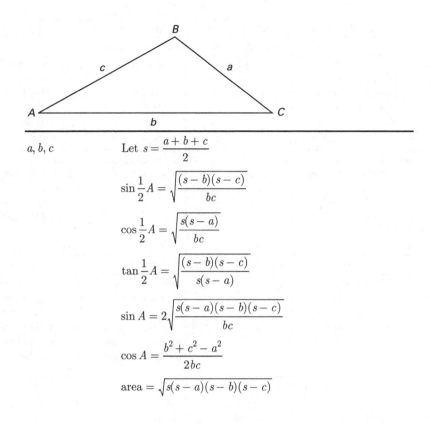

a, b, c Let $s = \dfrac{a+b+c}{2}$

$$\sin \frac{1}{2}A = \sqrt{\frac{(s-b)(s-c)}{bc}}$$

$$\cos \frac{1}{2}A = \sqrt{\frac{s(s-a)}{bc}}$$

$$\tan \frac{1}{2}A = \sqrt{\frac{(s-b)(s-c)}{s(s-a)}}$$

$$\sin A = 2\sqrt{\frac{s(s-a)(s-b)(s-c)}{bc}}$$

$$\cos A = \frac{b^2 + c^2 - a^2}{2bc}$$

$$\text{area} = \sqrt{s(s-a)(s-b)(s-c)}$$

APPENDIX 84.A
Polyphase Motor Classifications and Characteristics

speed regulations	speed control	starting torque	breakdown torque	application
general-purpose squirrel cage (NEMA design B)				
Drops about 3% for large to 5% for small sizes.	None, except multispeed types, designed for two to four fixed speeds.	100% for large; 275% for 1 hp 4 pole unit.	200% of full load.	Constant-speed service where starting is not excessive. Fans, blowers, rotary compressors, and centrifugal pumps.
high-torque squirrel cage (NEMA design C)				
Drops about 3% for large to 6% for small sizes.	None, except multispeed types, designed for two and four fixed speeds.	250% of full load for high-speed to 200% for low-speed designs.	200% of full load.	Constant-speed where fairly high starting torque is required infrequently with starting current about 550% of full load. Reciprocating pumps and compressors, crushers, etc.
high-slip squirrel cage (NEMA design D)				
Drops about 10% to 15% from no load to full load.	None, except multispeed types, designed for two to four fixed speeds.	225% to 300% full load, depending on speed with rotor resistance.	200%. Will usually not stall until loaded to maximum torque, which occurs at standstill.	Constant-speed and high starting torque, if starting is not too frequent, and for high-peak loads with or without flywheels. Punch presses, shears, elevators, etc.
low-torque squirrel cage (NEMA design F)				
Drops about 3% for large to 5% for small sizes.	None, except multispeed types, designed for two to four fixed speeds.	50% of full load for high-speed to 90% for low-speed designs.	135% to 170% of full load.	Constant-speed service where starting duty is light. Fans, blowers, centrifugal pumps, and similar loads.
wound rotor				
With rotor rings short circuited, drops about 3% for large to 5% for small sizes.	Speed can be reduced to 50% by rotor resistance. Speed varies inversely as load.	Up to 300% depending on external resistance in rotor circuit and how distributed.	300% when rotor slip rings are short circuited.	Where high starting torque with low starting current or where limited speed control is required. Fans, centrifugal and plunger pumps, compressors, conveyors, hoists, cranes, etc.
synchronous				
Constant.	None, except special motors designed for two fixed speeds.	40% for slow to 160% for medium-speed 80% pf. Specials develop higher.	Unity-pf motors 170%, 80%-pf motors 225%. Specials, up to 300%	For constant-speed service, direct connection to slow-speed machines and where pf correction is required.

Adapted from *Mechanical Engineering, Design Manual*, NAVFAC DM-3, Department of the Navy, copyright © 1972.

APPENDIX 84.B
DC and Single-Phase Motor Classifications and Characteristics

speed regulations	speed control	starting torque	breakdown torque	application
series				
Varies inversely as load. Races on light loads and full voltage.	Zero to maximum depending on control and load.	High. Varies as square of voltage. Limited by commutation, heating, and line capacity.	High. Limited by commutation, heating, and line capacity.	Where high starting torque is required and speed can be regulated. Traction, bridges, hoists, gates, car dumpers, car retarders.
shunt				
Drops 3% to 5% from no load to full load.	Any desired range depending on design, type of system.	Good. With constant field, varies directly as voltage applied to armature.	High. Limited by commutation, heating, and line capacity.	Where constant or adjustable speed is required and starting conditions are not severe. Fans, blowers, centrifugal pumps, conveyors, wood and metal-working machines, elevators.
compound				
Drops 7% to 20% from no load to full load depending on amount of compounding.	Any desired range, depending on design, type of control.	Higher than for shunt, depending on amount of compounding.	High. Limited by commutation, heating and line capacity.	Where high starting torque and fairly constant speed is required. Plunger pumps, punch presses, shears, bending rolls, geared elevators, conveyors, hoists.
split-phase				
Drops about 10% from no load to full load.	None.	75% for large to 175% for small sizes.	150% for large to 200% for small sizes.	Constant-speed service where starting is easy. Small fans, centrifugal pumps and light-running machines, where polyphase is not available.
capacitor start				
Drops about 5% for large to 10% for small sizes.	None.	150% to 350% of full load depending on design and size.	50% for large to 200% for small sizes.	Constant-speed service for any starting duty and quiet operation, where polyphase current cannot be used.
commutator type				
Drops about 5% for large to 10% for small sizes.	Repulsion induction, none. Brush-shifting types, four to one at full load.	250% for large to 350% for small sizes.	150% for large to 250%.	Constant-speed service for any starting duty where speed control is required and polyphase current cannot be used.

Adapted from *Mechanical Engineering, Design Manual*, NAVFAC DM-3, Department of the Navy, copyright © 1972.

Appendices

APPENDIX 85.A
Thermoelectric Constants for Thermocouples[*]
(mV, reference 32°F (0°C))

(a) chromel-alumel (type K)

°F	0	10	20	30	40	50	60	70	80	90
−300	−5.51	−5.60			millivolts					
−200	−4.29	−4.44	−4.58	−4.71	−4.84	−4.96	−5.08	−5.20	−5.30	−5.41
−100	−2.65	−2.83	−3.01	−3.19	−3.36	−3.52	−3.69	−3.84	−4.00	−4.15
−0	−0.68	−0.89	−1.10	−1.30	−1.50	−1.70	−1.90	−2.09	−2.28	−2.47
+0	−0.68	−0.49	−0.26	−0.04	0.18	0.40	0.62	0.84	1.06	1.29
100	1.52	1.74	1.97	2.20	2.43	2.66	2.89	3.12	3.36	3.59
200	3.82	4.05	4.28	4.51	4.74	4.97	5.20	5.42	5.65	5.87
300	6.09	6.31	6.53	6.76	6.98	7.20	7.42	7.64	7.87	8.09
400	8.31	8.54	8.76	8.98	9.21	9.43	9.66	9.88	10.11	10.34
500	10.57	10.79	11.02	11.25	11.48	11.71	11.94	12.17	12.40	12.63
600	12.86	13.09	13.32	13.55	13.78	14.02	14.25	14.48	14.71	14.95
700	15.18	15.41	15.65	15.88	16.12	16.35	16.59	16.82	17.06	17.29
800	17.53	17.76	18.00	18.23	18.47	18.70	18.94	19.18	19.41	19.65
900	19.89	20.13	20.36	20.60	20.84	21.07	21.31	21.54	21.78	22.02
1000	22.26	22.49	22.73	22.97	23.20	23.44	23.68	23.91	24.15	24.39
1100	24.63	24.86	25.10	25.34	25.57	25.81	26.05	26.28	26.52	26.75
1200	26.98	27.22	27.45	27.69	27.92	28.15	28.39	28.62	28.86	29.09
1300	29.32	29.56	29.79	30.02	30.25	30.49	30.72	30.95	31.18	31.42
1400	31.65	31.88	32.11	32.34	32.57	32.80	33.02	33.25	33.48	33.71
1500	33.93	34.16	34.39	34.62	34.84	35.07	35.29	35.52	35.75	35.97
1600	36.19	36.42	36.64	36.87	37.09	37.31	37.54	37.76	37.98	38.20
1700	38.43	38.65	38.87	39.09	39.31	39.53	39.75	39.96	40.18	40.40
1800	40.62	40.84	41.05	41.27	41.49	41.70	41.92	42.14	42.35	42.57
1900	42.78	42.99	43.21	43.42	43.63	43.85	44.06	44.27	44.49	44.70
2000	44.91	45.12	45.33	45.54	45.75	45.96	46.17	46.38	46.58	46.79

(b) iron-constantan (type J)

°F	0	10	20	30	40	50	60	70	80	90
−300	−7.52	−7.66			millivolts					
−200	−5.76	−5.96	−6.16	−6.35	−6.53	−6.71	−6.89	−7.06	−7.22	−7.38
−100	−3.49	−3.73	−3.97	−4.21	−4.44	−4.68	−4.90	−5.12	−5.34	−5.55
−0	−0.89	−1.16	−1.43	−1.70	−1.96	−2.22	−2.48	−2.74	−2.99	−3.24
+0	−0.89	−0.61	−0.34	−0.06	0.22	0.50	0.79	1.07	1.36	1.65
100	1.94	2.23	2.52	2.82	3.11	3.41	3.71	4.01	4.31	4.61
200	4.91	5.21	5.51	5.81	6.11	6.42	6.72	7.03	7.33	7.64
300	7.94	8.25	8.56	8.87	9.17	9.48	9.79	10.10	10.41	10.72
400	11.03	11.34	11.65	11.96	12.26	12.57	12.88	13.19	13.50	13.81
500	14.12	14.42	14.73	15.04	15.34	15.65	15.96	16.26	16.57	16.88
600	17.18	17.49	17.80	18.11	18.41	18.72	19.03	19.34	19.64	19.95
700	20.26	20.56	20.87	21.18	21.48	21.79	22.10	22.40	22.71	23.01
800	23.32	23.63	23.93	24.24	24.55	24.85	25.16	25.47	25.78	26.09
900	26.40	26.70	27.02	27.33	27.64	27.95	28.26	28.58	28.89	29.21
1000	29.52	29.84	30.16	30.48	30.80	31.12	31.44	31.76	32.08	32.40
1100	32.72	33.05	33.37	33.70	34.03	34.36	34.68	35.01	35.35	35.68
1200	36.01	36.35	36.69	37.02	37.36	37.71	38.05	38.39	38.74	39.08
1300	39.43	39.78	40.13	40.48	40.83	41.19	41.54	41.90	42.25	42.61

APPENDIX 85.A *(continued)*
Thermoelectric Constants for Thermocouples[*]
(mV, reference 32°F (0°C))

(c) copper-constantan (type T)

°F	0	10	20	30	40	50	60	70	80	90
−300	−5.284	−5.379			millivolts					
−200	−4.111	−4.246	−4.377	−4.504	−4.627	−4.747	−4.863	−4.974	−5.081	−5.185
−100	−2.559	−2.730	−2.897	−3.062	−3.223	−3.380	−3.533	−3.684	−3.829	−3.972
−0	−0.670	−0.872	−1.072	−1.270	−1.463	−1.654	−1.842	−2.026	−2.207	−2.385
+0	−0.670	−0.463	−0.254	−0.042	0.171	0.389	0.609	0.832	1.057	1.286
100	1.517	1.751	1.987	2.226	2.467	2.711	2.958	3.207	3.458	3.712
200	3.967	4.225	4.486	4.749	5.014	5.280	5.550	5.821	6.094	6.370
300	6.647	6.926	7.208	7.491	7.776	8.064	8.352	8.642	8.935	9.229
400	9.525	9.823	10.123	10.423	10.726	11.030	11.336	11.643	11.953	12.263
500	12.575	12.888	13.203	13.520	13.838	14.157	14.477	14.799	15.122	15.447

[*]This appendix is included to support general study and noncritical applications. *NBS Circular 561* has been superseded by *NBS Monograph 125,Thermocouple Reference Tables Based on the IPTS-68: Reference Tables in degrees Fahrenheit for Thermoelements versus Platinum,* presenting the same information as British Standards Institution standard *B.S. 4937.* In the spirit of globalization, these have been superseded in other countries by *IEC 584-1 (60584-1) Thermocouples Part 1: Reference Tables,* published in 1995. In addition to being based on modern correlations and the 1990 International Temperature Scale (ITS), the 160-page IEC 60584-1 duplicates all text in English and French, refers only to the Celsius temperature scale, and is available only by purchase or licensing.

APPENDIX 87.A
Standard Cash Flow Factors

	multiply	by	to obtain	

$$P = F(1+i)^{-n}$$
$$F \qquad (P/F, i\%, n) \qquad P$$

$$F = P(1+i)^n$$
$$P \qquad (F/P, i\%, n) \qquad F$$

$$P = A\left(\frac{(1+i)^n - 1}{i(1+i)^n}\right)$$
$$A \qquad (P/A, i\%, n) \qquad P$$

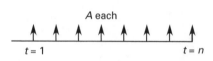

$$A = P\left(\frac{i(1+i)^n}{(1+i)^n - 1}\right)$$
$$P \qquad (A/P, i\%, n) \qquad A$$

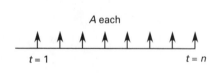

$$F = A\left(\frac{(1+i)^n - 1}{i}\right)$$
$$A \qquad (F/A, i\%, n) \qquad F$$

$$A = F\left(\frac{i}{(1+i)^n - 1}\right)$$
$$F \qquad (A/F, i\%, n) \qquad A$$

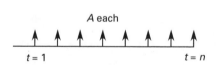

$$P = G\left(\frac{(1+i)^n - 1}{i^2(1+i)^n} - \frac{n}{i(1+i)^n}\right)$$
$$G \qquad (P/G, i\%, n) \qquad P$$

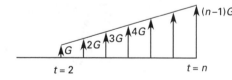

$$A = G\left(\frac{1}{i} - \frac{n}{(1+i)^n - 1}\right)$$
$$G \qquad (A/G, i\%, n) \qquad A$$

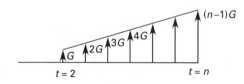

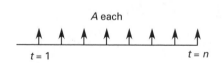

APPENDIX 87.B
Cash Flow Equivalent Factors

$i = 0.50\%$

n	P/F	P/A	P/G	F/P	F/A	A/P	A/F	A/G	n
1	0.9950	0.9950	0.0000	1.0050	1.0000	1.0050	1.0000	0.0000	1
2	0.9901	0.9851	0.9901	1.0100	2.0050	0.5038	0.4988	0.4988	2
3	0.9851	2.9702	2.9604	1.0151	3.0150	0.3367	0.3317	0.9967	3
4	0.9802	3.9505	5.9011	1.0202	4.0301	0.2531	0.2481	1.4938	4
5	0.9754	4.9259	9.8026	1.0253	5.0503	0.2030	0.1980	1.9900	5
6	0.9705	5.8964	14.6552	1.0304	6.0755	0.1696	0.1646	2.4855	6
7	0.9657	6.8621	20.4493	1.0355	7.1059	0.1457	0.1407	2.9801	7
8	0.9609	7.8230	27.1755	1.0407	8.1414	0.1278	0.1228	3.4738	8
9	0.9561	8.7791	34.8244	1.0459	9.1821	0.1139	0.1089	3.9668	9
10	0.9513	9.7304	43.3865	1.0511	10.2280	0.1028	0.0978	4.4589	10
11	0.9466	10.6770	52.8526	1.0564	11.2792	0.0937	0.0887	4.9501	11
12	0.9419	11.6189	63.2136	1.0617	12.3356	0.0861	0.0811	5.4406	12
13	0.9372	12.5562	74.4602	1.0670	13.3972	0.0796	0.0746	5.9302	13
14	0.9326	13.4887	86.5835	1.0723	14.4642	0.0741	0.0691	6.4190	14
15	0.9279	14.4166	99.5743	1.0777	15.5365	0.0694	0.0644	6.9069	15
16	0.9233	15.3399	113.4238	1.0831	16.6142	0.0652	0.0602	7.3940	16
17	0.9187	16.2586	128.1231	1.0885	17.6973	0.0615	0.0565	7.8803	17
18	0.9141	17.1728	143.6634	1.0939	18.7858	0.0582	0.0532	8.3658	18
19	0.9096	18.0824	160.0360	1.0994	19.8797	0.0553	0.0503	8.8504	19
20	0.9051	18.9874	177.2322	1.1049	20.9791	0.0527	0.0477	9.3342	20
21	0.9006	19.8880	195.2434	1.1104	22.0840	0.0503	0.0453	9.8172	21
22	0.8961	20.7841	214.0611	1.1160	23.1944	0.0481	0.0431	10.2993	22
23	0.8916	21.6757	233.6768	1.1216	24.3104	0.0461	0.0411	10.7806	23
24	0.8872	22.5629	254.0820	1.1272	25.4320	0.0443	0.0393	11.2611	24
25	0.8828	23.4456	275.2686	1.1328	26.5591	0.0427	0.0377	11.7407	25
26	0.8784	24.3240	297.2281	1.1385	27.6919	0.0411	0.0361	12.2195	26
27	0.8740	25.1980	319.9523	1.1442	28.8304	0.0397	0.0347	12.6975	27
28	0.8697	26.0677	343.4332	1.1499	29.9745	0.0384	0.0334	13.1747	28
29	0.8653	26.9330	367.6625	1.1556	31.1244	0.0371	0.0321	13.6510	29
30	0.8610	27.7941	392.6324	1.1614	32.2800	0.0360	0.0310	14.1265	30
31	0.8567	28.6508	418.3348	1.1672	33.4414	0.0349	0.0299	14.6012	31
32	0.8525	29.5033	444.7618	1.1730	34.6086	0.0339	0.0289	15.0750	32
33	0.8482	30.3515	471.9055	1.1789	35.7817	0.0329	0.0279	15.5480	33
34	0.8440	31.1955	499.7583	1.1848	36.9606	0.0321	0.0271	16.0202	34
35	0.8398	32.0354	528.3123	1.1907	38.1454	0.0312	0.0262	16.4915	35
36	0.8356	32.8710	557.5598	1.1967	39.3361	0.0304	0.0254	16.9621	36
37	0.8315	33.7025	587.4934	1.2027	40.5328	0.0297	0.0247	17.4317	37
38	0.8274	34.5299	618.1054	1.2087	41.7354	0.0290	0.0240	17.9006	38
39	0.8232	35.3531	649.3883	1.2147	42.9441	0.0283	0.0233	18.3686	39
40	0.8191	36.1722	681.3347	1.2208	44.1588	0.0276	0.0226	18.8359	40
41	0.8151	36.9873	713.9372	1.2269	45.3796	0.0270	0.0220	19.3022	41
42	0.8110	37.7983	747.1886	1.2330	46.6065	0.0265	0.0215	19.7678	42
43	0.8070	38.6053	781.0815	1.2392	47.8396	0.0259	0.0209	20.2325	43
44	0.8030	39.4082	815.6087	1.2454	49.0788	0.0254	0.0204	20.6964	44
45	0.7990	40.2072	850.7631	1.2516	50.3242	0.0249	0.0199	21.1595	45
46	0.7950	41.0022	886.5376	1.2579	51.5758	0.0244	0.0194	21.6217	46
47	0.7910	41.7932	922.9252	1.2642	52.8337	0.0239	0.0189	22.0831	47
48	0.7871	42.5803	959.9188	1.2705	54.0978	0.0235	0.0185	22.5437	48
49	0.7832	43.3635	997.5116	1.2768	55.3683	0.0231	0.0181	23.0035	49
50	0.7793	44.1428	1035.6966	1.2832	56.6452	0.0227	0.0177	23.4624	50
51	0.7754	44.9182	1074.4670	1.2896	57.9284	0.0223	0.0173	23.9205	51
52	0.7716	45.6897	1113.8162	1.2961	59.2180	0.0219	0.0169	24.3778	52
53	0.7677	46.4575	1153.7372	1.3026	60.5141	0.0215	0.0165	24.8343	53
54	0.7639	47.2214	1194.2236	1.3091	61.8167	0.0212	0.0162	25.2899	54
55	0.7601	47.9814	1235.2686	1.3156	63.1258	0.0208	0.0158	25.7447	55
60	0.7414	51.7256	1448.6458	1.3489	69.7700	0.0193	0.0143	28.0064	60
65	0.7231	55.3775	1675.0272	1.3829	76.5821	0.0181	0.0131	30.2475	65
70	0.7053	58.9394	1913.6427	1.4178	83.5661	0.0170	0.0120	32.4680	70
75	0.6879	62.4136	2163.7525	1.4536	90.7265	0.0160	0.0110	34.6679	75
80	0.6710	65.8023	2424.6455	1.4903	98.0677	0.0152	0.0102	36.8474	80
85	0.6545	69.1075	2695.6389	1.5280	105.5943	0.0145	0.0095	39.0065	85
90	0.6383	72.3313	2976.0769	1.5666	113.3109	0.0138	0.0088	41.1451	90
95	0.6226	75.4757	3265.3298	1.6061	121.2224	0.0132	0.0082	43.2633	95
100	0.6073	78.5426	3562.7934	1.6467	129.3337	0.0127	0.0077	45.3613	100

APPENDIX 87.B *(continued)*
Cash Flow Equivalent Factors

$i = 0.75\%$

n	P/F	P/A	P/G	F/P	F/A	A/P	A/F	A/G	n
1	0.9926	0.9926	0.0000	1.0075	1.0000	1.0075	1.0000	0.0000	1
2	0.9852	1.9777	0.9852	1.0151	2.0075	0.5056	0.4981	0.4981	2
3	0.9778	2.9556	2.9408	1.0227	3.0226	0.3383	0.3308	0.9950	3
4	0.9706	3.9261	5.8525	1.0303	4.0452	0.2547	0.2472	1.4907	4
5	0.9633	4.8894	9.7058	1.0381	5.0756	0.2045	0.1970	1.9851	5
6	0.9562	5.8456	14.4866	1.0459	6.1136	0.1711	0.1636	2.4782	6
7	0.9490	6.7946	20.1808	1.0537	7.1595	0.1472	0.1397	2.9701	7
8	0.9420	7.7366	26.7747	1.0616	8.2132	0.1293	0.1218	3.4608	8
9	0.9350	8.6716	34.2544	1.0696	9.2748	0.1153	0.1078	3.9502	9
10	0.9280	9.5996	42.6064	1.0776	10.3443	0.1042	0.0967	4.4384	10
11	0.9211	10.5207	51.8174	1.0857	11.4219	0.0951	0.0876	4.9253	11
12	0.9142	11.4349	61.8740	1.0938	12.5076	0.0875	0.0800	5.4110	12
13	0.9074	12.3423	72.7632	1.1020	13.6014	0.0810	0.0735	5.8954	13
14	0.9007	13.2430	84.4720	1.1103	14.7034	0.0755	0.0680	6.3786	14
15	0.8940	14.1370	96.9876	1.1186	15.8137	0.0707	0.0632	6.8606	15
16	0.8873	15.0243	110.2973	1.1270	16.9323	0.0666	0.0591	7.3413	16
17	0.8807	15.9050	124.3887	1.1354	18.0593	0.0629	0.0554	7.8207	17
18	0.8742	16.7792	139.2494	1.1440	19.1947	0.0596	0.0521	8.2989	18
19	0.8676	17.6468	154.8671	1.1525	20.3387	0.0567	0.0492	8.7759	19
20	0.8612	18.5080	171.2297	1.1612	21.4912	0.0540	0.0465	9.2516	20
21	0.8548	19.3628	188.3253	1.1699	22.6524	0.0516	0.0441	9.7261	21
22	0.8484	20.2112	206.1420	1.1787	23.8223	0.0495	0.0420	10.1994	22
23	0.8421	21.0533	224.6682	1.1875	25.0010	0.0475	0.0400	10.6714	23
24	0.8358	21.8891	243.8923	1.1964	26.1885	0.0457	0.0382	11.1422	24
25	0.8296	22.7188	263.8029	1.2054	27.3849	0.0440	0.0365	11.6117	25
26	0.8234	23.5422	284.3888	1.2144	28.5903	0.0425	0.0350	12.0800	26
27	0.8173	24.3595	305.6387	1.2235	29.8047	0.0411	0.0336	12.5470	27
28	0.8112	25.1707	327.5416	1.2327	31.0282	0.0397	0.0322	13.0128	28
29	0.8052	25.9759	350.0867	1.2420	32.2609	0.0385	0.0310	13.4774	29
30	0.7992	26.7751	373.2631	1.2513	33.5029	0.0373	0.0298	13.9407	30
31	0.7932	27.5683	397.0602	1.2607	34.7542	0.0363	0.0288	14.4028	31
32	0.7873	28.3557	421.4675	1.2701	36.0148	0.0353	0.0278	14.8636	32
33	0.7815	29.1371	446.4746	1.2796	37.2849	0.0343	0.0268	15.3232	33
34	0.7757	29.9128	472.0712	1.2892	38.5646	0.0334	0.0259	15.7816	34
35	0.7699	30.6827	498.2471	1.2989	39.8538	0.0326	0.0251	16.2387	35
36	0.7641	31.4468	524.9924	1.3086	41.1527	0.0318	0.0243	16.6946	36
37	0.7585	32.2053	552.2969	1.3185	42.4614	0.0311	0.0236	17.1493	37
38	0.7528	32.9581	580.1511	1.3283	43.7798	0.0303	0.0228	17.6027	38
39	0.7472	33.7053	608.5451	1.3383	45.1082	0.0297	0.0222	18.0549	39
40	0.7416	34.4469	637.4693	1.3483	46.4465	0.0290	0.0215	18.5058	40
41	0.7361	35.1831	666.9144	1.3585	47.7948	0.0284	0.0209	18.9556	41
42	0.7306	35.9137	696.8709	1.3686	49.1533	0.0278	0.0203	19.4040	42
43	0.7252	36.6389	727.3297	1.3789	50.5219	0.0273	0.0198	19.8513	43
44	0.7198	37.3587	758.2815	1.3893	51.9009	0.0268	0.0193	20.2973	44
45	0.7145	38.0732	789.7173	1.3997	53.2901	0.0263	0.0188	20.7421	45
46	0.7091	38.7823	821.6283	1.4102	54.6898	0.0258	0.0183	21.1856	46
47	0.7039	39.4862	854.0056	1.4207	56.1000	0.0253	0.0178	21.6280	47
48	0.6986	40.1848	886.8404	1.4314	57.5207	0.0249	0.0174	22.0691	48
49	0.6934	40.8782	920.1243	1.4421	58.9521	0.0245	0.0170	22.5089	49
50	0.6883	41.5664	953.8486	1.4530	60.3943	0.0241	0.0166	22.9476	50
51	0.6831	42.2496	988.0050	1.4639	61.8472	0.0237	0.0162	23.3850	51
52	0.6780	42.9276	1022.5852	1.4748	63.3111	0.0233	0.0158	23.8211	52
53	0.6730	43.6006	1057.5810	1.4859	64.7859	0.0229	0.0154	24.2561	53
54	0.6680	44.2686	1092.9842	1.4970	66.2718	0.0226	0.0151	24.6898	54
55	0.6630	44.9316	1128.7869	1.5083	67.7688	0.0223	0.0148	25.1223	55
60	0.6387	48.1734	1313.5189	1.5657	75.4241	0.0208	0.0133	27.2665	60
65	0.6153	51.2963	1507.0910	1.6253	83.3709	0.0195	0.0120	29.3801	65
70	0.5927	54.3046	1708.6065	1.6872	91.6201	0.0184	0.0109	31.4634	70
75	0.5710	57.2027	1917.2225	1.7514	100.1833	0.0175	0.0100	33.5163	75
80	0.5500	59.9944	2132.1472	1.8180	109.0725	0.0167	0.0092	35.5391	80
85	0.5299	62.6838	2352.6375	1.8873	118.3001	0.0160	0.0085	37.5318	85
90	0.5104	65.2746	2577.9961	1.9591	127.8790	0.0153	0.0078	39.4946	90
95	0.4917	67.7704	2807.5694	2.0337	137.8225	0.0148	0.0073	41.4277	95
100	0.4737	70.1746	3040.7453	2.1111	148.1445	0.0143	0.0068	43.3311	100

APPENDIX 87.B *(continued)*
Cash Flow Equivalent Factors

$i = 1.00\%$

n	P/F	P/A	P/G	F/P	F/A	A/P	A/F	A/G	n
1	0.9901	0.9901	0.0000	1.0100	1.0000	1.0100	1.0000	0.0000	1
2	0.9803	1.9704	0.9803	1.0201	2.0100	0.5075	0.4975	0.4975	2
3	0.9706	2.9410	2.9215	1.0303	3.0301	0.3400	0.3300	0.9934	3
4	0.9610	3.9020	5.8044	1.0406	4.0604	0.2563	0.2463	1.4876	4
5	0.9515	4.8534	9.6103	1.0510	5.1010	0.2060	0.1960	1.9801	5
6	0.9420	5.7955	14.3205	1.0615	6.1520	0.1725	0.1625	2.4710	6
7	0.9327	6.7282	19.9168	1.0721	7.2135	0.1486	0.1386	2.9602	7
8	0.9235	7.6517	26.3812	1.0829	8.2857	0.1307	0.1207	3.4478	8
9	0.9143	8.5660	33.6959	1.0937	9.3685	0.1167	0.1067	3.9337	9
10	0.9053	9.4713	41.8435	1.1046	10.4622	0.1056	0.0956	4.4179	10
11	0.8963	10.3676	50.8067	1.1157	11.5668	0.0965	0.0865	4.9005	11
12	0.8874	11.2551	60.5687	1.1268	12.6825	0.0888	0.0788	5.3815	12
13	0.8787	12.1337	71.1126	1.1381	13.8093	0.0824	0.0724	5.8607	13
14	0.8700	13.0037	82.4221	1.1495	14.9474	0.0769	0.0669	6.3384	14
15	0.8613	13.8651	94.4810	1.1610	16.0969	0.0721	0.0621	6.8143	15
16	0.8528	14.7179	107.2734	1.1726	17.2579	0.0679	0.0579	7.2886	16
17	0.8444	15.5623	120.7834	1.1843	18.4304	0.0643	0.0543	7.7613	17
18	0.8360	16.3983	134.9957	1.1961	19.6147	0.0610	0.0510	8.2323	18
19	0.8277	17.2260	149.8950	1.2081	20.8109	0.0581	0.0481	8.7017	19
20	0.8195	18.0456	165.4664	1.2202	22.0190	0.0554	0.0454	9.1694	20
21	0.8114	18.8570	181.6950	1.2324	23.2392	0.0530	0.0430	9.6354	21
22	0.8034	19.6604	198.5663	1.2447	24.4716	0.0509	0.0409	10.0998	22
23	0.7954	20.4558	216.0660	1.2572	25.7163	0.0489	0.0389	10.5626	23
24	0.7876	21.2434	234.1800	1.2697	26.9735	0.0471	0.0371	11.0237	24
25	0.7798	22.0232	252.8945	1.2824	28.2432	0.0454	0.0354	11.4831	25
26	0.7720	22.7952	272.1957	1.2953	29.5256	0.0439	0.0339	11.9409	26
27	0.7644	23.5596	292.0702	1.3082	30.8209	0.0424	0.0324	12.3971	27
28	0.7568	24.3164	312.5047	1.3213	32.1291	0.0411	0.0311	12.8516	28
29	0.7493	25.0658	333.4863	1.3345	33.4504	0.0399	0.0299	13.3044	29
30	0.7419	25.8077	355.0021	1.3478	34.7849	0.0387	0.0287	13.7557	30
31	0.7346	26.5423	377.0394	1.3613	36.1327	0.0377	0.0277	14.2052	31
32	0.7273	27.2696	399.5858	1.3749	37.4941	0.0367	0.0267	14.6532	32
33	0.7201	27.9897	422.6291	1.3887	38.8690	0.0357	0.0257	15.0995	33
34	0.7130	28.7027	446.1572	1.4026	40.2577	0.0348	0.0248	15.5441	34
35	0.7059	29.4086	470.1583	1.4166	41.6603	0.0340	0.0240	15.9871	35
36	0.6989	30.1075	494.6207	1.4308	43.0769	0.0332	0.0232	16.4285	36
37	0.6920	30.7995	519.5329	1.4451	44.5076	0.0325	0.0225	16.8682	37
38	0.6852	31.4847	544.8835	1.4595	45.9527	0.0318	0.0218	17.3063	38
39	0.6784	32.1630	570.6616	1.4741	47.4123	0.0311	0.0211	17.7428	39
40	0.6717	32.8347	596.8561	1.4889	48.8864	0.0305	0.0205	18.1776	40
41	0.6650	33.4997	623.4562	1.5038	50.3752	0.0299	0.0199	18.6108	41
42	0.6584	34.1581	650.4514	1.5188	51.8790	0.0293	0.0193	19.0424	42
43	0.6519	34.8100	677.8312	1.5340	53.3978	0.0287	0.0187	19.4723	43
44	0.6454	35.4555	705.5853	1.5493	54.9318	0.0282	0.0182	19.9006	44
45	0.6391	36.0945	733.7037	1.5648	56.4811	0.0277	0.0177	20.3273	45
46	0.6327	36.7272	762.1765	1.5805	58.0459	0.0272	0.0172	20.7524	46
47	0.6265	37.3537	790.9938	1.5963	59.6263	0.0268	0.0168	21.1758	47
48	0.6203	37.9740	820.1460	1.6122	61.2226	0.0263	0.0163	21.5976	48
49	0.6141	38.5881	849.6237	1.6283	62.8348	0.0259	0.0159	22.0178	49
50	0.6080	39.1961	879.4176	1.6446	64.4632	0.0255	0.0155	22.4363	50
51	0.6020	39.7981	909.5186	1.6611	66.1078	0.0251	0.0151	22.8533	51
52	0.5961	40.3942	939.9175	1.6777	67.7689	0.0248	0.0148	23.2686	52
53	0.5902	40.9844	970.6057	1.6945	69.4466	0.0244	0.0144	23.6823	53
54	0.5843	41.5687	1001.5743	1.7114	71.1410	0.0241	0.0141	24.0945	54
55	0.5785	42.1472	1032.8148	1.7285	72.8525	0.0237	0.0137	24.5049	55
60	0.5504	44.9550	1192.8061	1.8167	81.6697	0.0222	0.0122	26.5333	60
65	0.5237	47.6266	1358.3903	1.9094	90.9366	0.0210	0.0110	28.5217	65
70	0.4983	50.1685	1528.6474	2.0068	100.6763	0.0199	0.0099	30.4703	70
75	0.4741	52.5871	1702.7340	2.1091	110.9128	0.0190	0.0090	32.3793	75
80	0.4511	54.8882	1879.8771	2.2167	121.6715	0.0182	0.0082	34.2492	80
85	0.4292	57.0777	2059.3701	2.3298	132.9790	0.0175	0.0075	36.0801	85
90	0.4084	59.1609	2240.5675	2.4486	144.8633	0.0169	0.0069	37.8724	90
95	0.3886	61.1430	2422.8811	2.5735	157.3538	0.0164	0.0064	39.6265	95
100	0.3697	63.0289	2605.7758	2.7048	170.4814	0.0159	0.0059	41.3426	100

APPENDIX 87.B *(continued)*
Cash Flow Equivalent Factors

$i = 1.50\%$

n	P/F	P/A	P/G	F/P	F/A	A/P	A/F	A/G	n
1	0.9852	0.9852	0.0000	1.0150	1.0000	1.0150	1.0000	0.0000	1
2	0.9707	1.9559	0.9707	1.0302	2.0150	0.5113	0.4963	0.4963	2
3	0.9563	2.9122	2.8833	1.0457	3.0452	0.3434	0.3284	0.9901	3
4	0.9422	3.8544	5.7098	1.0614	4.0909	0.2594	0.2444	1.4814	4
5	0.9283	4.7826	9.4229	1.0773	5.1523	0.2091	0.1941	1.9702	5
6	0.9145	5.6972	13.9956	1.0934	6.2296	0.1755	0.1605	2.4566	6
7	0.9010	6.5982	19.4018	1.1098	7.3230	0.1516	0.1366	2.9405	7
8	0.8877	7.4859	25.6157	1.1265	8.4328	0.1336	0.1186	3.4219	8
9	0.8746	8.3605	32.6125	1.1434	9.5593	0.1196	0.1046	3.9008	9
10	0.8617	9.2222	40.3675	1.1605	10.7027	0.1084	0.0934	4.3772	10
11	0.8489	10.0711	48.8568	1.1779	11.8633	0.0993	0.0843	4.8512	11
12	0.8364	10.9075	58.0571	1.1956	13.0412	0.0917	0.0767	5.3227	12
13	0.8240	11.7315	67.9454	1.2136	14.2368	0.0852	0.0702	5.7917	13
14	0.8118	12.5434	78.4994	1.2318	15.4504	0.0797	0.0647	6.2582	14
15	0.7999	13.3432	89.6974	1.2502	16.6821	0.0749	0.0599	6.7223	15
16	0.7880	14.1313	101.5178	1.2690	17.9324	0.0708	0.0558	7.1839	16
17	0.7764	14.9076	113.9400	1.2880	19.2014	0.0671	0.0521	7.6431	17
18	0.7649	15.6726	126.9435	1.3073	20.4894	0.0638	0.0488	8.0997	18
19	0.7536	16.4262	140.5084	1.3270	21.7967	0.0609	0.0459	8.5539	19
20	0.7425	17.1686	154.6154	1.3469	23.1237	0.0582	0.0432	9.0057	20
21	0.7315	17.9001	169.2453	1.3671	24.4705	0.0559	0.0409	9.4550	21
22	0.7207	18.6208	184.3798	1.3876	25.8376	0.0537	0.0387	9.9018	22
23	0.7100	19.3309	200.0006	1.4084	27.2251	0.0517	0.0367	10.3462	23
24	0.6995	20.0304	216.0901	1.4295	28.6335	0.0499	0.0349	10.7881	24
25	0.6892	20.7196	232.6310	1.4509	30.0630	0.0483	0.0333	11.2276	25
26	0.6790	21.3986	249.6065	1.4727	31.5140	0.0467	0.0317	11.6646	26
27	0.6690	22.0676	267.0002	1.4948	32.9867	0.0453	0.0303	12.0992	27
28	0.6591	22.7267	284.7958	1.5172	34.4815	0.0440	0.0290	12.5313	28
29	0.6494	23.3761	302.9779	1.5400	35.9987	0.0428	0.0278	12.9610	29
30	0.6398	24.0158	321.5310	1.5631	37.5387	0.0416	0.0266	13.3883	30
31	0.6303	24.6461	340.4402	1.5865	39.1018	0.0406	0.0256	13.8131	31
32	0.6210	25.2671	359.6910	1.6103	40.6883	0.0396	0.0246	14.2355	32
33	0.6118	25.8790	379.2691	1.6345	42.2986	0.0386	0.0236	14.6555	33
34	0.6028	26.4817	399.1607	1.6590	43.9331	0.0378	0.0228	15.0731	34
35	0.5939	27.0756	419.3521	1.6839	45.5921	0.0369	0.0219	15.4882	35
36	0.5851	27.6607	439.8303	1.7091	47.2760	0.0362	0.0212	15.9009	36
37	0.5764	28.2371	460.5822	1.7348	48.9851	0.0354	0.0204	16.3112	37
38	0.5679	28.8051	481.5954	1.7608	50.7199	0.0347	0.0197	16.7191	38
39	0.5595	29.3646	502.8576	1.7872	52.4807	0.0341	0.0191	17.1246	39
40	0.5513	29.9158	524.3568	1.8140	54.2679	0.0334	0.0184	17.5277	40
41	0.5431	30.4590	546.0814	1.8412	56.0819	0.0328	0.0178	17.9284	41
42	0.5351	30.9941	568.0201	1.8688	57.9231	0.0323	0.0173	18.3267	42
43	0.5272	31.5212	590.1617	1.8969	59.7920	0.0317	0.0167	18.7227	43
44	0.5194	32.0406	612.4955	1.9253	61.6889	0.0312	0.0162	19.1162	44
45	0.5117	32.5523	635.0110	1.9542	63.6142	0.0307	0.0157	19.5074	45
46	0.5042	33.0565	657.6979	1.9835	65.5684	0.0303	0.0153	19.8962	46
47	0.4967	33.5532	680.5462	2.0133	67.5519	0.0298	0.0148	20.2826	47
48	0.4894	34.0426	703.5462	2.0435	69.5652	0.0294	0.0144	20.6667	48
49	0.4821	34.5247	726.6884	2.0741	71.6087	0.0290	0.0140	21.0484	49
50	0.4750	34.9997	749.9636	2.1052	73.6828	0.0286	0.0136	21.4277	50
51	0.4680	35.4677	773.3629	2.1368	75.7881	0.0282	0.0132	21.8047	51
52	0.4611	35.9287	796.8774	2.1689	77.9249	0.0278	0.0128	22.1794	52
53	0.4543	36.3830	820.4986	2.2014	80.0938	0.0275	0.0125	22.5517	53
54	0.4475	36.8305	844.2184	2.2344	82.2952	0.0272	0.0122	22.9217	54
55	0.4409	37.2715	868.0285	2.2679	84.5296	0.0268	0.0118	23.2894	55
60	0.4093	39.3803	988.1674	2.4432	96.2147	0.0254	0.0104	25.0930	60
65	0.3799	41.3378	1109.4752	2.6320	108.8028	0.0242	0.0092	26.8393	65
70	0.3527	43.1549	1231.1658	2.8355	122.3638	0.0232	0.0082	28.5290	70
75	0.3274	44.8416	1352.5600	3.0546	136.9728	0.0223	0.0073	30.1631	75
80	0.3039	46.4073	1473.0741	3.2907	152.7109	0.0215	0.0065	31.7423	80
85	0.2821	47.8607	1592.2095	3.5450	169.6652	0.0209	0.0059	33.2676	85
90	0.2619	49.2099	1709.5439	3.8189	187.9299	0.0203	0.0053	34.7399	90
95	0.2431	50.4622	1824.7224	4.1141	207.6061	0.0198	0.0048	36.1602	95
100	0.2256	51.6247	1937.4506	4.4320	228.8030	0.0194	0.0044	37.5295	100

APPENDIX 87.B (continued)
Cash Flow Equivalent Factors

$i = 2.00\%$

n	P/F	P/A	P/G	F/P	F/A	A/P	A/F	A/G	n
1	0.9804	0.9804	0.0000	1.0200	1.0000	1.0200	1.0000	0.0000	1
2	0.9612	1.9416	0.9612	1.0404	2.0200	0.5150	0.4950	0.4950	2
3	0.9423	2.8839	2.8458	1.0612	3.0604	0.3468	0.3268	0.9868	3
4	0.9238	3.8077	5.6173	1.0824	4.1216	0.2626	0.2426	1.4752	4
5	0.9057	4.7135	9.2403	1.1041	5.2040	0.2122	0.1922	1.9604	5
6	0.8880	5.6014	13.6801	1.1262	6.3081	0.1785	0.1585	2.4423	6
7	0.8706	6.4720	18.9035	1.1487	7.4343	0.1545	0.1345	2.9208	7
8	0.8535	7.3255	24.8779	1.1717	8.5830	0.1365	0.1165	3.3961	8
9	0.8368	8.1622	31.5720	1.1951	9.7546	0.1225	0.1025	3.8681	9
10	0.8203	8.9826	38.9551	1.2190	10.9497	0.1113	0.0913	4.3367	10
11	0.8043	9.7868	46.9977	1.2434	12.1687	0.1022	0.0822	4.8021	11
12	0.7885	10.5753	55.6712	1.2682	13.4121	0.0946	0.0746	5.2642	12
13	0.7730	11.3484	64.9475	1.2936	14.6803	0.0881	0.0681	5.7231	13
14	0.7579	12.1062	74.7999	1.3195	15.9739	0.0826	0.0626	6.1786	14
15	0.7430	12.8493	85.2021	1.3459	17.2934	0.0778	0.0578	6.6309	15
16	0.7284	13.5777	96.1288	1.3728	18.6393	0.0737	0.0537	7.0799	16
17	0.7142	14.2919	107.5554	1.4002	20.0121	0.0700	0.0500	7.5256	17
18	0.7002	14.9920	119.4581	1.4282	21.4123	0.0667	0.0467	7.9681	18
19	0.6864	15.6785	131.8139	1.4568	22.8406	0.0638	0.0438	8.4073	19
20	0.6730	16.3514	144.6003	1.4859	24.2974	0.0612	0.0412	8.8433	20
21	0.6598	17.0112	157.7959	1.5157	25.7833	0.0588	0.0388	9.2760	21
22	0.6468	17.6580	171.3795	1.5460	27.2990	0.0566	0.0366	9.7055	22
23	0.6342	18.2922	185.3309	1.5769	28.8450	0.0547	0.0347	10.1317	23
24	0.6217	18.9139	199.6305	1.6084	30.4219	0.0529	0.0329	10.5547	24
25	0.6095	19.5235	214.2592	1.6406	32.0303	0.0512	0.0312	10.9745	25
26	0.5976	20.1210	229.1987	1.6734	33.6709	0.0497	0.0297	11.3910	26
27	0.5859	20.7069	244.4311	1.7069	35.3443	0.0483	0.0283	11.8043	27
28	0.5744	21.2813	259.9392	1.7410	37.0512	0.0470	0.0270	12.2145	28
29	0.5631	21.8444	275.7064	1.7758	38.7922	0.0458	0.0258	12.6214	29
30	0.5521	22.3965	291.7164	1.8114	40.5681	0.0446	0.0246	13.0251	30
31	0.5412	22.9377	307.9538	1.8476	42.3794	0.0436	0.0236	13.4257	31
32	0.5306	23.4683	324.4035	1.8845	44.2270	0.0426	0.0226	13.8230	32
33	0.5202	23.9886	341.0508	1.9222	46.1116	0.0417	0.0217	14.2172	33
34	0.5100	24.4986	357.8817	1.9607	48.0338	0.0408	0.0208	14.6083	34
35	0.5000	24.9986	374.8826	1.9999	49.9945	0.0400	0.0200	14.9961	35
36	0.4902	25.4888	392.0405	2.0399	51.9944	0.0392	0.0192	15.3809	36
37	0.4806	25.9695	409.3424	2.0807	54.0343	0.0385	0.0185	15.7625	37
38	0.4712	26.4406	426.7764	2.1223	56.1149	0.0378	0.0178	16.1409	38
39	0.4619	26.9026	444.3304	2.1647	58.2372	0.0372	0.0172	16.5163	39
40	0.4529	27.3555	461.9931	2.2080	60.4020	0.0366	0.0166	16.8885	40
41	0.4440	27.7995	479.7535	2.2522	62.6100	0.0360	0.0160	17.2576	41
42	0.4353	28.2348	497.6010	2.2972	64.8622	0.0354	0.0154	17.6237	42
43	0.4268	28.6616	515.5253	2.3432	67.1595	0.0349	0.0149	17.9866	43
44	0.4184	29.0800	533.5165	2.3901	69.5027	0.0344	0.0144	18.3465	44
45	0.4102	29.4902	551.5652	2.4379	71.8927	0.0339	0.0139	18.7034	45
46	0.4022	29.8923	569.6621	2.4866	74.3306	0.0335	0.0135	19.0571	46
47	0.3943	30.2866	587.7985	2.5363	76.8172	0.0330	0.0130	19.4079	47
48	0.3865	30.6731	605.9657	2.5871	79.3535	0.0326	0.0126	19.7556	48
49	0.3790	31.0521	624.1557	2.6388	81.9406	0.0322	0.0122	20.1003	49
50	0.3715	31.4236	642.3606	2.6916	84.5794	0.0318	0.0118	20.4420	50
51	0.3642	31.7878	660.5727	2.7454	87.2710	0.0315	0.0115	20.7807	51
52	0.3571	32.1449	678.7849	2.8003	90.0164	0.0311	0.0111	21.1164	52
53	0.3501	32.4950	696.9900	2.8563	92.8167	0.0308	0.0108	21.4491	53
54	0.3432	32.8383	715.1815	2.9135	95.6731	0.0305	0.0105	21.7789	54
55	0.3365	33.1748	733.3527	2.9717	98.5865	0.0301	0.0101	22.1057	55
60	0.3048	34.7609	823.6975	3.2810	114.0515	0.0288	0.0088	23.6961	60
65	0.2761	36.1975	912.7085	3.6225	131.1262	0.0276	0.0076	25.2147	65
70	0.2500	37.4986	999.8343	3.9996	149.9779	0.0267	0.0067	26.6632	70
75	0.2265	38.6771	1084.6393	4.4158	170.7918	0.0259	0.0059	28.0434	75
80	0.2051	39.7445	1166.7868	4.8754	193.7720	0.0252	0.0052	29.3572	80
85	0.1858	40.7113	1246.0241	5.3829	219.1439	0.0246	0.0046	30.6064	85
90	0.1683	41.5869	1322.1701	5.9431	247.1567	0.0240	0.0040	31.7929	90
95	0.1524	42.3800	1395.1033	6.5617	278.0850	0.0236	0.0036	32.9189	95
100	0.1380	43.0984	1464.7527	7.2446	312.2323	0.0232	0.0032	33.9863	100

APPENDIX 87.B *(continued)*
Cash Flow Equivalent Factors

$i = 3.00\%$

n	P/F	P/A	P/G	F/P	F/A	A/P	A/F	A/G	n
1	0.9709	0.9709	0.0000	1.0300	1.0000	1.0300	1.0000	0.0000	1
2	0.9426	1.9135	0.9426	1.0609	2.0300	0.5226	0.4926	0.4926	2
3	0.9151	2.8286	2.7729	1.0927	3.0909	0.3535	0.3235	0.9803	3
4	0.8885	3.7171	5.4383	1.1255	4.1836	0.2690	0.2390	1.4631	4
5	0.8626	4.5797	8.8888	1.1593	5.3091	0.2184	0.1884	1.9409	5
6	0.8375	5.4172	13.0762	1.1941	6.4684	0.1846	0.1546	2.4138	6
7	0.8131	6.2303	17.9547	1.2299	7.6625	0.1605	0.1305	2.8819	7
8	0.7894	7.0197	23.4806	1.2668	8.8923	0.1425	0.1125	3.3450	8
9	0.7664	7.7861	29.6119	1.3048	10.1591	0.1284	0.0984	3.8032	9
10	0.7441	8.5302	36.3088	1.3439	11.4639	0.1172	0.0872	4.2565	10
11	0.7224	9.2526	43.5330	1.3842	12.8078	0.1081	0.0781	4.7049	11
12	0.7014	9.9540	51.2482	1.4258	14.1920	0.1005	0.0705	5.1485	12
13	0.6810	10.6350	59.4196	1.4685	15.6178	0.0940	0.0640	5.5872	13
14	0.6611	11.2961	68.0141	1.5126	17.0863	0.0885	0.0585	6.0210	14
15	0.6419	11.9379	77.0002	1.5580	18.5989	0.0838	0.0538	6.4500	15
16	0.6232	12.5611	86.3477	1.6047	20.1569	0.0796	0.0496	6.8742	16
17	0.6050	13.1661	96.0280	1.6528	21.7616	0.0760	0.0460	7.2936	17
18	0.5874	13.7535	106.0137	1.7024	23.4144	0.0727	0.0427	7.7081	18
19	0.5703	14.3238	116.2788	1.7535	25.1169	0.0698	0.0398	8.1179	19
20	0.5537	14.8775	126.7987	1.8061	26.8704	0.0672	0.0372	8.5229	20
21	0.5375	15.4150	137.5496	1.8603	28.6765	0.0649	0.0349	8.9231	21
22	0.5219	15.9369	148.5094	1.9161	30.5368	0.0627	0.0327	9.3186	22
23	0.5067	16.4436	159.6566	1.9736	32.4529	0.0608	0.0308	9.7093	23
24	0.4919	16.9355	170.9711	2.0328	34.4265	0.0590	0.0290	10.0954	24
25	0.4776	17.4131	182.4336	2.0938	36.4593	0.0574	0.0274	10.4768	25
26	0.4637	17.8768	194.0260	2.1566	38.5530	0.0559	0.0259	10.8535	26
27	0.4502	18.3270	205.7309	2.2213	40.7096	0.0546	0.0246	11.2255	27
28	0.4371	18.7641	217.5320	2.2879	42.9309	0.0533	0.0233	11.5930	28
29	0.4243	19.1885	229.4137	2.3566	45.2189	0.0521	0.0221	11.9558	29
30	0.4120	19.6004	241.3613	2.4273	47.5754	0.0510	0.0210	12.3141	30
31	0.4000	20.0004	253.3609	2.5001	50.0027	0.0500	0.0200	12.6678	31
32	0.3883	20.3888	265.3993	2.5751	52.5028	0.0490	0.0190	13.0169	32
33	0.3770	20.7658	277.4642	2.6523	55.0778	0.0482	0.0182	13.3616	33
34	0.3660	21.1318	289.5437	2.7319	57.7302	0.0473	0.0173	13.7018	34
35	0.3554	21.4872	301.6267	2.8139	60.4621	0.0465	0.0165	14.0375	35
36	0.3450	21.8323	313.7028	2.8983	63.2759	0.0458	0.0158	14.3688	36
37	0.3350	22.1672	325.7622	2.9852	66.1742	0.0451	0.0151	14.6957	37
38	0.3252	22.4925	337.7956	3.0748	69.1594	0.0445	0.0145	15.0182	38
39	0.3158	22.8082	349.7942	3.1670	72.2342	0.0438	0.0138	15.3363	39
40	0.3066	23.1148	361.7499	3.2620	75.4013	0.0433	0.0133	15.6502	40
41	0.2976	23.4124	373.6551	3.3599	78.6633	0.0427	0.0127	15.9597	41
42	0.2890	23.7014	385.5024	3.4607	82.0232	0.0422	0.0122	16.2650	42
43	0.2805	23.9819	397.2852	3.5645	85.4839	0.0417	0.0117	16.5660	43
44	0.2724	24.2543	408.9972	3.6715	89.0484	0.0412	0.0112	16.8629	44
45	0.2644	24.5187	420.6325	3.7816	92.7199	0.0408	0.0108	17.1556	45
46	0.2567	24.7754	432.1856	3.8950	96.5015	0.0404	0.0104	17.4441	46
47	0.2493	25.0247	443.6515	4.0119	100.3965	0.0400	0.0100	17.7285	47
48	0.2420	25.2667	455.0255	4.1323	104.4084	0.0396	0.0096	18.0089	48
49	0.2350	25.5017	466.3031	4.2562	108.5406	0.0392	0.0092	18.2852	49
50	0.2281	25.7298	477.4803	4.3839	112.7969	0.0389	0.0089	18.5575	50
51	0.2215	25.9512	488.5535	4.5154	117.1808	0.0385	0.0085	18.8258	51
52	0.2150	26.1662	499.5191	4.6509	121.6962	0.0382	0.0082	19.0902	52
53	0.2088	26.3750	510.3742	4.7904	126.3471	0.0379	0.0079	19.3507	53
54	0.2027	26.5777	521.1157	4.9341	131.1375	0.0376	0.0076	19.6073	54
55	0.1968	26.7744	531.7411	5.0821	136.0716	0.0373	0.0073	19.8600	55
60	0.1697	27.6756	583.0526	5.8916	163.0534	0.0361	0.0061	21.0674	60
65	0.1464	28.4529	631.2010	6.8300	194.3328	0.0351	0.0051	22.1841	65
70	0.1263	29.1234	676.0869	7.9178	230.5941	0.0343	0.0043	23.2145	70
75	0.1089	29.7018	717.6978	9.1789	272.6309	0.0337	0.0037	24.1634	75
80	0.0940	30.2008	756.0865	10.6409	321.3630	0.0331	0.0031	25.0353	80
85	0.0811	30.6312	791.3529	12.3357	377.8570	0.0326	0.0026	25.8349	85
90	0.0699	31.0024	823.6302	14.3005	443.3489	0.0323	0.0023	26.5667	90
95	0.0603	31.3227	853.0742	16.5782	519.2720	0.0319	0.0019	27.2351	95
100	0.0520	31.5989	879.8540	19.2186	607.2877	0.0316	0.0016	27.8444	100

APPENDIX 87.B *(continued)*
Cash Flow Equivalent Factors

$i = 4.00\%$

n	P/F	P/A	P/G	F/P	F/A	A/P	A/F	A/G	n
1	0.9615	0.9615	0.0000	1.0400	1.0000	1.0400	1.0000	0.0000	1
2	0.9246	1.8861	0.9246	1.0816	2.0400	0.5302	0.4902	0.4902	2
3	0.8890	2.7751	2.7025	1.1249	3.1216	0.3603	0.3203	0.9739	3
4	0.8548	3.6299	5.2670	1.1699	4.2465	0.2755	0.2355	1.4510	4
5	0.8219	4.4518	8.5547	1.2167	5.4163	0.2246	0.1846	1.9216	5
6	0.7903	5.2421	12.5062	1.2653	6.6330	0.1908	0.1508	2.3857	6
7	0.7599	6.0021	17.0657	1.3159	7.8983	0.1666	0.1266	2.8433	7
8	0.7307	6.7327	22.1806	1.3686	9.2142	0.1485	0.1085	3.2944	8
9	0.7026	7.4353	27.8013	1.4233	10.5828	0.1345	0.0945	3.7391	9
10	0.6756	8.1109	33.8814	1.4802	12.0061	0.1233	0.0833	4.1773	10
11	0.6496	8.7605	40.3772	1.5395	13.4864	0.1141	0.0741	4.6090	11
12	0.6246	9.3851	47.2477	1.6010	15.0258	0.1066	0.0666	5.0343	12
13	0.6006	9.9856	54.4546	1.6651	16.6268	0.1001	0.0601	5.4533	13
14	0.5775	10.5631	61.9618	1.7317	18.2919	0.0947	0.0547	5.8659	14
15	0.5553	11.1184	69.7355	1.8009	20.0236	0.0899	0.0499	6.2721	15
16	0.5339	11.6523	77.7441	1.8730	21.8245	0.0858	0.0458	6.6720	16
17	0.5134	12.1657	85.9581	1.9479	23.6975	0.0822	0.0422	7.0656	17
18	0.4936	12.6593	94.3498	2.0258	25.6454	0.0790	0.0390	7.4530	18
19	0.4746	13.1339	102.8933	2.1068	27.6712	0.0761	0.0361	7.8342	19
20	0.4564	13.5903	111.5647	2.1911	29.7781	0.0736	0.0336	8.2091	20
21	0.4388	14.0292	120.3414	2.2788	31.9692	0.0713	0.0313	8.5779	21
22	0.4220	14.4511	129.2024	2.3699	34.2480	0.0692	0.0292	8.9407	22
23	0.4057	14.8568	138.1284	2.4647	36.6179	0.0673	0.0273	9.2973	23
24	0.3901	15.2470	147.1012	2.5633	39.0826	0.0656	0.0256	9.6479	24
25	0.3751	15.6221	156.1040	2.6658	41.6459	0.0640	0.0240	9.9925	25
26	0.3607	15.9828	165.1212	2.7725	44.3117	0.0626	0.0226	10.3312	26
27	0.3468	16.3296	174.1385	2.8834	47.0842	0.0612	0.0212	10.6640	27
28	0.3335	16.6631	183.1424	2.9987	49.9676	0.0600	0.0200	10.9909	28
29	0.3207	16.9837	192.1206	3.1187	52.9663	0.0589	0.0189	11.3120	29
30	0.3083	17.2920	201.0618	3.2434	56.0849	0.0578	0.0178	11.6274	30
31	0.2965	17.5885	209.9556	3.3731	59.3283	0.0569	0.0169	11.9371	31
32	0.2851	17.8736	218.7924	3.5081	62.7015	0.0559	0.0159	12.2411	32
33	0.2741	18.1476	227.5634	3.6484	66.2095	0.0551	0.0151	12.5396	33
34	0.2636	18.4112	236.2607	3.7943	69.8579	0.0543	0.0143	12.8324	34
35	0.2534	18.6646	244.8768	3.9461	73.6522	0.0536	0.0136	13.1198	35
36	0.2437	18.9083	253.4052	4.1039	77.5983	0.0529	0.0129	13.4018	36
37	0.2343	19.1426	261.8399	4.2681	81.7022	0.0522	0.0122	13.6784	37
38	0.2253	19.3679	270.1754	4.4388	85.9703	0.0516	0.0116	13.9497	38
39	0.2166	19.5845	278.4070	4.6164	90.4091	0.0511	0.0111	14.2157	39
40	0.2083	19.7928	286.5303	4.8010	95.0255	0.0505	0.0105	14.4765	40
41	0.2003	19.9931	294.5414	4.9931	99.8265	0.0500	0.0100	14.7322	41
42	0.1926	20.1856	302.4370	5.1928	104.8196	0.0495	0.0095	14.9828	42
43	0.1852	20.3708	310.2141	5.4005	110.0124	0.0491	0.0091	15.2284	43
44	0.1780	20.5488	317.8700	5.6165	115.4129	0.0487	0.0087	15.4690	44
45	0.1712	20.7200	325.4028	5.8412	121.0294	0.0483	0.0083	15.7047	45
46	0.1646	20.8847	332.8104	6.0748	126.8706	0.0479	0.0079	15.9356	46
47	0.1583	21.0429	340.0914	6.3178	132.9454	0.0475	0.0075	16.1618	47
48	0.1522	21.1951	347.2446	6.5705	139.2632	0.0472	0.0072	16.3832	48
49	0.1463	21.3415	354.2689	6.8333	145.8337	0.0469	0.0069	16.6000	49
50	0.1407	21.4822	361.1638	7.1067	152.6671	0.0466	0.0066	16.8122	50
51	0.1353	21.6175	367.9289	7.3910	159.7738	0.0463	0.0063	17.0200	51
52	0.1301	21.7476	374.5638	7.6866	167.1647	0.0460	0.0060	17.2232	52
53	0.1251	21.8727	381.0686	7.9941	174.8513	0.0457	0.0057	17.4221	53
54	0.1203	21.9930	387.4436	8.3138	182.8454	0.0455	0.0055	17.6167	54
55	0.1157	22.1086	393.6890	8.6464	191.1592	0.0452	0.0052	17.8070	55
60	0.0951	22.6235	422.9966	10.5196	237.9907	0.0442	0.0042	18.6972	60
65	0.0781	23.0467	449.2014	12.7987	294.9684	0.0434	0.0034	19.4909	65
70	0.0642	23.3945	472.4789	15.5716	364.2905	0.0427	0.0027	20.1961	70
75	0.0528	23.6804	493.0408	18.9453	448.6314	0.0422	0.0022	20.8206	75
80	0.0434	23.9154	511.1161	23.0498	551.2450	0.0418	0.0018	21.3718	80
85	0.0357	24.1085	526.9384	28.0436	676.0901	0.0415	0.0015	21.8569	85
90	0.0293	24.2673	540.7369	34.1193	827.9833	0.0412	0.0012	22.2826	90
95	0.0241	24.3978	552.7307	41.5114	1012.7846	0.0410	0.0010	22.6550	95
100	0.0198	24.5050	563.1249	50.5049	1237.6237	0.0408	0.0008	22.9800	100

APPENDIX 87.B *(continued)*
Cash Flow Equivalent Factors

$i = 5.00\%$

n	P/F	P/A	P/G	F/P	F/A	A/P	A/F	A/G	n
1	0.9524	0.9524	0.0000	1.0500	1.0000	1.0500	1.0000	0.0000	1
2	0.9070	1.8594	0.9070	1.1025	2.0500	0.5378	0.4878	0.4878	2
3	0.8638	2.7232	2.6347	1.1576	3.1525	0.3672	0.3172	0.9675	3
4	0.8227	3.5460	5.1028	1.2155	4.3101	0.2820	0.2320	1.4391	4
5	0.7835	4.3295	8.2369	1.2763	5.5256	0.2310	0.1810	1.9025	5
6	0.7462	5.0757	11.9680	1.3401	6.8019	0.1970	0.1470	2.3579	6
7	0.7107	5.7864	16.2321	1.4071	8.1420	0.1728	0.1228	2.8052	7
8	0.6768	6.4632	20.9700	1.4775	9.5491	0.1547	0.1047	3.2445	8
9	0.6446	7.1078	26.1268	1.5513	11.0266	0.1407	0.0907	3.6758	9
10	0.6139	7.7217	31.6520	1.6289	12.5779	0.1295	0.0795	4.0991	10
11	0.5847	8.3064	37.4988	1.7103	14.2068	0.1204	0.0704	4.5144	11
12	0.5568	8.8633	43.6241	1.7959	15.9171	0.1128	0.0628	4.9219	12
13	0.5303	9.3936	49.9879	1.8856	17.7130	0.1065	0.0565	5.3215	13
14	0.5051	9.8986	56.5538	1.9799	19.5986	0.1010	0.0510	5.7133	14
15	0.4810	10.3797	63.2880	2.0789	21.5786	0.0963	0.0463	6.0973	15
16	0.4581	10.8378	70.1597	2.1829	23.6575	0.0923	0.0423	6.4736	16
17	0.4363	11.2741	77.1405	2.2920	25.8404	0.0887	0.0387	6.8423	17
18	0.4155	11.6896	84.2043	2.4066	28.1324	0.0855	0.0355	7.2034	18
19	0.3957	12.0853	91.3275	2.5270	30.5390	0.0827	0.0327	7.5569	19
20	0.3769	12.4622	98.4884	2.6533	33.0660	0.0802	0.0302	7.9030	20
21	0.3589	12.8212	105.6673	2.7860	35.7193	0.0780	0.0280	8.2416	21
22	0.3418	13.1630	112.8461	2.9253	38.5052	0.0760	0.0260	8.5730	22
23	0.3256	13.4886	120.0087	3.0715	41.4305	0.0741	0.0241	8.8971	23
24	0.3101	13.7986	127.1402	3.2251	44.5020	0.0725	0.0225	9.2140	24
25	0.2953	14.0939	134.2275	3.3864	47.7271	0.0710	0.0210	9.5238	25
26	0.2812	14.3752	141.2585	3.5557	51.1135	0.0696	0.0196	9.8266	26
27	0.2678	14.6430	148.2226	3.7335	54.6691	0.0683	0.0183	10.1224	27
28	0.2551	14.8981	155.1101	3.9201	58.4026	0.0671	0.0171	10.4114	28
29	0.2429	15.1411	161.9126	4.1161	62.3227	0.0660	0.0160	10.6936	29
30	0.2314	15.3725	168.6226	4.3219	66.4388	0.0651	0.0151	10.9691	30
31	0.2204	15.5928	175.2333	4.5380	70.7608	0.0641	0.0141	11.2381	31
32	0.2099	15.8027	181.7392	4.7649	75.2988	0.0633	0.0133	11.5005	32
33	0.1999	16.0025	188.1351	5.0032	80.0638	0.0625	0.0125	11.7566	33
34	0.1904	16.1929	194.4168	5.2533	85.0670	0.0618	0.0118	12.0063	34
35	0.1813	16.3742	200.5807	5.5160	90.3203	0.0611	0.0111	12.2498	35
36	0.1727	16.5469	206.6237	5.7918	95.8363	0.0604	0.0104	12.4872	36
37	0.1644	16.7113	212.5434	6.0814	101.6281	0.0598	0.0098	12.7186	37
38	0.1566	16.8679	218.3378	6.3855	107.7095	0.0593	0.0093	12.9440	38
39	0.1491	17.0170	224.0054	6.7048	114.0950	0.0588	0.0088	13.1636	39
40	0.1420	17.1591	229.5452	7.0400	120.7998	0.0583	0.0083	13.3775	40
41	0.1353	17.2944	234.9564	7.3920	127.8398	0.0578	0.0078	13.5857	41
42	0.1288	17.4232	240.2389	7.7616	135.2318	0.0574	0.0074	13.7884	42
43	0.1227	17.5459	245.3925	8.1497	142.9933	0.0570	0.0070	13.9857	43
44	0.1169	17.6628	250.4175	8.5572	151.1430	0.0566	0.0066	14.1777	44
45	0.1113	17.7741	255.3145	8.9850	159.7002	0.0563	0.0063	14.3644	45
46	0.1060	17.8801	260.0844	9.4343	168.6852	0.0559	0.0059	14.5461	46
47	0.1009	17.9810	264.7281	9.9060	178.1194	0.0556	0.0056	14.7226	47
48	0.0961	18.0772	269.2467	10.4013	188.0254	0.0553	0.0053	14.8943	48
49	0.0916	18.1687	273.6418	10.9213	198.4267	0.0550	0.0050	15.0611	49
50	0.0872	18.2559	277.9148	11.4674	209.3480	0.0548	0.0048	15.2233	50
51	0.0831	18.3390	282.0673	12.0408	220.8154	0.0545	0.0045	15.3808	51
52	0.0791	18.4181	286.1013	12.6428	232.8562	0.0543	0.0043	15.5337	52
53	0.0753	18.4934	290.0184	13.2749	245.4990	0.0541	0.0041	15.6823	53
54	0.0717	18.5651	293.8208	13.9387	258.7739	0.0539	0.0039	15.8265	54
55	0.0683	18.6335	297.5104	14.6356	272.7126	0.0537	0.0037	15.9664	55
60	0.0535	18.9293	314.3432	18.6792	353.5837	0.0528	0.0028	16.6062	60
65	0.0419	19.1611	328.6910	23.8399	456.7980	0.0522	0.0022	17.1541	65
70	0.0329	19.3427	340.8409	30.4264	588.5285	0.0517	0.0017	17.6212	70
75	0.0258	19.4850	351.0721	38.8327	756.6537	0.0513	0.0013	18.0176	75
80	0.0202	19.5965	359.6460	49.5614	971.2288	0.0510	0.0010	18.3526	80
85	0.0158	19.6838	366.8007	63.2544	1245.0871	0.0508	0.0008	18.6346	85
90	0.0124	19.7523	372.7488	80.7304	1597.6073	0.0506	0.0006	18.8712	90
95	0.0097	19.8059	377.6774	103.0347	2040.6935	0.0505	0.0005	19.0689	95
100	0.0076	19.8479	381.7492	131.5013	2610.0252	0.0504	0.0004	19.2337	100

APPENDIX 87.B (continued)
Cash Flow Equivalent Factors

$i = 6.00\%$

n	P/F	P/A	P/G	F/P	F/A	A/P	A/F	A/G	n
1	0.9434	0.9434	0.0000	1.0600	1.0000	1.0600	1.0000	0.0000	1
2	0.8900	1.8334	0.8900	1.1236	2.0600	0.5454	0.4854	0.4854	2
3	0.8396	2.6730	2.5692	1.1910	3.1836	0.3741	0.3141	0.9612	3
4	0.7921	3.4651	4.9455	1.2625	4.3746	0.2886	0.2286	1.4272	4
5	0.7473	4.2124	7.9345	1.3382	5.6371	0.2374	0.1774	1.8836	5
6	0.7050	4.9173	11.4594	1.4185	6.9753	0.2034	0.1434	2.3304	6
7	0.6651	5.5824	15.4497	1.5036	8.3938	0.1791	0.1191	2.7676	7
8	0.6274	6.2098	19.8416	1.5938	9.8975	0.1610	0.1010	3.1952	8
9	0.5919	6.8017	24.5768	1.6895	11.4913	0.1470	0.0870	3.6133	9
10	0.5584	7.3601	29.6023	1.7908	13.1808	0.1359	0.0759	4.0220	10
11	0.5268	7.8869	34.8702	1.8983	14.9716	0.1268	0.0668	4.4213	11
12	0.4970	8.3838	40.3369	2.0122	16.8699	0.1193	0.0593	4.8113	12
13	0.4688	8.8527	45.9629	2.1329	18.8821	0.1130	0.0530	5.1920	13
14	0.4423	9.2950	51.7128	2.2609	21.0151	0.1076	0.0476	5.5635	14
15	0.4173	9.7122	57.5546	2.3966	23.2760	0.1030	0.0430	5.9260	15
16	0.3936	10.1059	63.4592	2.5404	25.6725	0.0990	0.0390	6.2794	16
17	0.3714	10.4773	69.4011	2.6928	28.2129	0.0954	0.0354	6.6240	17
18	0.3503	10.8276	75.3569	2.8543	30.9057	0.0924	0.0324	6.9597	18
19	0.3305	11.1581	81.3062	3.0256	33.7600	0.0896	0.0296	7.2867	19
20	0.3118	11.4699	87.2304	3.2071	36.7856	0.0872	0.0272	7.6051	20
21	0.2942	11.7641	93.1136	3.3996	39.9927	0.0850	0.0250	7.9151	21
22	0.2775	12.0416	98.9412	3.6035	43.3923	0.0830	0.0230	8.2166	22
23	0.2618	12.3034	104.7007	3.8197	46.9958	0.0813	0.0213	8.5099	23
24	0.2470	12.5504	110.3812	4.0489	50.8156	0.0797	0.0197	8.7951	24
25	0.2330	12.7834	115.9732	4.2919	54.8645	0.0782	0.0182	9.0722	25
26	0.2198	13.0032	121.4684	4.5494	59.1564	0.0769	0.0169	9.3414	26
27	0.2074	13.2105	126.8600	4.8223	63.7058	0.0757	0.0157	9.6029	27
28	0.1956	13.4062	132.1420	5.1117	68.5281	0.0746	0.0146	9.8568	28
29	0.1846	13.5907	137.3096	5.4184	73.6398	0.0736	0.0136	10.1032	29
30	0.1741	13.7648	142.3588	5.7435	79.0582	0.0726	0.0126	10.3422	30
31	0.1643	13.9291	147.2864	6.0881	84.8017	0.0718	0.0118	10.5740	31
32	0.1550	14.0840	152.0901	6.4534	90.8898	0.0710	0.0110	10.7988	32
33	0.1462	14.2302	156.7681	6.8406	97.3432	0.0703	0.0103	11.0166	33
34	0.1379	14.3681	161.3192	7.2510	104.1838	0.0696	0.0096	11.2276	34
35	0.1301	14.4982	165.7427	7.6861	111.4348	0.0690	0.0090	11.4319	35
36	0.1227	14.6210	170.0387	8.1473	119.1209	0.0684	0.0084	11.6298	36
37	0.1158	14.7368	174.2072	8.6361	127.2681	0.0679	0.0079	11.8213	37
38	0.1092	14.8460	178.2490	9.1543	135.9042	0.0674	0.0074	12.0065	38
39	0.1031	14.9491	182.1652	9.7035	145.0585	0.0669	0.0069	12.1857	39
40	0.0972	15.0463	185.9568	10.2857	154.7620	0.0665	0.0065	12.3590	40
41	0.0917	15.1380	189.6256	10.9029	165.0477	0.0661	0.0061	12.5264	41
42	0.0865	15.2245	193.1732	11.5570	175.9505	0.0657	0.0057	12.6883	42
43	0.0816	15.3062	196.6017	12.2505	187.5076	0.0653	0.0053	12.8446	43
44	0.0770	15.3832	199.9130	12.9855	199.7580	0.0650	0.0050	12.9956	44
45	0.0727	15.4558	203.1096	13.7646	212.7435	0.0647	0.0047	13.1413	45
46	0.0685	15.5244	206.1938	14.5905	226.5081	0.0644	0.0044	13.2819	46
47	0.0647	15.5890	209.1681	15.4659	241.0986	0.0641	0.0041	13.4177	47
48	0.0610	15.6500	212.0351	16.3939	256.5645	0.0639	0.0039	13.5485	48
49	0.0575	15.7076	214.7972	17.3775	272.9584	0.0637	0.0037	13.6748	49
50	0.0543	15.7619	217.4574	18.4202	290.3359	0.0634	0.0034	13.7964	50
51	0.0512	15.8131	220.0181	19.5254	308.7561	0.0632	0.0032	13.9137	51
52	0.0483	15.8614	222.4823	20.6969	328.2814	0.0630	0.0030	14.0267	52
53	0.0456	15.9070	224.8525	21.9387	348.9783	0.0629	0.0029	14.1355	53
54	0.0430	15.9500	227.1316	23.2550	370.9170	0.0627	0.0027	14.2402	54
55	0.0406	15.9905	229.3222	24.6503	394.1720	0.0625	0.0025	14.3411	55
60	0.0303	16.1614	239.0428	32.9877	533.1282	0.0619	0.0019	14.7909	60
65	0.0227	16.2891	246.9450	44.1450	719.0829	0.0614	0.0014	15.1601	65
70	0.0169	16.3845	253.3271	59.0759	967.9322	0.0610	0.0010	15.4613	70
75	0.0126	16.4558	258.4527	79.0569	1300.9487	0.0608	0.0008	15.7058	75
80	0.0095	16.5091	262.5493	105.7960	1746.5999	0.0606	0.0006	15.9033	80
85	0.0071	16.5489	265.8096	141.5789	2342.9817	0.0604	0.0004	16.0620	85
90	0.0053	16.5787	268.3946	189.4645	3141.0752	0.0603	0.0003	16.1891	90
95	0.0039	16.6009	270.4375	253.5463	4209.1042	0.0602	0.0002	16.2905	95
100	0.0029	16.6175	272.0471	339.3021	5638.3681	0.0602	0.0002	16.3711	100

APPENDIX 87.B *(continued)*
Cash Flow Equivalent Factors

$i = 7.00\%$

n	P/F	P/A	P/G	F/P	F/A	A/P	A/F	A/G	n
1	0.9346	0.9346	0.0000	1.0700	1.0000	1.0700	1.0000	0.0000	1
2	0.8734	1.8080	0.8734	1.1449	2.0700	0.5531	0.4831	0.4831	2
3	0.8163	2.6243	2.5060	1.2250	3.2149	0.3811	0.3111	0.9549	3
4	0.7629	3.3872	4.7947	1.3108	4.4399	0.2952	0.2252	1.4155	4
5	0.7130	4.1002	7.6467	1.4026	5.7507	0.2439	0.1739	1.8650	5
6	0.6663	4.7665	10.9784	1.5007	7.1533	0.2098	0.1398	2.3032	6
7	0.6227	5.3893	14.7149	1.6058	8.6540	0.1856	0.1156	2.7304	7
8	0.5820	5.9713	18.7889	1.7182	10.2598	0.1675	0.0975	3.1465	8
9	0.5439	6.5152	23.1404	1.8385	11.9780	0.1535	0.0835	3.5517	9
10	0.5083	7.0236	27.7156	1.9672	13.8164	0.1424	0.0724	3.9461	10
11	0.4751	7.4987	32.4665	2.1049	15.7836	0.1334	0.0634	4.3296	11
12	0.4440	7.9427	37.3506	2.2522	17.8885	0.1259	0.0559	4.7025	12
13	0.4150	8.3577	42.3302	2.4098	20.1406	0.1197	0.0497	5.0648	13
14	0.3878	8.7455	47.3718	2.5785	22.5505	0.1143	0.0443	5.4167	14
15	0.3624	9.1079	52.4461	2.7590	25.1290	0.1098	0.0398	5.7583	15
16	0.3387	9.4466	57.5271	2.9522	27.8881	0.1059	0.0359	6.0897	16
17	0.3166	9.7632	62.5923	3.1588	30.8402	0.1024	0.0324	6.4110	17
18	0.2959	10.0591	67.6219	3.3799	33.9990	0.0994	0.0294	6.7225	18
19	0.2765	10.3356	72.5991	3.6165	37.3790	0.0968	0.0268	7.0242	19
20	0.2584	10.5940	77.5091	3.8697	40.9955	0.0944	0.0244	7.3163	20
21	0.2415	10.8355	82.3393	4.1406	44.8652	0.0923	0.0223	7.5990	21
22	0.2257	11.0612	87.0793	4.4304	49.0057	0.0904	0.0204	7.8725	22
23	0.2109	11.2722	91.7201	4.7405	53.4361	0.0887	0.0187	8.1369	23
24	0.1971	11.4693	96.2545	5.0724	58.1767	0.0872	0.0172	8.3923	24
25	0.1842	11.6536	100.6765	5.4274	63.2490	0.0858	0.0158	8.6391	25
26	0.1722	11.8258	104.9814	5.8074	68.6765	0.0846	0.0146	8.8773	26
27	0.1609	11.9867	109.1656	6.2139	74.4838	0.0834	0.0134	9.1072	27
28	0.1504	12.1371	113.2264	6.6488	80.6977	0.0824	0.0124	9.3289	28
29	0.1406	12.2777	117.1622	7.1143	87.3465	0.0814	0.0114	9.5427	29
30	0.1314	12.4090	120.9718	7.6123	94.4608	0.0806	0.0106	9.7487	30
31	0.1228	12.5318	124.6550	8.1451	102.0730	0.0798	0.0098	9.9471	31
32	0.1147	12.6466	128.2120	8.7153	110.2182	0.0791	0.0091	10.1381	32
33	0.1072	12.7538	131.6435	9.3253	118.9334	0.0784	0.0084	10.3219	33
34	0.1002	12.8540	134.9507	9.9781	128.2588	0.0778	0.0078	10.4987	34
35	0.0937	12.9477	138.1353	10.6766	138.2369	0.0772	0.0072	10.6687	35
36	0.0875	13.0352	141.1990	11.4239	148.9135	0.0767	0.0067	10.8321	36
37	0.0818	13.1170	144.1441	12.2236	160.3374	0.0762	0.0062	10.9891	37
38	0.0765	13.1935	146.9730	13.0793	172.5610	0.0758	0.0058	11.1398	38
39	0.0715	13.2649	149.6883	13.9948	185.6403	0.0754	0.0054	11.2845	39
40	0.0668	13.3317	152.2928	14.9745	199.6351	0.0750	0.0050	11.4233	40
41	0.0624	13.3941	154.7892	16.0227	214.6096	0.0747	0.0047	11.5565	41
42	0.0583	13.4524	157.1807	17.1443	230.6322	0.0743	0.0043	11.6842	42
43	0.0545	13.5070	159.4702	18.3444	247.7765	0.0740	0.0040	11.8065	43
44	0.0509	13.5579	161.6609	19.6285	266.1209	0.0738	0.0038	11.9237	44
45	0.0476	13.6055	163.7559	21.0025	285.7493	0.0735	0.0035	12.0360	45
46	0.0445	13.6500	165.7584	22.4726	306.7518	0.0733	0.0033	12.1435	46
47	0.0416	13.6916	167.6714	24.0457	329.2244	0.0730	0.0030	12.2463	47
48	0.0389	13.7305	169.4981	25.7289	353.2701	0.0728	0.0028	12.3447	48
49	0.0363	13.7668	171.2417	27.5299	378.9990	0.0726	0.0026	12.4387	49
50	0.0339	13.8007	172.9051	29.4570	406.5289	0.0725	0.0025	12.5287	50
51	0.0317	13.8325	174.4915	31.5190	435.9860	0.0723	0.0023	12.6146	51
52	0.0297	13.8621	176.0037	33.7253	467.5050	0.0721	0.0021	12.6967	52
53	0.0277	13.8898	177.4447	36.0861	501.2303	0.0720	0.0020	12.7751	53
54	0.0259	13.9157	178.8173	38.6122	537.3164	0.0719	0.0019	12.8500	54
55	0.0242	13.9399	180.1243	41.3150	575.9286	0.0717	0.0017	12.9215	55
60	0.0173	14.0392	185.7677	57.9464	813.5204	0.0712	0.0012	13.2321	60
65	0.0123	14.1099	190.1452	81.2729	1146.7552	0.0709	0.0009	13.4760	65
70	0.0088	14.1604	193.5185	113.9894	1614.1342	0.0706	0.0006	13.6662	70
75	0.0063	14.1964	196.1035	159.8760	2269.6574	0.0704	0.0004	13.8136	75
80	0.0045	14.2220	198.0748	224.2344	3189.0627	0.0703	0.0003	13.9273	80
85	0.0032	14.2403	199.5717	314.5003	4478.5761	0.0702	0.0002	14.0146	85
90	0.0023	14.2533	200.7042	441.1030	6287.1854	0.0702	0.0002	14.0812	90
95	0.0016	14.2626	201.5581	618.6697	8823.8535	0.0701	0.0001	14.1319	95
100	0.0012	14.2693	202.2001	867.7163	12381.6618	0.0701	0.0001	14.1703	100

APPENDIX 87.B *(continued)*
Cash Flow Equivalent Factors

$i = 8.00\%$

n	P/F	P/A	P/G	F/P	F/A	A/P	A/F	A/G	n
1	0.9259	0.9259	0.0000	1.0800	1.0000	1.0800	1.0000	0.0000	1
2	0.8573	1.7833	0.8573	1.1664	2.0800	0.5608	0.4808	0.4808	2
3	0.7938	2.5771	2.4450	1.2597	3.2464	0.3880	0.3080	0.9487	3
4	0.7350	3.3121	4.6501	1.3605	4.5061	0.3019	0.2219	1.4040	4
5	0.6806	3.9927	7.3724	1.4693	5.8666	0.2505	0.1705	1.8465	5
6	0.6302	4.6229	10.5233	1.5869	7.3359	0.2163	0.1363	2.2763	6
7	0.5835	5.2064	14.0242	1.7138	8.9228	0.1921	0.1121	2.6937	7
8	0.5403	5.7466	17.8061	1.8509	10.6366	0.1740	0.0940	3.0985	8
9	0.5002	6.2469	21.8081	1.9990	12.4876	0.1601	0.0801	3.4910	9
10	0.4632	6.7101	25.9768	2.1589	14.4866	0.1490	0.0690	3.8713	10
11	0.4289	7.1390	30.2657	2.3316	16.6455	0.1401	0.0601	4.2395	11
12	0.3971	7.5361	34.6339	2.5182	18.9771	0.1327	0.0527	4.5957	12
13	0.3677	7.9038	39.0463	2.7196	21.4953	0.1265	0.0465	4.9402	13
14	0.3405	8.2442	43.4723	2.9372	24.2149	0.1213	0.0413	5.2731	14
15	0.3152	8.5595	47.8857	3.1722	27.1521	0.1168	0.0368	5.5945	15
16	0.2919	8.8514	52.2640	3.4259	30.3243	0.1130	0.0330	5.9046	16
17	0.2703	9.1216	56.5883	3.7000	33.7502	0.1096	0.0296	6.2037	17
18	0.2502	9.3719	60.8426	3.9960	37.4502	0.1067	0.0267	6.4920	18
19	0.2317	9.6036	65.0134	4.3157	41.4463	0.1041	0.0241	6.7697	19
20	0.2145	9.8181	69.0898	4.6610	45.7620	0.1019	0.0219	7.0369	20
21	0.1987	10.0168	73.0629	5.0338	50.4229	0.0998	0.0198	7.2940	21
22	0.1839	10.2007	76.9257	5.4365	55.4568	0.0980	0.0180	7.5412	22
23	0.1703	10.3711	80.6726	5.8715	60.8933	0.0964	0.0164	7.7786	23
24	0.1577	10.5288	84.2997	6.3412	66.7648	0.0950	0.0150	8.0066	24
25	0.1460	10.6748	87.8041	6.8485	73.1059	0.0937	0.0137	8.2254	25
26	0.1352	10.8100	91.1842	7.3964	79.9544	0.0925	0.0125	8.4352	26
27	0.1252	10.9352	94.4390	7.9881	87.3508	0.0914	0.0114	8.6363	27
28	0.1159	11.0511	97.5687	8.6271	95.3388	0.0905	0.0105	8.8289	28
29	0.1073	11.1584	100.5738	9.3173	103.9659	0.0896	0.0096	9.0133	29
30	0.0994	11.2578	103.4558	10.0627	113.2832	0.0888	0.0088	9.1897	30
31	0.0920	11.3498	106.2163	10.8677	123.3459	0.0881	0.0081	9.3584	31
32	0.0852	11.4350	108.8575	11.7371	134.2135	0.0875	0.0075	9.5197	32
33	0.0789	11.5139	111.3819	12.6760	145.9506	0.0869	0.0069	9.6737	33
34	0.0730	11.5869	113.7924	13.6901	158.6267	0.0863	0.0063	9.8208	34
35	0.0676	11.6546	116.0920	14.7853	172.3168	0.0858	0.0058	9.9611	35
36	0.0626	11.7172	118.2839	15.9682	187.1021	0.0853	0.0053	10.0949	36
37	0.0580	11.7752	120.3713	17.2456	203.0703	0.0849	0.0049	10.2225	37
38	0.0537	11.8289	122.3579	18.6253	220.3159	0.0845	0.0045	10.3440	38
39	0.0497	11.8786	124.2470	20.1153	238.9412	0.0842	0.0042	10.4597	39
40	0.0460	11.9246	126.0422	21.7245	259.0565	0.0839	0.0039	10.5699	40
41	0.0426	11.9672	127.7470	23.4625	280.7810	0.0836	0.0036	10.6747	41
42	0.0395	12.0067	129.3651	25.3395	304.2435	0.0833	0.0033	10.7744	42
43	0.0365	12.0432	130.8998	27.3666	329.5830	0.0830	0.0030	10.8692	43
44	0.0338	12.0771	132.3547	29.5560	356.9496	0.0828	0.0028	10.9592	44
45	0.0313	12.1084	133.7331	31.9204	386.5056	0.0826	0.0026	11.0447	45
46	0.0290	12.1374	135.0384	34.4741	418.4261	0.0824	0.0024	11.1258	46
47	0.0269	12.1643	136.2739	37.2320	452.9002	0.0822	0.0022	11.2028	47
48	0.0249	12.1891	137.4428	40.2106	490.1322	0.0820	0.0020	11.2758	48
49	0.0230	12.2122	138.5480	43.4274	530.3427	0.0819	0.0019	11.3451	49
50	0.0213	12.2335	139.5928	46.9016	573.7702	0.0817	0.0017	11.4107	50
51	0.0197	12.2532	140.5799	50.6537	620.6718	0.0816	0.0016	11.4729	51
52	0.0183	12.2715	141.5121	54.7060	671.3255	0.0815	0.0015	11.5318	52
53	0.0169	12.2884	142.3923	59.0825	726.0316	0.0814	0.0014	11.5875	53
54	0.0157	12.3041	143.2229	63.8091	785.1141	0.0813	0.0013	11.6403	54
55	0.0145	12.3186	144.0065	68.9139	848.9232	0.0812	0.0012	11.6902	55
60	0.0099	12.3766	147.3000	101.2571	1253.2133	0.0808	0.0008	11.9015	60
65	0.0067	12.4160	149.7387	148.7798	1847.2481	0.0805	0.0005	12.0602	65
70	0.0046	12.4428	151.5326	218.6064	2720.0801	0.0804	0.0004	12.1783	70
75	0.0031	12.4611	152.8448	321.2045	4002.5566	0.0802	0.0002	12.2658	75
80	0.0021	12.4735	153.8001	471.9548	5886.9354	0.0802	0.0002	12.3301	80
85	0.0014	12.4820	154.4925	693.4565	8655.7061	0.0801	0.0001	12.3772	85
90	0.0010	12.4877	154.9925	1018.9151	12723.9386	0.0801	0.0001	12.4116	90
95	0.0007	12.4917	155.3524	1497.1205	18701.5069	0.0801	0.0001	12.4365	95
100	0.0005	12.4943	155.6107	2199.7613	27484.5157	0.0800	0.0000	12.4545	100

APPENDIX 87.B *(continued)*
Cash Flow Equivalent Factors

$i = 9.00\%$

n	P/F	P/A	P/G	F/P	F/A	A/P	A/F	A/G	n
1	0.9174	0.9174	0.0000	1.0900	1.0000	1.0900	1.0000	0.0000	1
2	0.8417	1.7591	0.8417	1.1881	2.0900	0.5685	0.4785	0.4785	2
3	0.7722	2.5313	2.3860	1.2950	3.2781	0.3951	0.3051	0.9426	3
4	0.7084	3.2397	4.5113	1.4116	4.5731	0.3087	0.2187	1.3925	4
5	0.6499	3.8897	7.1110	1.5386	5.9847	0.2571	0.1671	1.8282	5
6	0.5963	4.4859	10.0924	1.6771	7.5233	0.2229	0.1329	2.2498	6
7	0.5470	5.0330	13.3746	1.8280	9.2004	0.1987	0.1087	2.6574	7
8	0.5019	5.5348	16.8877	1.9926	11.0285	0.1807	0.0907	3.0512	8
9	0.4604	5.9952	20.5711	2.1719	13.0210	0.1668	0.0768	3.4312	9
10	0.4224	6.4177	24.3728	2.3674	15.1929	0.1558	0.0658	3.7978	10
11	0.3875	6.8052	28.2481	2.5804	17.5603	0.1469	0.0569	4.1510	11
12	0.3555	7.1607	32.1590	2.8127	20.1407	0.1397	0.0497	4.4910	12
13	0.3262	7.4869	36.0731	3.0658	22.9534	0.1336	0.0436	4.8182	13
14	0.2992	7.7862	39.9633	3.3417	26.0192	0.1284	0.0384	5.1326	14
15	0.2745	8.0607	43.8069	3.6425	29.3609	0.1241	0.0341	5.4346	15
16	0.2519	8.3126	47.5849	3.9703	33.0034	0.1203	0.0303	5.7245	16
17	0.2311	8.5436	51.2821	4.3276	36.9737	0.1170	0.0270	6.0024	17
18	0.2120	8.7556	54.8860	4.7171	41.3013	0.1142	0.0242	6.2687	18
19	0.1945	8.9501	58.3868	5.1417	46.0185	0.1117	0.0217	6.5236	19
20	0.1784	9.1285	61.7770	5.6044	51.1601	0.1095	0.0195	6.7674	20
21	0.1637	9.2922	65.0509	6.1088	56.7645	0.1076	0.0176	7.0006	21
22	0.1502	9.4424	68.2048	6.6586	62.8733	0.1059	0.0159	7.2232	22
23	0.1378	9.5802	71.2359	7.2579	69.5319	0.1044	0.0144	7.4357	23
24	0.1264	9.7066	74.1433	7.9111	76.7898	0.1030	0.0130	7.6384	24
25	0.1160	9.8226	76.9265	8.6231	84.7009	0.1018	0.0118	7.8316	25
26	0.1064	9.9290	79.5863	9.3992	93.3240	0.1007	0.0107	8.0156	26
27	0.0976	10.0266	82.1241	10.2451	102.7231	0.0997	0.0097	8.1906	27
28	0.0895	10.1161	84.5419	11.1671	112.9682	0.0989	0.0089	8.3571	28
29	0.0822	10.1983	86.8422	12.1722	124.1354	0.0981	0.0081	8.5154	29
30	0.0754	10.2737	89.0280	13.2677	136.3076	0.0973	0.0073	8.6657	30
31	0.0691	10.3428	91.1024	14.4618	149.5752	0.0967	0.0067	8.8083	31
32	0.0634	10.4062	93.0690	15.7633	164.0370	0.0961	0.0061	8.9436	32
33	0.0582	10.4644	94.9314	17.1820	179.8003	0.0956	0.0056	9.0718	33
34	0.0534	10.5178	96.6935	18.7284	196.9823	0.0951	0.0051	9.1933	34
35	0.0490	10.5668	98.3590	20.4140	215.7108	0.0946	0.0046	9.3083	35
36	0.0449	10.6118	99.9319	22.2512	236.1247	0.0942	0.0042	9.4171	36
37	0.0412	10.6530	101.4162	24.2538	258.3759	0.0939	0.0039	9.5200	37
38	0.0378	10.6908	102.8158	26.4367	282.6298	0.0935	0.0035	9.6172	38
39	0.0347	10.7255	104.1345	28.8160	309.0665	0.0932	0.0032	9.7090	39
40	0.0318	10.7574	105.3762	31.4094	337.8824	0.0930	0.0030	9.7957	40
41	0.0292	10.7866	106.5445	34.2363	369.2919	0.0927	0.0027	9.8775	41
42	0.0268	10.8134	107.6432	37.3175	403.5281	0.0925	0.0025	9.9546	42
43	0.0246	10.8380	108.6758	40.6761	440.8457	0.0923	0.0023	10.0273	43
44	0.0226	10.8605	109.6456	44.3370	481.5218	0.0921	0.0021	10.0958	44
45	0.0207	10.8812	110.5561	48.3273	525.8587	0.0919	0.0019	10.1603	45
46	0.0190	10.9002	111.4103	52.6767	574.1860	0.0917	0.0017	10.2210	46
47	0.0174	10.9176	112.2115	57.4176	626.8628	0.0916	0.0016	10.2780	47
48	0.0160	10.9336	112.9625	62.5852	684.2804	0.0915	0.0015	10.3317	48
49	0.0147	10.9482	113.6661	68.2179	746.8656	0.0913	0.0013	10.3821	49
50	0.0134	10.9617	114.3251	74.3575	815.0836	0.0912	0.0012	10.4295	50
51	0.0123	10.9740	114.9420	81.0497	889.4411	0.0911	0.0011	10.4740	51
52	0.0113	10.9853	115.5193	88.3442	970.4908	0.0910	0.0010	10.5158	52
53	0.0104	10.9957	116.0593	96.2951	1058.8349	0.0909	0.0009	10.5549	53
54	0.0095	11.0053	116.5642	104.9617	1155.1301	0.0909	0.0009	10.5917	54
55	0.0087	11.0140	117.0362	114.4083	1260.0918	0.0908	0.0008	10.6261	55
60	0.0057	11.0480	118.9683	176.0313	1944.7921	0.0905	0.0005	10.7683	60
65	0.0037	11.0701	120.3344	270.8460	2998.2885	0.0903	0.0003	10.8702	65
70	0.0024	11.0844	121.2942	416.7301	4619.2232	0.0902	0.0002	10.9427	70
75	0.0016	11.0938	121.9646	641.1909	7113.2321	0.0901	0.0001	10.9940	75
80	0.0010	11.0998	122.4306	986.5517	10950.5741	0.0901	0.0001	11.0299	80
85	0.0007	11.1038	122.7533	1517.9320	16854.8003	0.0901	0.0001	11.0551	85
90	0.0004	11.1064	122.9758	2335.5266	25939.1842	0.0900	0.0000	11.0726	90
95	0.0003	11.1080	123.1287	3593.4971	39916.6350	0.0900	0.0000	11.0847	95
100	0.0002	11.1091	123.2335	5529.0408	61422.6755	0.0900	0.0000	11.0930	100

APPENDIX 87.B (continued)
Cash Flow Equivalent Factors

$i = 10.00\%$

n	P/F	P/A	P/G	F/P	F/A	A/P	A/F	A/G	n
1	0.9091	0.9091	0.0000	1.1000	1.0000	1.1000	1.0000	0.0000	1
2	0.8264	1.7355	0.8264	1.2100	2.1000	0.5762	0.4762	0.4762	2
3	0.7513	2.4869	2.3291	1.3310	3.3100	0.4021	0.3021	0.9366	3
4	0.6830	3.1699	4.3781	1.4641	4.6410	0.3155	0.2155	1.3812	4
5	0.6209	3.7908	6.8618	1.6105	6.1051	0.2638	0.1638	1.8101	5
6	0.5645	4.3553	9.6842	1.7716	7.7156	0.2296	0.1296	2.2236	6
7	0.5132	4.8684	12.7631	1.9487	9.4872	0.2054	0.1054	2.6216	7
8	0.4665	5.3349	16.0287	2.1436	11.4359	0.1874	0.0874	3.0045	8
9	0.4241	5.7590	19.4215	2.3579	13.5795	0.1736	0.0736	3.3724	9
10	0.3855	6.1446	22.8913	2.5937	15.9374	0.1627	0.0627	3.7255	10
11	0.3505	6.4951	26.3963	2.8531	18.5312	0.1540	0.0540	4.0641	11
12	0.3186	6.8137	29.9012	3.1384	21.3843	0.1468	0.0468	4.3884	12
13	0.2897	7.1034	33.3772	3.4523	24.5227	0.1408	0.0408	4.6988	13
14	0.2633	7.3667	36.8005	3.7975	27.9750	0.1357	0.0357	4.9955	14
15	0.2394	7.6061	40.1520	4.1772	31.7725	0.1315	0.0315	5.2789	15
16	0.2176	7.8237	43.4164	4.5950	35.9497	0.1278	0.0278	5.5493	16
17	0.1978	8.0216	46.5819	5.0545	40.5447	0.1247	0.0247	5.8071	17
18	0.1799	8.2014	49.6395	5.5599	45.5992	0.1219	0.0219	6.0526	18
19	0.1635	8.3649	52.5827	6.1159	51.1591	0.1195	0.0195	6.2861	19
20	0.1486	8.5136	55.4069	6.7275	57.2750	0.1175	0.0175	6.5081	20
21	0.1351	8.6487	58.1095	7.4002	64.0025	0.1156	0.0156	6.7189	21
22	0.1228	8.7715	60.6893	8.1403	71.4027	0.1140	0.0140	6.9189	22
23	0.1117	8.8832	63.1462	8.9543	79.5430	0.1126	0.0126	7.1085	23
24	0.1015	8.9847	65.4813	9.8497	88.4973	0.1113	0.0113	7.2881	24
25	0.0923	9.0770	67.6964	10.8347	98.3471	0.1102	0.0102	7.4580	25
26	0.0839	9.1609	69.7940	11.9182	109.1818	0.1092	0.0092	7.6186	26
27	0.0763	9.2372	71.7773	13.1100	121.0999	0.1083	0.0083	7.7704	27
28	0.0693	9.3066	73.6495	14.4210	134.2099	0.1075	0.0075	7.9137	28
29	0.0630	9.3696	75.4146	15.8631	148.6309	0.1067	0.0067	8.0489	29
30	0.0573	9.4269	77.0766	17.4494	164.4940	0.1061	0.0061	8.1762	30
31	0.0521	9.4790	78.6395	19.1943	181.9434	0.1055	0.0055	8.2962	31
32	0.0474	9.5264	80.1078	21.1138	201.1378	0.1050	0.0050	8.4091	32
33	0.0431	9.5694	81.4856	23.2252	222.2515	0.1045	0.0045	8.5152	33
34	0.0391	9.6086	82.7773	25.5477	245.4767	0.1041	0.0041	8.6149	34
35	0.0356	9.6442	83.9872	28.1024	271.0244	0.1037	0.0037	8.7086	35
36	0.0323	9.6765	85.1194	30.9127	299.1268	0.1033	0.0033	8.7965	36
37	0.0294	9.7059	86.1781	34.0039	330.0395	0.1030	0.0030	8.8789	37
38	0.0267	9.7327	87.1673	37.4043	364.0434	0.1027	0.0027	8.9562	38
39	0.0243	9.7570	88.0908	41.1448	401.4478	0.1025	0.0025	9.0285	39
40	0.0221	9.7791	88.9525	45.2593	442.5926	0.1023	0.0023	9.0962	40
41	0.0201	9.7991	89.7560	49.7852	487.8518	0.1020	0.0020	9.1596	41
42	0.0183	9.8174	90.5047	54.7637	537.6370	0.1019	0.0019	9.2188	42
43	0.0166	9.8340	91.2019	60.2401	592.4007	0.1017	0.0017	9.2741	43
44	0.0151	9.8491	91.8508	66.2641	652.6408	0.1015	0.0015	9.3258	44
45	0.0137	9.8628	92.4544	72.8905	718.9048	0.1014	0.0014	9.3740	45
46	0.0125	9.8753	93.0157	80.1795	791.7953	0.1013	0.0013	9.4190	46
47	0.0113	9.8866	93.5372	88.1975	871.9749	0.1011	0.0011	9.4610	47
48	0.0103	9.8969	94.0217	97.0172	960.1723	0.1010	0.0010	9.5001	48
49	0.0094	9.9063	94.4715	106.7190	1057.1896	0.1009	0.0009	9.5365	49
50	0.0085	9.9148	94.8889	117.3909	1163.9085	0.1009	0.0009	9.5704	50
51	0.0077	9.9226	95.2761	129.1299	1281.2994	0.1008	0.0008	9.6020	51
52	0.0070	9.9296	95.6351	142.0429	1410.4293	0.1007	0.0007	9.6313	52
53	0.0064	9.9360	95.9679	156.2472	1552.4723	0.1006	0.0006	9.6586	53
54	0.0058	9.9418	96.2763	171.8719	1708.7195	0.1006	0.0006	9.6840	54
55	0.0053	9.9471	96.5619	189.0591	1880.5914	0.1005	0.0005	9.7075	55
60	0.0033	9.9672	97.7010	304.4816	3034.8164	0.1003	0.0003	9.8023	60
65	0.0020	9.9796	98.4705	490.3707	4893.7073	0.1002	0.0002	9.8672	65
70	0.0013	9.9873	98.9870	789.7470	7887.4696	0.1001	0.0001	9.9113	70
75	0.0008	9.9921	99.3317	1271.8954	12708.9537	0.1001	0.0001	9.9410	75
80	0.0005	9.9951	99.5606	2048.4002	20474.0021	0.1000	0.0000	9.9609	80
85	0.0003	9.9970	99.7120	3298.9690	32979.6903	0.1000	0.0000	9.9742	85
90	0.0002	9.9981	99.8118	5313.0226	53120.2261	0.1000	0.0000	9.9831	90
95	0.0001	9.9988	99.8773	8556.6760	85556.7605	0.1000	0.0000	9.9889	95
100	0.0001	9.9993	99.9202	13780.6123	137796.1234	0.1000	0.0000	9.9927	100

APPENDIX 87.B *(continued)*
Cash Flow Equivalent Factors

$i = 12.00\%$

n	P/F	P/A	P/G	F/P	F/A	A/P	A/F	A/G	n
1	0.8929	0.8929	0.0000	1.1200	1.0000	1.1200	1.0000	0.0000	1
2	0.7972	1.6901	0.7972	1.2544	2.1200	0.5917	0.4717	0.4717	2
3	0.7118	2.4018	2.2208	1.4049	3.3744	0.4163	0.2963	0.9246	3
4	0.6355	3.0373	4.1273	1.5735	4.7793	0.3292	0.2092	1.3589	4
5	0.5674	3.6048	6.3970	1.7623	6.3528	0.2774	0.1574	1.7746	5
6	0.5066	4.1114	8.9302	1.9738	8.1152	0.2432	0.1232	2.1720	6
7	0.4523	4.5638	11.6443	2.2107	10.0890	0.2191	0.0991	2.5515	7
8	0.4039	4.9676	14.4714	2.4760	12.2997	0.2013	0.0813	2.9131	8
9	0.3606	5.3282	17.3563	2.7731	14.7757	0.1877	0.0677	3.2574	9
10	0.3220	5.6502	20.2541	3.1058	17.5487	0.1770	0.0570	3.5847	10
11	0.2875	5.9377	23.1288	3.4785	20.6546	0.1684	0.0484	3.8953	11
12	0.2567	6.1944	25.9523	3.8960	24.1331	0.1614	0.0414	4.1897	12
13	0.2292	6.4235	28.7024	4.3635	28.0291	0.1557	0.0357	4.4683	13
14	0.2046	6.6282	31.3624	4.8871	32.3926	0.1509	0.0309	4.7317	14
15	0.1827	6.8109	33.9202	5.4736	37.2797	0.1468	0.0268	4.9803	15
16	0.1631	6.9740	36.3670	6.1304	42.7533	0.1434	0.0234	5.2147	16
17	0.1456	7.1196	38.6973	6.8660	48.8837	0.1405	0.0205	5.4353	17
18	0.1300	7.2497	40.9080	7.6900	55.7497	0.1379	0.0179	5.6427	18
19	0.1161	7.3658	42.9979	8.6128	63.4397	0.1358	0.0158	6.8375	19
20	0.1037	7.4694	44.9676	9.6463	72.0524	0.1339	0.0139	6.0202	20
21	0.0926	7.5620	46.8188	10.8038	81.6987	0.1322	0.0122	6.1913	21
22	0.0826	7.6446	48.5543	12.1003	92.5026	0.1308	0.0108	6.3514	22
23	0.0738	7.7184	50.1776	13.5523	104.6029	0.1296	0.0096	6.5010	23
24	0.0659	7.7843	51.6929	15.1786	118.1552	0.1285	0.0085	6.6406	24
25	0.0588	7.8431	53.1046	17.0001	133.3339	0.1275	0.0075	6.7708	25
26	0.0525	7.8957	54.4177	19.0401	150.3339	0.1267	0.0067	6.8921	26
27	0.0469	7.9426	55.6369	21.3249	169.3740	0.1259	0.0059	7.0049	27
28	0.0419	7.9844	56.7674	23.8839	190.6989	0.1252	0.0052	7.1098	28
29	0.0374	8.0218	57.8141	26.7499	214.5828	0.1247	0.0047	7.2071	29
30	0.0334	8.0552	58.7821	29.9599	241.3327	0.1241	0.0041	7.2974	30
31	0.0298	8.0850	59.6761	33.5551	271.2926	0.1237	0.0037	7.3811	31
32	0.0266	8.1116	60.5010	37.5817	304.8477	0.1233	0.0033	7.4586	32
33	0.0238	8.1354	61.2612	42.0915	342.4294	0.1229	0.0029	7.5302	33
34	0.0212	8.1566	61.9612	47.1425	384.5210	0.1226	0.0026	7.5965	34
35	0.0189	8.1755	62.6052	52.7996	431.6635	0.1223	0.0023	7.6577	35
36	0.0169	8.1924	63.1970	59.1356	484.4631	0.1221	0.0021	7.7141	36
37	0.0151	8.2075	63.7406	66.2318	543.5987	0.1218	0.0018	7.7661	37
38	0.0135	8.2210	64.2394	74.1797	609.8305	0.1216	0.0016	7.8141	38
39	0.0120	8.2330	64.6967	83.0812	684.0102	0.1215	0.0015	7.8582	39
40	0.0107	8.2438	65.1159	93.0510	767.0914	0.1213	0.0013	7.8988	40
41	0.0096	8.2534	65.4997	104.2171	860.1424	0.1212	0.0012	7.9361	41
42	0.0086	8.2619	65.8509	116.7231	964.3595	0.1210	0.0010	7.9704	42
43	0.0076	8.2696	66.1722	130.7299	1081.0826	0.1209	0.0009	8.0019	43
44	0.0068	8.2764	66.4659	146.4175	1211.8125	0.1208	0.0008	8.0308	44
45	0.0061	8.2825	66.7342	163.9876	1358.2300	0.1207	0.0007	8.0572	45
46	0.0054	8.2880	66.9792	183.6661	1522.2176	0.1207	0.0007	8.0815	46
47	0.0049	8.2928	67.2028	205.7061	1705.8838	0.1206	0.0006	8.1037	47
48	0.0043	8.2972	67.4068	230.3908	1911.5898	0.1205	0.0005	8.1241	48
49	0.0039	8.3010	67.5929	258.0377	2141.9806	0.1205	0.0005	8.1427	49
50	0.0035	8.3045	67.7624	289.0022	2400.0182	0.1204	0.0004	8.1597	50
51	0.0031	8.3076	67.9169	323.6825	2689.0204	0.1204	0.0004	8.1753	51
52	0.0028	8.3103	68.0576	362.5243	3012.7029	0.1203	0.0003	8.1895	52
53	0.0025	8.3128	68.1856	406.0273	3375.2272	0.1203	0.0003	8.2025	53
54	0.0022	8.3150	68.3022	454.7505	3781.2545	0.1203	0.0003	8.2143	54
55	0.0020	8.3170	68.4082	509.3206	4236.0050	0.1202	0.0002	8.2251	55
60	0.0011	8.3240	68.8100	897.5969	7471.6411	0.1201	0.0001	8.2664	60
65	0.0006	8.3281	69.0581	1581.8725	13173.9374	0.1201	0.0001	8.2922	65
70	0.0004	8.3303	69.2103	2787.7998	23223.3319	0.1200	0.0000	8.3082	70
75	0.0002	8.3316	69.3031	4913.0558	40933.7987	0.1200	0.0000	8.3181	75
80	0.0001	8.3324	69.3594	8658.4831	72145.6925	0.1200	0.0000	8.3241	80
85	0.0001	8.3328	69.3935	15259.2057	127151.7140	0.1200	0.0000	8.3278	85
90	0.0000	8.3330	69.4140	26891.9342	224091.1185	0.1200	0.0000	8.3300	90
95	0.0000	8.3332	69.4263	47392.7766	394931.4719	0.1200	0.0000	8.3313	95
100	0.0000	8.3332	69.4336	83522.2657	696010.5477	0.1200	0.0000	8.3321	100

APPENDIX 87.B *(continued)*
Cash Flow Equivalent Factors

$i = 15.00\%$

n	P/F	P/A	P/G	F/P	F/A	A/P	A/F	A/G	n
1	0.8696	0.8696	0.0000	1.1500	1.0000	1.1500	1.0000	0.0000	1
2	0.7561	1.6257	0.7561	1.3225	2.1500	0.6151	0.4651	0.4651	2
3	0.6575	2.2832	2.0712	1.5209	3.4725	0.4380	0.2880	0.9071	3
4	0.5718	2.8550	3.7864	1.7490	4.9934	0.3503	0.2003	1.3263	4
5	0.4972	3.3522	5.7751	2.0114	6.7424	0.2983	0.1483	1.7228	5
6	0.4323	3.7845	7.9368	2.3131	8.7537	0.2642	0.1142	2.0972	6
7	0.3759	4.1604	10.1924	2.6600	11.0668	0.2404	0.0904	2.4498	7
8	0.3269	4.4873	12.4807	3.0590	13.7268	0.2229	0.0729	2.7813	8
9	0.2843	4.7716	14.7548	3.5179	16.7858	0.2096	0.0596	3.0922	9
10	0.2472	5.0188	16.9795	4.0456	20.3037	0.1993	0.0493	3.3832	10
11	0.2149	5.2337	19.1289	4.6524	24.3493	0.1911	0.0411	3.6549	11
12	0.1869	5.4206	21.1849	5.3503	29.0017	0.1845	0.0345	3.9082	12
13	0.1625	5.5831	23.1352	6.1528	34.3519	0.1791	0.0291	4.1438	13
14	0.1413	5.7245	24.9725	7.0757	40.5047	0.1747	0.0247	4.3624	14
15	0.1229	5.8474	26.9630	8.1371	47.5804	0.1710	0.0210	4.5650	15
16	0.1069	5.9542	28.2960	9.3576	55.7175	0.1679	0.0179	4.7522	16
17	0.0929	6.0472	29.7828	10.7613	65.0751	0.1654	0.0154	4.9251	17
18	0.0808	6.1280	31.1565	12.3755	75.8364	0.1632	0.0132	5.0843	18
19	0.0703	6.1982	32.4213	14.2318	88.2118	0.1613	0.0113	5.2307	19
20	0.0611	6.2593	33.5822	16.3665	102.4436	0.1598	0.0098	5.3651	20
21	0.0531	6.3125	34.6448	18.8215	118.8101	0.1584	0.0084	5.4883	21
22	0.0462	6.3587	35.6150	21.6447	137.6316	0.1573	0.0073	5.6010	22
23	0.0402	6.3988	36.4988	24.8915	159.2764	0.1563	0.0063	5.7040	23
24	0.0349	6.4338	37.3023	28.6252	184.1678	0.1554	0.0054	5.7979	24
25	0.0304	6.4641	38.0314	32.9190	212.7930	0.1547	0.0047	5.8834	25
26	0.0264	6.4906	38.6918	37.8568	245.7120	0.1541	0.0041	5.9612	26
27	0.0230	6.5135	39.2890	43.5353	283.5688	0.1535	0.0035	6.0319	27
28	0.0200	6.5335	39.8283	50.0656	327.1041	0.1531	0.0031	6.0960	28
29	0.0174	6.5509	40.3146	57.5755	377.1697	0.1527	0.0027	6.1541	29
30	0.0151	6.5660	40.7526	66.2118	434.7451	0.1523	0.0023	6.2066	30
31	0.0131	6.5791	41.1466	76.1435	500.9569	0.1520	0.0020	6.2541	31
32	0.0114	6.5905	41.5006	87.5651	577.1005	0.1517	0.0017	6.2970	32
33	0.0099	6.6005	41.8184	100.6998	664.6655	0.1515	0.0015	6.3357	33
34	0.0086	6.6091	42.1033	115.8048	765.3654	0.1513	0.0013	6.3705	34
35	0.0075	6.6166	42.3586	133.1755	881.1702	0.1511	0.0011	6.4019	35
36	0.0065	6.6231	42.5872	153.1519	1014.3457	0.1510	0.0010	6.4301	36
37	0.0057	6.6288	42.7916	176.1246	1167.4975	0.1509	0.0009	6.4554	37
38	0.0049	6.6338	42.9743	202.5433	1343.6222	0.1507	0.0007	6.4781	38
39	0.0043	6.6380	43.1374	232.9248	1546.1655	0.1506	0.0006	6.4985	39
40	0.0037	6.6418	43.2830	267.8635	1779.0903	0.1506	0.0006	6.5168	40
41	0.0032	6.6450	43.4128	308.0431	2046.9539	0.1505	0.0005	6.5331	41
42	0.0028	6.6478	43.5286	354.2495	2354.9969	0.1504	0.0004	6.5478	42
43	0.0025	6.6503	43.6317	407.3870	2709.2465	0.1504	0.0004	6.5609	43
44	0.0021	6.6524	43.7235	468.4950	3116.6334	0.1503	0.0003	6.5725	44
45	0.0019	6.6543	43.8051	538.7693	3585.1285	0.1503	0.0003	6.5830	45
46	0.0016	6.6559	43.8778	619.5847	4123.8977	0.1502	0.0002	6.5923	46
47	0.0014	6.6573	43.9423	712.5224	4743.4824	0.1502	0.0002	6.6006	47
48	0.0012	6.6585	43.9997	819.4007	5456.0047	0.1502	0.0002	6.6080	48
49	0.0011	6.6596	44.0506	942.3108	6275.4055	0.1502	0.0002	6.6146	49
50	0.0009	6.6605	44.0958	1083.6574	7217.7163	0.1501	0.0001	6.6205	50
51	0.0008	6.6613	44.1360	1246.2061	8301.3737	0.1501	0.0001	6.6257	51
52	0.0007	6.6620	44.1715	1433.1370	9547.5798	0.1501	0.0001	6.6304	52
53	0.0006	6.6626	44.2031	1648.1075	10980.7167	0.1501	0.0001	6.6345	53
54	0.0005	6.6631	44.2311	1895.3236	12628.8243	0.1501	0.0001	6.6382	54
55	0.0005	6.6636	44.2558	2179.6222	14524.1479	0.1501	0.0001	6.6414	55
60	0.0002	6.6651	44.3431	4383.9987	29219.9916	0.1500	0.0000	6.6530	60
65	0.0001	6.6659	44.3903	8817.7874	58778.5826	0.1500	0.0000	6.6593	65
70	0.0001	6.6663	44.4156	17735.7200	118231.4669	0.1500	0.0000	6.6627	70
75	0.0000	6.6665	44.4292	35672.8680	237812.4532	0.1500	0.0000	6.6646	75
80	0.0000	6.6666	44.4364	71750.8794	478332.5293	0.1500	0.0000	6.6656	80
85	0.0000	6.6666	44.4402	144316.6470	962104.3133	0.1500	0.0000	6.6661	85
90	0.0000	6.6666	44.4422	290272.3252	1935142.1680	0.1500	0.0000	6.6664	90
95	0.0000	6.6667	44.4433	583841.3276	3892268.8509	0.1500	0.0000	6.6665	95
100	0.0000	6.6667	44.4438	1174313.4507	7828749.6713	0.1500	0.0000	6.6666	100

Appendices

APPENDIX 87.B *(continued)*
Cash Flow Equivalent Factors

$i = 20.00\%$

n	P/F	P/A	P/G	F/P	F/A	A/P	A/F	A/G	n
1	0.8333	0.8333	0.0000	1.2000	1.0000	1.2000	1.0000	0.0000	1
2	0.6944	1.5278	0.6944	1.4400	2.2000	0.6545	0.4545	0.4545	2
3	0.5787	2.1065	1.8519	1.7280	3.6400	0.4747	0.2747	0.8791	3
4	0.4823	2.5887	3.2986	2.0736	5.3680	0.3863	0.1863	1.2742	4
5	0.4019	2.9906	4.9061	2.4883	7.4416	0.3344	0.1344	1.6405	5
6	0.3349	3.3255	6.5806	2.9860	9.9299	0.3007	0.1007	1.9788	6
7	0.2791	3.6046	8.2551	3.5832	12.9159	0.2774	0.0774	2.2902	7
8	0.2326	3.8372	9.8831	4.2998	16.4991	0.2606	0.0606	2.5756	8
9	0.1938	4.0310	11.4335	5.1598	20.7989	0.2481	0.0481	2.8364	9
10	0.1615	4.1925	12.8871	6.1917	25.9587	0.2385	0.0385	3.0739	10
11	0.1346	4.3271	14.2330	7.4301	32.1504	0.2311	0.0311	3.2893	11
12	0.1122	4.4392	15.4667	8.9161	39.5805	0.2253	0.0253	3.4841	12
13	0.0935	4.5327	16.5883	10.6993	48.4966	0.2206	0.0206	3.6597	13
14	0.0779	4.6106	17.6008	12.8392	59.1959	0.2169	0.0169	3.8175	14
15	0.0649	4.6755	18.5095	15.4070	72.0351	0.2139	0.0139	3.9588	15
16	0.0541	4.7296	19.3208	18.4884	87.4421	0.2114	0.0114	4.0851	16
17	0.0451	4.7746	20.0419	22.1861	105.9306	0.2094	0.0094	4.1976	17
18	0.0376	4.8122	20.6805	26.6233	128.1167	0.2078	0.0078	4.2975	18
19	0.0313	4.8435	21.2439	31.9480	154.7400	0.2065	0.0065	4.3861	19
20	0.0261	4.8696	21.7395	38.3376	186.6880	0.2054	0.0054	4.4643	20
21	0.0217	4.8913	22.1742	46.0051	225.0256	0.2044	0.0044	4.5334	21
22	0.0181	4.9094	22.5546	55.2061	271.0307	0.2037	0.0037	4.5941	22
23	0.0151	4.9245	22.8867	66.2474	326.2369	0.2031	0.0031	4.6475	23
24	0.0126	4.9371	23.1760	79.4968	392.4842	0.2025	0.0025	4.6943	24
25	0.0105	4.9476	23.4276	95.3962	471.9811	0.2021	0.0021	4.7352	25
26	0.0087	4.9563	23.6460	114.4755	567.3773	0.2018	0.0018	4.7709	26
27	0.0073	4.9636	23.8353	137.3706	681.8528	0.2015	0.0015	4.8020	27
28	0.0061	4.9697	23.9991	164.8447	819.2233	0.2012	0.0012	4.8291	28
29	0.0051	4.9747	24.1406	197.8136	984.0680	0.2010	0.0010	4.8527	29
30	0.0042	4.9789	24.2628	237.3763	1181.8816	0.2008	0.0008	4.8731	30
31	0.0035	4.9824	24.3681	284.8516	1419.2579	0.2007	0.0007	4.8908	31
32	0.0029	4.9854	24.4588	341.8219	1704.1095	0.2006	0.0006	4.9061	32
33	0.0024	4.9878	24.5368	410.1863	2045.9314	0.2005	0.0005	4.9194	33
34	0.0020	4.9898	24.6038	492.2235	2456.1176	0.2004	0.0004	4.9308	34
35	0.0017	4.9915	24.6614	590.6682	2948.3411	0.2003	0.0003	4.9406	35
36	0.0014	4.9929	24.7108	708.8019	3539.0094	0.2003	0.0003	4.9491	36
37	0.0012	4.9941	24.7531	850.5622	4247.8112	0.2002	0.0002	4.9564	37
38	0.0010	4.9951	24.7894	1020.6747	5098.3735	0.2002	0.0002	4.9627	38
39	0.0008	4.9959	24.8204	1224.8096	6119.0482	0.2002	0.0002	4.9681	39
40	0.0007	4.9966	24.8469	1469.7716	7343.8578	0.2001	0.0001	4.9728	40
41	0.0006	4.9972	24.8696	1763.7259	8813.6294	0.2001	0.0001	4.9767	41
42	0.0005	4.9976	24.8890	2116.4711	10577.3553	0.2001	0.0001	4.9801	42
43	0.0004	4.9980	24.9055	2539.7653	12693.8263	0.2001	0.0001	4.9831	43
44	0.0003	4.9984	24.9196	3047.7183	15233.5916	0.2001	0.0001	4.9856	44
45	0.0003	4.9986	24.9316	3657.2620	18281.3099	0.2001	0.0001	4.9877	45
46	0.0002	4.9989	24.9419	4388.7144	21938.5719	0.2000	0.0000	4.9895	46
47	0.0002	4.9991	24.9506	5266.4573	26327.2863	0.2000	0.0000	4.9911	47
48	0.0002	4.9992	24.9581	6319.7487	31593.7436	0.2000	0.0000	4.9924	48
49	0.0001	4.9993	24.9644	7583.6985	37913.4923	0.2000	0.0000	4.9935	49
50	0.0001	4.9995	24.9698	9100.4382	45497.1908	0.2000	0.0000	4.9945	50
51	0.0001	4.9995	24.9744	10920.5258	54597.6289	0.2000	0.0000	4.9953	51
52	0.0001	4.9996	24.9783	13104.6309	65518.1547	0.2000	0.0000	4.9960	52
53	0.0001	4.9997	24.9816	15725.5571	78622.7856	0.2000	0.0000	4.9966	53
54	0.0001	4.9997	24.9844	18870.6685	94348.3427	0.2000	0.0000	4.9971	54
55	0.0000	4.9998	24.9868	22644.8023	113219.0113	0.2000	0.0000	4.9976	55
60	0.0000	4.9999	24.9942	56347.5144	281732.5718	0.2000	0.0000	4.9989	60
65	0.0000	5.0000	24.9975	140210.6469	701048.2346	0.2000	0.0000	4.9995	65
70	0.0000	5.0000	24.9989	348888.9569	1744439.7847	0.2000	0.0000	4.9998	70
75	0.0000	5.0000	24.9995	868147.3693	4340731.8466	0.2000	0.0000	4.9999	75

APPENDIX 87.B *(continued)*
Cash Flow Equivalent Factors

$i = 25.00\%$

n	P/F	P/A	P/G	F/P	F/A	A/P	A/F	A/G	n
1	0.8000	0.8000	0.0000	1.2500	1.0000	1.2500	1.0000	0.0000	1
2	0.6400	1.4400	0.6400	1.5625	2.2500	0.6944	0.0444	0.4444	2
3	0.5120	1.9520	1.6640	1.9531	3.8125	0.5123	0.2623	0.8525	3
4	0.4096	2.3616	2.8928	2.4414	5.7656	0.4234	0.1734	1.2249	4
5	0.3277	2.6893	4.2035	3.0518	8.2070	0.3718	0.1218	1.5631	5
6	0.2621	2.9514	5.5142	3.8147	11.2588	0.3383	0.0888	1.8683	6
7	0.2097	3.1611	6.7725	4.7684	15.0735	0.3163	0.0663	2.1424	7
8	0.1678	3.3289	7.9469	5.9605	19.8419	0.3004	0.0504	2.3872	8
9	0.1342	3.4631	9.0207	7.4506	25.8023	0.2888	0.0388	2.6048	9
10	0.1074	3.5705	9.9870	9.3132	33.2529	0.2801	0.0301	2.7971	10
11	0.0859	3.6564	10.8460	11.6415	42.5661	0.2735	0.0235	2.9663	11
12	0.0687	3.7251	11.6020	14.5519	54.2077	0.2684	0.0184	3.1145	12
13	0.0550	3.7801	12.2617	18.1899	68.7596	0.2645	0.0145	3.2437	13
14	0.0440	3.8241	12.8334	22.7374	86.9495	0.2615	0.0115	3.3559	14
15	0.0352	3.8593	13.3260	28.4217	109.6868	0.2591	0.0091	3.4530	15
16	0.0281	3.8874	13.7482	35.5271	138.1085	0.2572	0.0072	3.5366	16
17	0.0225	3.9099	14.1085	44.4089	173.6357	0.2558	0.0058	3.6084	17
18	0.0180	3.9279	14.4147	55.5112	218.0446	0.2546	0.0046	3.6698	18
19	0.0144	3.9424	14.6741	69.3889	273.5558	0.2537	0.0037	3.7222	19
20	0.0115	3.9539	14.8932	86.7362	342.9447	0.2529	0.0029	3.7667	20
21	0.0092	3.9631	15.0777	108.4202	429.6809	0.2523	0.0023	3.8045	21
22	0.0074	3.9705	15.2326	135.5253	538.1011	0.2519	0.0019	3.8365	22
23	0.0059	3.9764	15.3625	169.4066	673.6264	0.2515	0.0015	3.8634	23
24	0.0047	3.9811	15.4711	211.7582	843.0329	0.2512	0.0012	3.8861	24
25	0.0038	3.9849	15.5618	264.6978	1054.7912	0.2509	0.0009	3.9052	25
26	0.0030	3.9879	15.6373	330.8722	1319.4890	0.2508	0.0008	3.9212	26
27	0.0024	3.9903	15.7002	413.5903	1650.3612	0.2506	0.0006	3.9346	27
28	0.0019	3.9923	15.7524	516.9879	2063.9515	0.2505	0.0005	3.9457	28
29	0.0015	3.9938	15.7957	646.2349	2580.9394	0.2504	0.0004	3.9551	29
30	0.0012	3.9950	15.8316	807.7936	3227.1743	0.2503	0.0003	3.9628	30
31	0.0010	3.9960	15.8614	1009.7420	4034.9678	0.2502	0.0002	3.9693	31
32	0.0008	3.9968	15.8859	1262.1774	5044.7098	0.2502	0.0002	3.9746	32
33	0.0006	3.9975	15.9062	1577.7218	6306.8872	0.2502	0.0002	3.9791	33
34	0.0005	3.9980	15.9229	1972.1523	7884.6091	0.2501	0.0001	3.9828	34
35	0.0004	3.9984	15.9367	2465.1903	9856.7613	0.2501	0.0001	3.9858	35
36	0.0003	3.9987	15.9481	3081.4879	12321.9516	0.2501	0.0001	3.9883	36
37	0.0003	3.9990	15.9574	3851.8599	15403.4396	0.2501	0.0001	3.9904	37
38	0.0002	3.9992	15.9651	4814.8249	19255.2994	0.2501	0.0001	3.9921	38
39	0.0002	3.9993	15.9714	6018.5311	24070.1243	0.2500	0.0000	3.9935	39
40	0.0001	3.9995	15.9766	7523.1638	30088.6554	0.2500	0.0000	3.9947	40
41	0.0001	3.9996	15.9809	9403.9548	37611.8192	0.2500	0.0000	3.9956	41
42	0.0001	3.9997	15.9843	11754.9435	47015.7740	0.2500	0.0000	3.9964	42
43	0.0001	3.9997	15.9872	14693.6794	58770.7175	0.2500	0.0000	3.9971	43
44	0.0001	3.9998	15.9895	18367.0992	73464.3969	0.2500	0.0000	3.9976	44
45	0.0000	3.9998	15.9915	22958.8740	91831.4962	0.2500	0.0000	3.9980	45
46	0.0000	3.9999	15.9930	28698.5925	114790.3702	0.2500	0.0000	3.9984	46
47	0.0000	3.9999	15.9943	35873.2407	143488.9627	0.2500	0.0000	3.9987	47
48	0.0000	3.9999	15.9954	44841.5509	179362.2034	0.2500	0.0000	3.9989	48
49	0.0000	3.9999	15.9962	56051.9386	224203.7543	0.2500	0.0000	3.9991	49
50	0.0000	3.9999	15.9969	70064.9232	280255.6929	0.2500	0.0000	3.9993	50
51	0.0000	4.0000	15.9975	87581.1540	350320.6161	0.2500	0.0000	3.9994	51
52	0.0000	4.0000	15.9980	109476.4425	437901.7701	0.2500	0.0000	3.9995	52
53	0.0000	4.0000	15.9983	136845.5532	547378.2126	0.2500	0.0000	3.9996	53
54	0.0000	4.0000	15.9986	171056.9414	684223.7658	0.2500	0.0000	3.9997	54
55	0.0000	4.0000	15.9989	213821.1768	855280.7072	0.2500	0.0000	3.9997	55
60	0.0000	4.0000	15.9996	652530.4468	2610117.7872	0.2500	0.0000	3.9999	60

APPENDIX 87.B *(continued)*
Cash Flow Equivalent Factors

$i = 30.00\%$

n	P/F	P/A	P/G	F/P	F/A	A/P	A/F	A/G	n
1	0.7692	0.7692	0.0000	1.3000	1.0000	1.3000	1.0000	0.000	1
2	0.5917	1.3609	0.5917	1.6900	2.3000	0.7348	0.4348	0.434	2
3	0.4552	1.8161	1.5020	2.1970	3.9900	0.5506	0.2506	0.827	3
4	0.3501	2.1662	2.5524	2.8561	6.1870	0.4616	0.1616	1.178	4
5	0.2693	2.4356	3.6297	3.7129	9.0431	0.4106	0.1106	1.490	5
6	0.2072	2.6427	4.6656	4.8268	12.7560	0.3784	0.0784	1.765	6
7	0.1594	2.8021	5.6218	6.2749	17.5828	0.3569	0.0569	2.006	7
8	0.1226	2.9247	6.4800	8.1573	23.8577	0.3419	0.0419	2.215	8
9	0.0943	3.0190	7.2343	10.6045	32.0150	0.3312	0.0312	2.396	9
10	0.0725	3.0915	7.8872	13.7858	42.6195	0.3235	0.0235	2.551	10
11	0.0558	3.1473	8.4452	17.9216	56.4053	0.3177	0.0177	2.683	11
12	0.0429	3.1903	8.9173	23.2981	74.3270	0.3135	0.0135	2.795	12
13	0.0330	3.2233	9.3135	30.2875	97.6250	0.3102	0.0102	2.889	13
14	0.0254	3.2487	9.6437	39.3738	127.9125	0.3078	0.0078	2.968	14
15	0.0195	3.2682	9.9172	51.1859	167.2863	0.3060	0.0060	3.034	15
16	0.0150	3.2832	10.1426	66.5417	218.4722	0.3046	0.0046	3.089	16
17	0.0116	3.2948	10.3276	86.5042	285.0139	0.3035	0.0035	3.134	17
18	0.0089	3.3037	10.4788	112.4554	371.5180	0.3027	0.0027	3.171	18
19	0.0068	3.3105	10.6019	146.1920	483.9734	0.3021	0.0021	3.202	19
20	0.0053	3.3158	10.7019	190.0496	630.1655	0.3016	0.0016	3.227	20
21	0.0040	3.3198	10.7828	247.0645	820.2151	0.3012	0.0012	3.248	21
22	0.0031	3.3230	10.8482	321.1839	1067.2796	0.3009	0.0009	3.264	22
23	0.0024	3.3254	10.9009	417.5391	1388.4635	0.3007	0.0007	3.278	23
24	0.0018	3.3272	10.9433	542.8008	1806.0026	0.3006	0.0006	3.289	24
25	0.0014	3.3286	10.9773	705.6410	2348.8033	0.3004	0.0004	3.297	25
26	0.0011	3.3297	11.0045	917.3333	3054.4443	0.3003	0.0003	3.305	26
27	0.0008	3.3305	11.0263	1192.5333	3971.7776	0.3003	0.0003	3.310	27
28	0.0006	3.3312	11.0437	1550.2933	5164.3109	0.3002	0.0002	3.315	28
29	0.0005	3.3317	11.0576	2015.3813	6714.6042	0.3001	0.0001	3.318	29
30	0.0004	3.3321	11.0687	2619.9956	8729.9855	0.3001	0.0001	3.321	30
31	0.0003	3.3324	11.0775	3405.9943	11349.9811	0.3001	0.0001	3.324	31
32	0.0002	3.3326	11.0845	4427.7926	14755.9755	0.3001	0.0001	3.326	32
33	0.0002	3.3328	11.0901	5756.1304	19183.7681	0.3001	0.0001	3.327	33
34	0.0001	3.3329	11.0945	7482.9696	24939.8985	0.3000	0.0000	3.328	34
35	0.0001	3.3330	11.0980	9727.8604	32422.8681	0.3000	0.0000	3.329	35
36	0.0001	3.3331	11.1007	12646.2186	42150.7285	0.3000	0.0000	3.330	36
37	0.0001	3.3331	11.1029	16440.0841	54796.9471	0.3000	0.0000	3.331	37
38	0.0000	3.3332	11.1047	21372.1094	71237.0312	0.3000	0.0000	3.331	38
39	0.0000	3.3332	11.1060	27783.7422	92609.1405	0.3000	0.0000	3.331	39
40	0.0000	3.3332	11.1071	36118.8648	120392.8827	0.3000	0.0000	3.332	40
41	0.0000	3.3333	11.1080	46954.5243	156511.7475	0.3000	0.0000	3.332	41
42	0.0000	3.3333	11.1086	61040.8815	203466.2718	0.3000	0.0000	3.332	42
43	0.0000	3.3333	11.1092	79353.1460	264507.1533	0.3000	0.0000	3.332	43
44	0.0000	3.3333	11.1096	103159.0898	343860.2993	0.3000	0.0000	3.332	44
45	0.0000	3.3333	11.1099	134106.8167	447019.3890	0.3000	0.0000	3.333	45
46	0.0000	3.3333	11.1102	174338.8617	581126.2058	0.3000	0.0000	3.333	46
47	0.0000	3.3333	11.1104	226640.5202	755465.0675	0.3000	0.0000	3.333	47
48	0.0000	3.3333	11.1105	294632.6763	982105.5877	0.3000	0.0000	3.333	48
49	0.0000	3.3333	11.1107	383022.4792	1276738.2640	0.3000	0.0000	3.333	49
50	0.0000	3.3333	11.1108	497929.2230	1659760.7433	0.3000	0.0000	3.333	50

APPENDIX 87.B (continued)
Cash Flow Equivalent Factors

$i = 40.00\%$

n	P/F	P/A	P/G	F/P	F/A	A/P	A/F	A/G	n
1	0.7143	0.7143	0.0000	1.4000	1.0000	1.4000	1.0000	0.000	1
2	0.5102	1.2245	0.5102	1.9600	2.4000	0.8167	0.4167	0.416	2
3	0.3644	1.5889	1.2391	2.7440	4.3600	0.6294	0.2294	0.779	3
4	0.2603	1.8492	2.0200	3.8416	7.1040	0.5408	0.1408	1.092	4
5	0.1859	2.0352	2.7637	5.3782	10.9456	0.4914	0.0914	1.358	5
6	0.1328	2.1680	3.4278	7.5295	16.3238	0.4613	0.0613	1.581	6
7	0.0949	2.2628	3.9970	10.5414	23.8534	0.4419	0.0419	1.766	7
8	0.0678	2.3306	4.4713	14.7579	34.3947	0.4291	0.0291	1.918	8
9	0.0484	2.3790	4.8585	20.6610	49.1526	0.4203	0.0203	2.042	9
10	0.0346	2.4136	5.1696	28.9255	69.8137	0.4143	0.0143	2.141	10
11	0.0247	2.4383	5.4166	40.4957	98.7391	0.4101	0.0101	2.221	11
12	0.0176	2.4559	5.6106	56.6939	139.2348	0.4072	0.0072	2.284	12
13	0.0126	2.4685	5.7618	79.3715	195.9287	0.4051	0.0051	2.334	13
14	0.0090	2.4775	5.8788	111.1201	275.3002	0.4036	0.0036	2.372	14
15	0.0064	2.4839	5.9688	155.5681	386.4202	0.4026	0.0026	2.403	15
16	0.0046	2.4885	6.0376	217.7953	541.9883	0.4018	0.0018	2.426	16
17	0.0033	2.4918	6.0901	304.9135	759.7837	0.4013	0.0013	2.444	17
18	0.0023	2.4941	6.1299	426.8789	1064.6971	0.4009	0.0009	2.457	18
19	0.0017	2.4958	6.1601	597.6304	1491.5760	0.4007	0.0007	2.468	19
20	0.0012	2.4970	6.1828	836.6826	2089.2064	0.4005	0.0005	2.476	20
21	0.0009	2.4979	6.1998	1171.3556	2925.8889	0.4003	0.0003	2.482	21
22	0.0006	2.4985	6.2127	1639.8978	4097.2445	0.4002	0.0002	2.486	22
23	0.0004	2.4989	6.2222	2295.8569	5737.1423	0.4002	0.0002	2.490	23
24	0.0003	2.4992	6.2294	3214.1997	8032.9993	0.4001	0.0001	2.492	24
25	0.0002	2.4994	6.2347	4499.8796	11247.1990	0.4001	0.0001	2.494	25
26	0.0002	2.4996	6.2387	6299.8314	15747.0785	0.4001	0.0001	2.495	26
27	0.0001	2.4997	6.2416	8819.7640	22046.9099	0.4000	0.0000	2.496	27
28	0.0001	2.4998	6.2438	12347.6696	30866.6739	0.4000	0.0000	2.497	28
29	0.0001	2.4999	6.2454	17286.7374	43214.3435	0.4000	0.0000	2.498	29
30	0.0000	2.4999	6.2466	24201.4324	60501.0809	0.4000	0.0000	2.498	30
31	0.0000	2.4999	6.2475	33882.0053	84702.5132	0.4000	0.0000	2.499	31
32	0.0000	2.4999	6.2482	47434.8074	118584.5185	0.4000	0.0000	2.499	32
33	0.0000	2.5000	6.2487	66408.7304	166019.3260	0.4000	0.0000	2.499	33
34	0.0000	2.5000	6.2490	92972.2225	232428.0563	0.4000	0.0000	2.499	34
35	0.0000	2.5000	6.2493	130161.1116	325400.2789	0.4000	0.0000	2.499	35
36	0.0000	2.5000	6.2495	182225.5562	455561.3904	0.4000	0.0000	2.499	36
37	0.0000	2.5000	6.2496	255115.7786	637786.9466	0.4000	0.0000	2.499	37
38	0.0000	2.5000	6.2497	357162.0901	892902.7252	0.4000	0.0000	2.499	38
39	0.0000	2.5000	6.2498	500026.9261	1250064.8153	0.4000	0.0000	2.499	39
40	0.0000	2.5000	6.2498	700037.6966	1750091.7415	0.4000	0.0000	2.499	40
41	0.0000	2.5000	6.2499	980052.7752	2450129.4381	0.4000	0.0000	2.500	41
42	0.0000	2.5000	6.2499	1372073.8853	3430182.2133	0.4000	0.0000	2.500	42
43	0.0000	2.5000	6.2499	1920903.4394	4802256.0986	0.4000	0.0000	2.500	43
44	0.0000	2.5000	6.2500	2689264.8152	6723159.5381	0.4000	0.0000	2.500	44
45	0.0000	2.5000	6.2500	3764970.7413	9412424.3533	0.4000	0.0000	2.500	45

Glossary

Glossary

A

AASHTO: American Association of State and Highway Transportation Officials.

Abandonment: The reversion of title to the owner of the underlying fee where an easement for highway purposes is no longer needed.

Absorbed dose: The energy deposited by radiation as it passes through a material.

Absorption: The process by which a liquid is drawn into and tends to fill permeable pores in a porous body. Also, the increase in weight of a porous solid body resulting from the penetration of liquid into its permeable pores.

Accelerated flow: A form of varied flow in which the velocity is increasing and the depth is decreasing.

Acid: Any compound that dissociates in water into H^+ ions. (The combination of H^+ and water, H_3O^+, is known as the hydronium ion.) Acids conduct electricity in aqueous solutions, have a sour taste, turn blue litmus paper red, have a pH between 0 and 7, and neutralize bases, forming salts and water.

Active pressure: Pressure causing a wall to move away from the soil.

Adenosine triphosphate (ATP): The macromolecule that functions as an energy carrier in cells. The energy is stored in a high-energy bond between the second and third phosphates.

Adjudication: A court proceeding to determine rights to the use of water on a particular stream or aquifer.

Admixture: Material added to a concrete mixture to increase its workability, strength, or imperviousness, or to lower its freezing point.

Adsorbed water: Water held near the surface of a material by electrochemical forces.

Adsorption edge: The pH range where solute adsorption changes sharply.

Advection: The transport of solutes along stream lines at the average linear seepage flow velocity.

Aeration: Mixing water with air, either by spraying water or diffusing air through water.

Aerobe: A microorganism whose growth requires free oxygen.

Aerobic: Requiring oxygen. Descriptive of a bacterial class that functions in the presence of free dissolved oxygen.

Aggregate, coarse: Aggregate retained on a no. 4 sieve.

Aggregate, fine: Aggregate passing the no. 4 sieve and retained on the no. 200 sieve.

Aggregate, lightweight: Aggregate having a dry density of 70 lbm/ft^3 (32 kg/m^3) or less.

Agonic line: A line with no magnetic declination.

Air change (Air flush): A complete replacement of the air in a room or other closed space.

Algae: Simple photosynthetic plants having neither roots, stems, nor leaves.

Alidade: A tachometric instrument consisting of a telescope similar to a transit, an upright post that supports the standards of the horizontal axis of the telescope, and a straightedge whose edges are essentially in the same direction as the line of sight. An alidade is used in the field in conjunction with a plane table.

Alkalinity: A measure of the capacity of a water to neutralize acid without significant pH change. It is usually associated with the presence of hydroxyl, carbonate, and/or bicarbonate radicals in the water.

Alluvial deposit: A material deposited within the alluvium.

Alluvium: Sand, silt, clay, gravel, etc., deposited by running water.

Alternate depths: For a particular channel geometry and discharge, two depths at which water flows with the same specific energy in uniform flow. One depth corresponds to subcritical flow; the other corresponds to supercritical flow.

Amictic: Experiencing no overturns or mixing. Typical of polar lakes.

Amphoteric behavior: Ability of an aqueous complex or solid material to have a negative, neutral, or positive charge.

Anabolism: The phase of metabolism involving the formation of organic compounds; usually an energy-utilizing process.

Anabranch: The intertwining channels of a braided stream.

Anaerobe: A microorganism that grows only or best in the absence of free oxygen.

Anaerobic: Not requiring oxygen. Descriptive of a bacterial class that functions in the absence of free dissolved oxygen.

Anion: Negative ion that migrates to the positive electrode (anode) in an electrolytic solution.

Anticlinal spring: A portion of an exposed aquifer (usually on a slope) between two impervious layers.

Apparent specific gravity of asphalt mixture: A ratio of the unit weight of an asphalt mixture (excluding voids permeable to water) to the unit weight of water.

Appurtenance: That which belongs with or is designed to complement something else. For example, a manhole is a sewer appurtenance.

Apron: An underwater "floor" constructed along the channel bottom to prevent scour. Aprons are almost always extensions of spillways and culverts.

Aquiclude: A saturated geologic formation with insufficient porosity to support any significant removal rate or contribute to the overall groundwater regime. In groundwater analysis, an aquiclude is considered to confine an aquifer at its boundaries.

Aquifer: Rock or sediment in a formation, group of formations, or part of a formation that is saturated and sufficiently permeable to transmit economic quantities of water to wells and springs.

Aquifuge: An underground geological formation that has absolutely no porosity or interconnected openings through which water can enter or be removed.

Aquitard: A saturated geologic unit that is permeable enough to contribute to the regional groundwater flow regime, but not permeable enough to supply a water well or other economic use.

Arterial highway: A general term denoting a highway primarily for through traffic, usually on a continuous route.

Artesian formation: An aquifer in which the piezometric height is greater than the aquifer height. In an artesian formation, the aquifer is confined and the water is under hydrostatic pressure.

Artesian spring: Water from an artesian formation that flows to the ground surface naturally, under hydrostatic pressure, due to a crack or other opening in the formation's confining layer.

Asphalt emulsion: A mixture of asphalt cement with water. Asphalt emulsions are produced by adding a small amount of emulsifying soap to asphalt and water. The asphalt sets when the water evaporates.

Atomic mass unit (AMU): One AMU is one twelfth the atomic weight of carbon.

Atomic number: The number of protons in the nucleus of an atom.

Atomic weight: Approximately, the sum of the numbers of protons and neutrons in the nucleus of an atom.

Autotroph: An organism than can synthesize all of its organic components from inorganic sources.

Auxiliary lane: The portion of a roadway adjoining the traveled way for truck climbing, speed change, or other purposes supplementary to through traffic movement.

Avogadro's law: A gram-mole of any substance contains 6.022×10^{23} molecules.

Azimuth: The horizontal angle measured from the plane of the meridian to the vertical plane containing the line. The azimuth gives the direction of the line with respect to the meridian and is usually measured in a clockwise direction with respect to either the north or south meridian.

B

Backward pass: The steps in a critical path analysis in which the latest start times are determined, usually after the earliest finish times have been determined in the forward pass.

Backwater: Water upstream from a dam or other obstruction that is deeper than it would normally be without the obstruction.

Backwater curve: A plot of depth versus location along the channel containing backwater.

Base: (a) A layer of selected, processed, or treated aggregate material of planned thickness and quality placed immediately below the pavement and above the subbase or subgrade soil. (b) Any compound that dissociates in water into OH$^-$ ions. Bases conduct electricity in aqueous solutions, have a bitter taste, turn red litmus paper blue, have a pH between 7 and 14, and neutralize acids, forming salts and water.

Base course: The bottom portion of a pavement where the top and bottom portions are not the same composition.

Base flow: Component of stream discharge that comes from groundwater flow. Water infiltrates and moves through the ground very slowly; up to 2 years may elapse between precipitation and discharge.

Batter pile: A pile inclined from the vertical.

Bed: A layer of rock in the earth. Also the bottom of a body of water such as a river, lake, or sea.

Bell: An enlarged section at the base of a pile or pier used as an anchor.

Belt highway: An arterial highway carrying traffic partially or entirely around an urban area.

Bent: A supporting structure (usually of a bridge) consisting of a beam or girder transverse to the supported roadway and that is supported, in turn, by columns at each end, making an inverted "U" shape.

Benthic zone: The bottom zone of a lake, where oxygen levels are low.

Benthos: Organisms (typically anaerobic) occupying the benthic zone.

Bentonite: A volcanic clay that exhibits extremely large volume changes with moisture content changes.

Berm: A shelf, ledge, or pile.

Bifurcation ratio: The average number of streams feeding into the next side (order) waterway. The range is usually 2 to 4.

Binary fission: An asexual reproductive process in which one cell splits into two independent daughter cells.

Bioaccumulation: The process by which chemical substances are ingested and retained by organisms, either from the environment directly or through consumption of food containing the chemicals.

Bioaccumulation factor: *See* Bioconcentration factor.

Bioactivation process: A process using sedimentation, trickling filter, and secondary sedimentation before adding activated sludge. Aeration and final sedimentation are the follow-up processes.

Bioassay: The determination of kinds, quantities, or concentrations, and in some cases, the locations of material in the body, whether by direct measurement (in vivo counting) or by analysis and evaluation of materials excreted or removed (in vitro) from the body.

Bioavailability: A measure of what fraction, how much, or the rate that a substance (ingested, breathed in, dermal contact) is actually absorbed biologically. Bioavailability measurements are typically based on absorption into the blood or liver tissue.

Biochemical oxygen demand (BOD): The quantity of oxygen needed by microorganisms in a body of water to decompose the organic matter present. An index of water pollution.

Bioconcentration: The increase in concentration of a chemical in an organism resulting from tissue absorption (bioaccumulation) levels exceeding the rate of metabolism and excretion (biomagnification).

Bioconcentration factor: The ratio of chemical concentration in an organism to chemical concentration in the surrounding environment.

Biodegradation: The use of microorganisms to degrade contaminants.

Biogas: A mixture of approximately 55% methane and 45% carbon dioxide that results from the digestion of animal dung.

Biomagnification: The cumulative increase in the concentration of a persistent substance in successively higher levels of the food chain.

Biomagnification factor: *See* Bioconcentration factor.

Biomass: Renewable organic plant and animal material such as plant residue, sawdust, tree trimmings, rice straw, poultry litter and other animal wastes, some industrial wastes, and the paper component of municipal solid waste that can be converted to energy.

Biosorption process: A process that mixes raw sewage and sludge that have been pre-aerated in a separate tank.

Biosphere: The part of the world in which life exists.

Biota: All of the species of plants and animals indigenous to an area.

Bleeding: A form of segregation in which some of the water in the mix tends to rise to the surface of freshly placed concrete.

Blind drainage: Geographically large (with respect to the drainage basin) depressions that store water during a storm and therefore stop it from contributing to surface runoff.

Bloom: A phenomenon whereby excessive nutrients within a body of water results in an explosion of plant life, resulting in a depletion of oxygen and fish kill. Usually caused by urban runoff containing fertilizers.

Bluff: A high and steep bank or cliff.

Body burden: The total amount of a particular chemical in the body.

Braided stream: A wide, shallow stream with many anabranches.

Branch sewer: A sewer off the main sewer.

Breaking chain: A technique used when the slope is too steep to permit bringing the full length of the chain or tape to a horizontal position. When breaking chain, the distance is measured in partial tape lengths.

Breakpoint chlorination: Application of chlorine that results in a minimum of chloramine residuals. No significant free chlorine residual is produced unless the breakpoint is reached.

Bulking: *See* Sludge bulking.

Bulk specific gravity of asphalt mixture: Ratio of the unit weight of an asphalt mixture (including permeable and impermeable voids) to the unit weight of water.

Butte: A hill with steep sides that usually stands away from other hills.

C

Caisson: An air- and watertight chamber used as a foundation and/or used to work or excavate below the water level.

Capillary water: Water just above the water table that is drawn up out of an aquifer due to capillary action of the soil.

Carbonaceous demand: Oxygen demand due to biological activity in a water sample, exclusive of nitrogenous demand.

Carbonate hardness: Hardness associated with the presence of bicarbonate radicals in the water.

Carcinogen: A cancer-causing agent.

Carrier (biological): An individual harboring a disease agent without apparent symptoms.

Cascade impactor: *See* Impactor.

Cased hole: An excavation whose sides are lined or sheeted.

Catabolism: The chemical reactions by which food materials or nutrients are converted into simpler substances for the production of energy and cell materials.

Catena: A group of soils of similar origin occurring in the same general locale but that differ slightly in properties.

Cation: Positive ion that migrates to the negative electrode (cathode) in an electrolytic solution.

Cell wall: The cell structure exterior to the cell membrane of plants, algae, bacteria, and fungi. It gives cells form and shape.

Cement-treated base: A base layer constructed with good-quality, well-graded aggregate mixed with up to 6% cement.

CFR: The Code of (U.S.) Federal Regulations, a compilation of all federal documents that have general applicability and legal effect, as published by the Office of the Federal Register.

Channelization: The separation or regulation of conflicting traffic movements into definite paths of travel by use of pavement markings, raised islands, or other means.

Chat: Small pieces of crushed rock and gravel. May be used for paving roads and roofs.

Check: A short section of built-up channel placed in a canal or irrigation ditch and provided with gates or flashboards to control flow or raise upstream level for diversion.

Chelate: A chemical compound into which a metallic ion (usually divalent) is tightly bound.

Chelation: (a) The process in which a compound or organic material attracts, combines with, and removes a metallic ion. (b) The process of removing metallic contaminants by having them combine with special added substances.

Chemical precipitation: Settling out of suspended solids caused by adding coagulating chemicals.

Chemocline: A steep chemical (saline) gradient separating layers in meromictic lakes.

Chloramine: Compounds of chlorine and ammonia (e.g., NH_2Cl, $NHCl_2$, or NCl_3).

Chlorine demand: The difference between applied chlorine and the chlorine residual. Chlorine demand is chlorine that has been reduced in chemical reactions and is no longer available for disinfection.

Class: A major taxonomic subdivision of a phylum. Each class is composed of one or more related orders.

Clean-out: A pipe through which snakes can be pushed to unplug a sewer.

Clearance: Distance between successive vehicles as measured between the vehicles, back bumper to front bumper.

Coliform: Gram-negative, lactose-fermenting rods, including escherichieae coli and similar species that normally inhabit the colon (large intestine). Commonly included in the coliform are Enterobacteria aerogenes, Klebsiella species, and other related bacteria.

Colloid: A fine particle ranging in size from 1 to 500 millimicrons. Colloids cause turbidity because they do not easily settle out.

Combined residuals: Compounds of an additive (such as chlorine) that have combined with something else. Chloramines are examples of combined residuals.

Combined system: A system using a single sewer for disposal of domestic waste and storm water.

Comminutor: A device that cuts solid waste into small pieces.

Compaction: Densification of soil by mechanical means, involving the expulsion of excess air.

Compensation level: In a lake, the depth of the limnetic area where oxygen production from light and photosynthesis are exactly balanced by depletion.

Complete mixing: Mixing accomplished by mechanical means (stirring).

Compound: A homogeneous substance composed of two or more elements that can be decomposed by chemical means only.

Concrete: A mixture of portland cement, fine aggregate, coarse aggregate, and water, with or without admixtures.

Concrete, normal weight: Concrete having a hardened density of approximately 150 lbm/ft^3.

Concrete, plain: Concrete that is not reinforced with steel.

Concrete, structural lightweight: A concrete containing lightweight aggregate.

Condemnation: The process by which property is acquired for public purposes through legal proceedings under power of eminent domain.

Cone of depression: The shape of the water table around a well during and immediately after use. The cone's water surface level differs from the original water table by the well's drawdown.

Confined water: Artesian water overlaid with an impervious layer, usually under pressure.

Confirmed test: A follow-up test used if the presumptive test for coliforms is positive.

Conflagration: Total involvement or engulfment (as in a fire).

Conjugate depths: The depths on either side of a hydraulic jump.

Connate water: Pressurized water (usually, high in mineral content) trapped in the pore spaces of sedimentary rock at the time it was formed.

Consolidation: Densification of soil by mechanical means, involving expulsion of excess water.

Contraction: A decrease in the width or depth of flow caused by the geometry of a weir, orifice, or obstruction.

Control of access: The condition where the right of owners or occupants of abutting land or other persons to access in connection with a highway is fully or partially controlled by public authority.

Critical depth: The depth that minimizes the specific energy of flow.

Critical flow: Flow at the critical depth and velocity. Critical flow minimizes the specific energy and maximizes discharge.

Critical slope: The slope that produces critical flow.

Critical velocity: The velocity that minimizes specific energy. When water is moving at its critical velocity, a disturbance wave cannot move upstream since the wave moves at the critical velocity.

Cuesta: (From the Spanish word for cliff); a hill with a steep slope on one side and a gentle slope on the other.

Cunette: A small channel in the invert of a large combined sewer for dry weather flow.

Curing: The process and procedures used for promoting the hydration of cement. It consists of controlling the temperature and moisture from and into the concrete.

Cyclone impactor: *See* Impactor.

D

Dead load: An inert, inactive load, primarily due to the structure's own weight.

Decision sight distance: Sight distance allowing for additional decision time in cases of complex conditions.

Delta: A deposit of sand and other sediment, usually triangular in shape. Deltas form at the mouths of rivers where the water flows into the sea.

Deoxygenation: The act of removing dissolved oxygen from water.

Deposition: The laying down of sediment such as sand, soil, clay, or gravel by wind or water. It may later be compacted into hard rock and buried by other sediment.

Depression storage: Initial storage of rain in small surface puddles.

Depth-area-duration analysis: A study made to determine the maximum amounts of rain within a given time period over a given area.

Detrial mineral: Mineral grain resulting from the mechanical disintegration of a parent rock.

Dewatering: Removal of excess moisture from sludge waste.

Digestion: Conversion of sludge solids to gas.

Dilatancy: The tendency of a material to increase in volume when undergoing shear.

Dilution disposal: Relying on a large water volume (lake or stream) to dilute waste to an acceptable concentration.

Dimictic: Experiencing two overturns per year. Dimictic lakes are usually found in temperate climates.

Dimiper lake: The freely circulating surface water with a small but variable temperature gradient.

Dimple spring: A depression in the earth below the water table.

Distribution coefficient: *See* Partition coefficient.

Divided highway: A highway with separated roadbeds for traffic in opposing directions.

Domestic waste: Waste that originates from households.

Downpull: A force on a gate, typically less at lower depths than at upper depths due to increased velocity, when the gate is partially open.

Drainage density: The total length of streams in a watershed divided by the drainage area.

Drawdown: The lowering of the water table level of an unconfined aquifer (or of the potentiometric surface of a confined aquifer) by pumping of wells.

Drawdown curve: *See* Cone of depression.

Dredge line: *See* Mud line.

Dry weather flow: *See* Base flow.

Dystrophic: Receiving large amounts of organic matter from surrounding watersheds, particularly humic materials from wetlands that stain the water brown. Dystrophic lakes have low plankton productivity, except in highly productive littoral zones.

E

Easement: A right to use or control the property of another for designated purposes.

Effective specific gravity of an asphalt mixture: Ratio of the unit weight of an asphalt mixture (excluding voids permeable to asphalt) to the unit weight of water.

Effluent: That which flows out of a process.

Effluent stream: A stream that intersects the water table and receives groundwater. Effluent streams seldom go completely dry during rainless periods.

Element: A pure substance that cannot be decomposed by chemical means.

Elutriation: A counter-current sludge washing process used to remove dissolved salts.

Elutriator: A device that purifies, separates, or washes material passing through it.

Embankment: A raised structure constructed of natural soil from excavation or borrow sources.

Eminent domain: The power to take private property for public use without the owner's consent upon payment of just compensation.

Emulsion: *See* Asphalt emulsion.

Encroachment: Use of the highway right-of-way for non-highway structures or other purposes.

Energy gradient: The slope of the specific energy line (i.e., the sum of the potential and velocity heads).

Enteric: Intestinal.

Enzyme: An organic (protein) catalyst that causes changes in other substances without undergoing any alteration itself.

Ephemeral stream: A stream that goes dry during rainless periods.

Epilimnion: (Gr. for "upper lake"); the freely circulating surface water with a small but variable temperature gradient.

Equivalent weight: The amount of substance (in grams) that supplies one mole of reacting units. It is calculated as the molecular weight divided by the change in oxidation number experienced in a chemical reaction. An alternative calculation is the atomic weight of an element divided by its valence or the molecular weight of a radical or compound divided by its valence.

Escarpment: A steep slope or cliff.

Escherichieae coli (E. coli): *See* Coliform.

Estuary: An area where fresh water meets salt water.

Eutrophic: Nutrient-rich; a eutrophic lake typically has a high surface area-to-volume ratio.

Eutrophication: The enrichment of water bodies by nutrients (e.g., phosphorus). Eutroficaction of a lake normally contributes to a slow evolution into a bog, marsh, and ultimately, dry land.

Evaporite: Sediment deposited when sea water evaporates. Gypsum, salt, and anhydrite are evaporites.

Evapotranspiration: Evaporation of water from a study area due to all sources including water, soil, snow, ice, vegetation, and transpiration.

F

Facultative: Able to live under different or changing conditions. Descriptive of a bacterial class that functions either in the presence or absence of free dissolved oxygen.

Fecal coliform: Coliform bacterium present in the intestinal tracts and feces of warm-blooded animals.

Fines: Silt- and/or clay-sized particles.

First-stage demand: *See* Carbonaceous demand.

Flexible pavement: A pavement having sufficiently low bending resistance to maintain intimate contact with the underlying structure, yet having the required stability furnished by aggregate interlock, internal friction, and cohesion to support traffic.

Float: The amount of time that an activity can be delayed without delaying any succeeding activities.

Floc: Agglomerated colloidal particles.

Flotation: Addition of chemicals and bubbled air to liquid waste in order to get solids to float to the top as scum.

Flow regime: (a) Subcritical, critical, or supercritical. (b) Entrance control or exit control.

Flowing well: A well that flows under hydrostatic pressure to the surface. Also called an Artesian well.

Flume: In general, any open channel for carrying water. More specifically, an open channel constructed above the earth's surface, usually supported on a trestle or on piers.

Force main: A sewer line that is pressured.

Forebay: A reservoir holding water for subsequent use after it has been discharged from a dam.

Forward pass: The steps in a critical path analysis in which the earliest finish times are determined, usually before the latest start times are determined in the backward pass.

Free residuals: Ions or compounds not combined or reduced. The presence of free residuals signifies excess dosage.

Freeboard distance: The vertical distance between the water surface and the crest of a dam or top of a channel side. The distance the water surface can rise before it overflows.

Freehaul: Pertaining to hauling "for free" (i.e., without being able to bill an extra amount over the contract charge).

Freeway: A divided arterial highway with full control of access.

Freeze (in piles): A large increase in the ultimate capacity (and required driving energy) of a pile after it has been driven some distance.

Friable: Easily crumbled.

Frontage road: A local street or road auxiliary to and located on the side of an arterial highway for service to abutting property and adjacent areas, and for control of access.

Frost susceptibility: Susceptible to having water continually drawn up from the water table by capillary action, forming ice crystals below the surface (but above the frost line).

Fulvic acid: The alkaline-soluble portion of organic material (i.e., humus) that remains in solution at low pH and is of lower molecular weight. A breakdown product of cellulose from vascular plants.

Fungi: Aerobic, multicellular, nonphotosynthetic heterotrophic, eucaryote protists that degrade dead organic matter, releasing carbon dioxide and nitrogen.

G

Gap: Corresponding time between successive vehicles (back bumper to front bumper) as they pass a point on a roadway.

Gap-graded: A soil with a discontinuous range of soil particle sizes; for example, containing large particles and small particles but no medium-sized particles.

Geobar: A polymeric material in the form of a bar.

Geocell: A three-dimensional, permeable, polymeric (synthetic or natural) honeycomb or web structure, made of alternating strips of geotextiles, geogrids, or geomembranes.

Geocomposite: A manufactured or assembled material using at least one geosynthetic product among the components.

Geofoam: A polymeric material that has been formed by the application of the polymer in semiliquid form through the use of a foaming agent. Results in a lightweight material with high void content.

Geographic Information System (GIS): A digital database containing geographic information.

Geogrid: A planar, polymeric (synthetic or natural) structure consisting of a regular open network of integrally connected tensile elements that may be linked or formed by extrusion, bonding, or interlacing (knitting or lacing).

Geomat: A three-dimensional, permeable, polymeric (synthetic or natural) structure made of bonded filaments, used for soil protection and to bind roots and small plants in erosion control applications.

Geomembrane: A planar, relatively impermeable, polymeric (synthetic or natural) sheet. May be bituminous, elastomeric, or plastomeric.

Geonet: A planar, polymeric structure consisting of a regular dense network whose constituent elements are linked by knots or extrusions and whose openings are much larger than the constituents.

Geopipe: A polymeric pipe.

Geospacer: A three-dimensional polymeric structure with large void spaces.

Geostrip: A polymeric material in the form of a strip, with a width less than approximately 200 mm.

Geosynthetic: A planar, polymeric (synthetic or natural) material.

Geotextile: A planar, permeable, polymeric (synthetic or natural) textile material that may be woven, nonwoven, or knitted.

Glacial till: Soil resulting from a receding glacier, consisting of mixed clay, sand, gravel, and boulders.

GMT: *See* UTC.

Gobar gas: *See* Biogas.

Gore: The area immediately beyond the divergence of two roadways bounded by the edges of those roadways.

Gradient: The energy (head) loss per unit distance. *See also* Slope.

Gravel: Granular material retained on a no. 4 sieve.

Gravitational water: Free water in transit downward through the vadose (unsaturated) zone.

Grillage: A footing or part of a footing consisting of horizontally laid timbers or steel beams.

Groundwater: Loosely, all water that is underground as opposed to on the surface of the ground. Usually refers to water in the saturated zone below the water table.

Gumbo: Silty soil that becomes soapy, sticky, or waxy when wet.

H

Hard water: Water containing dissolved salts of calcium and magnesium, typically associated with bicarbonates, sulfates, and chlorides.

Hardpan: A shallow layer of earth material that has become relatively hard and impermeable, usually through the decomposition of minerals.

Head (Total hydraulic): the sum of the elevation head, pressure head, and velocity head at a given point in an aquifer.

Headwall: Entrance to a culvert or sluiceway.

Headway: The time between successive vehicles as they pass a common point.

Heat of hydration: The exothermic heat given off by concrete as it cures.

Horizon: A layer of soil with different color or composition than the layers above and below it.

HOV: High-occupancy vehicle (e.g., bus).

Humic acid: The alkaline-soluble portion of organic material (i.e., humus) that precipitates from solution at low pH and is of higher molecular weight. A breakdown product of cellulose from vascular plants.

Humus: A grayish-brown sludge consisting of relatively large particle biological debris, such as the material sloughed off from a trickling filter.

Hydration: The chemical reaction between water and cement.

Hydraulic depth: Ratio of area in flow to the width of the channel at the fluid surface.

Hydraulic jump: An abrupt increase in flow depth that occurs when the velocity changes from supercritical to subcritical.

Hydraulic radius: Ratio of area in flow to wetted perimeter.

Hydrogen ion: The hydrogen atom stripped of its one orbital electron (H^+). It associates with a water molecule to form the hydronium ion (H_3O^+).

Hydrological cycle: The cycle experienced by water in its travel from the ocean, through evaporation and precipitation, percolation, runoff, and return to the ocean.

Hydrometeor: Any form of water falling from the sky.

Hydronium ion: *See* Hydrogen ion.

Hydrophilic: Seeking or liking water.

Hydrophobic: Avoiding or disliking water.

Hygroscopic: Absorbing moisture from the air.

Hygroscopic water: Moisture tightly adhering in a thin film to soil grains that is not removed by gravity or capillary forces.

Hypolimnion: (Gr. for "lower lake"); the deep, cold layer in a lake, below the epiliminion and metalimnion, cut off from the air above.

I

Igneous rock: Rock that forms when molten rock (magma or lava) cools and hardens.

Impactor: An environmental device that removes and measures micron-sized dusts, particles, and aerosols from an air stream.

Impervious layer: A geologic layer through which no water can pass.

Independent float: The amount of time that an activity can be delayed without affecting the float on any preceding or succeeding activities.

Infiltration: (a) Groundwater that enters sewer pipes through cracks and joints. (b) The movement of water downward from the ground surface through the upper soil.

Influent: Flow entering a process.

Influent stream: A stream above the water table that contributes to groundwater recharge. Influent streams may go dry during the rainless season.

Initial loss: The sum of interception and depression loss, excluding blind drainage.

In situ: "In place"; without removal; in original location.

Interception: The process by which precipitation is captured on the surfaces of vegetation and other impervious surfaces and evaporates before it reaches the land surface.

Interflow: Infiltrated subsurface water that travels to a stream without percolating down to the water level.

Intrusion: An igneous rock formed from magma that pushed its way through other rock layers. Magma often moves through rock fractures, where it cools and hardens.

Inverse condemnation: The legal process that may be initiated by a property owner to compel the payment of fair compensation when the owner's property has been taken or damaged for a public purpose.

Inversion layer: An extremely stable layer in the atmosphere in which temperature increases with elevation and mobility of airborne particles is restricted.

Inverted siphon: A sewer line that drops below the hydraulic grade line.

In vitro: Removed or obtained from an organism.

In vivo: Within an organism.

Ion: An atom that has either lost or gained one or more electrons, becoming an electrically charged particle.

Isogonic line: A line representing the magnetic declination.

Isotopes: Atoms of the same atomic number but having different atomic weights due to a variation in the number of neutrons.

J

Jam density: The density at which vehicles or pedestrians come to a halt.

Juvenile water: Water formed chemically within the earth from magma that has not participated in the hydrologic cycle.

K

Kingdom: A major taxonomic category consisting of several phyla or divisions.

Krause process: Mixing raw sewage, activated sludge, and material from sludge digesters.

L

Lagging: Heavy planking used to construct walls in excavations and braced cuts.

Lamp holes: Sewer inspection holes large enough to lower a lamp into but too small for a person.

Lane occupancy (ratio): The ratio of a lane's occupied time to the total observation time. Typically measured by a lane detector.

Lapse rate, dry: The rate that the atmospheric temperature decreases with altitude for a dry, adiabatic air mass.

Lapse rate, wet: The rate that the atmospheric temperature decreases with altitude for a moist, adiabatic air mass. The exact rate is a function of the moisture content.

Lateral: A sewer line that branches off from another.

Lava: Hot, liquid rock above ground. Also called lava once it has cooled and hardened.

Limnetic: Open water; extending down to the compensation level. Limnetic lake areas are occupied by suspended organisms (plankton) and free swimming fish.

Limnology: The branch of hydrology that pertains to the study of lakes.

Lipids: A group of organic compounds composed of carbon and hydrogen (e.g., fats, phospholipids, waxes, and steroids) that are soluble in a nonpolar, organic liquid (e.g., ether or chloroform); a constituent of living cells.

Lipiphilic: Having an affinity for lipids.

Littoral: Shallow; heavily oxygenated. In littoral lake areas, light penetrates all the way through to the bottom, and the zone is usually occupied by a diversity of rooted plants and animals.

Live load: The weight of all nonpermanent objects in a structure, including people and furniture. Live load does not include seismic or wind loading.

Loess: A deposit of wind-blown silt.

Lysimeter: A container used to observe and measure percolation and mineral leaching losses due to water percolating through the soil in it.

M

Magma: Hot, liquid rock under the earth's surface.

Main: A large sewer at which all other branches terminate.

Malodorous: Offensive smelling.

Marl: An earthy substance containing 35% to 65% clay and 65% to 35% carbonate formed under marine or freshwater conditions.

Meander corner: A survey point set where boundaries intersect the bank of a navigable stream, wide river, or large lake.

Meandering stream: A stream with large curving changes of direction.

Median: The portion of a divided highway separating traffic traveling in opposite directions.

Median lane: A lane within the median to accommodate left-turning vehicles.

Meridian: A great circle of the earth passing through the poles.

Meromictic lake: A lake with a permanent hypolimnion layer that never mixes with the epilimnion. The hypolimnion layer is perennially stagnant and saline.

Mesa: A flat-topped hill with steep sides.

Mesophyllic bacteria: Bacteria growing between 10°C and 40°C, with an optimum temperature of 37°C. 40°C is, therefore, the upper limit for most wastewater processes.

Metabolism: All cellular chemical reactions by which energy is provided for vital processes and new cell substances are assimilated.

Metalimnion: A middle portion of a lake, between the epilimnion and hypolimnion, characterized by a steep and rapid decline in temperature (e.g., 1°C for each meter of depth).

Metamorphic rock: Rock that has changed from one form to another by heat or pressure.

Meteoric water: *See* Hydrometeor.

Methylmercury: A form of mercury that is readily absorbed through the gills of fish, resulting in large bioconcentration factors. Methymercury is passed on to organisms higher in the food chain.

Microorganism: A microscopic form of life.

Mixture: A heterogeneous physical combination of two or more substances, each retaining its identity and specific properties.

Mohlman index: *See* Sludge volume index.

Mole: A quantity of substance equal to its molecular weight in grams (gmole or gram-mole) or in pounds (pmole or pound-mole).

Molecular weight: The sum of the atomic weights of all atoms in a molecule.

Monomictic: Experiencing one overturn per year. Monomictic lakes are typically very large and/or deep.

Mud line: The lower surface of an excavation or braced cut.

N

Nephelometric turbidity unit: The unit of measurement for visual turbidity in water and other solutions.

Net rain: That portion of rain that contributes to surface runoff.

Nitrogen fixation: The formation of nitrogen compounds (NH_3, organic nitrogen) from free atmospheric nitrogen (N_2).

Nitrogenous demand: Oxygen demand from nitrogen consuming bacteria.

Node: An activity in a precedence (critical path) diagram.

Nonpathogenic: Not capable of causing disease.

Nonpoint source: A pollution source caused by sediment, nutrients, and organic and toxic substances originating from land-use activities, usually carried to lakes and streams by surface runoff from rain or snowmelt. Nonpoint pollutants include fertilizers, herbicides, insecticides, oil and grease, sediment, salt, and bacteria. Nonpoint sources do not generally require NPDES permits.

Nonstriping sight distance: Nonstriping sight distances are in between stopping and passing distances, and they exceed the minimum sight distances required for marking no-passing zones. They provide a practical distance to complete the passing maneuver in a reasonably safe manner, eliminating the need for a no-passing zone pavement marking.

Normal depth: The depth of uniform flow. This is a unique depth of flow for any combination of channel conditions. Normal depth can be determined from the Manning equation.

Normally consolidated soil: Soil that has never been consolidated by a greater stress than presently existing.

NPDES: National Pollutant Discharge Elimination System.

NTU: *See* Nephelometric turbidity unit.

O

Observation well: A nonpumping well used to observe the elevation of the water table or the potentiometric surface. An observation well is generally of larger diameter than a piezometer well and typically is screened or slotted throughout the thickness of the aquifer.

Odor number: *See* Threshold odor number.

Oligotrophic: Nutrient-poor. Oligotrophic lakes typically have low surface area-to-volume ratios and largely inorganic sediments and are surrounded by nutrient-poor soil.

Order: In taxonomy, a major subdivision of a class. Each order consists of one or more related families.

Orthotropic bridge deck: A bridge deck, usually steel plate covered with a wearing surface, reinforced in one direction with integral cast-in-place concrete ribs. Used to reduce the bridge deck mass.

Orthotropic material: A material with different strengths (stiffnesses) along different axes.

Osmosis: The flow of a solvent through a semipermeable membrane separating two solutions of different concentrations.

Outcrop: A natural exposure of a rock bed at the earth's surface.

Outfall: A pipe that discharges treated wastewater into a lake, stream, or ocean.

Overchute: A flume passing over a canal to carry flood-waters away without contaminating the canal water below. An elevated culvert.

Overhaul: Pertaining to billable hauling (i.e., with being able to bill an extra amount over the contract charge).

Overland flow: Water that travels over the ground surface to a stream.

Overturn: (a) The seasonal (fall) increase in epilimnion depth to include the entire lake depth, generally aided by unstable temperature/density gradients. (b) The seasonal (spring) mixing of lake layers, generally aided by wind.

Oxidation: The loss of electrons in a chemical reaction. Opposite of *reduction*.

Oxidation number: An electrical charge assigned by a set of prescribed rules, used in predicting the formation of compounds in chemical reactions.

P

Pan: A container used to measure surface evaporation rates.

Parkway: An arterial highway for noncommercial traffic, with full or partial control of access, usually located within a park or a ribbon of park-like development.

Partial treatment: Primary treatment only.

Partition coefficient: The ratio of the contaminant concentration in the solid (unabsorbed) phase to the contaminant concentration in the liquid (absorbed) phase when the system is in equilibrium; typically represented as K_d.

Passing sight distance: The length of roadway ahead required to pass another vehicle without meeting an oncoming vehicle.

Passive pressure: A pressure acting to counteract active pressure.

Pathogenic: Capable of causing disease.

Pathway: An environmental route by which chemicals can reach receptors.

Pay as you throw: An administration scheme by which individuals are charged based on the volume of municipal waste discarded.

Pedology: The study of the formation, development, and classification of natural soils.

Penetration treatment: Application of light liquid asphalt to the roadbed material. Used primarily to reduce dust.

Perched spring: A localized saturated area that occurs above an impervious layer.

Percolation: The movement of water through the subsurface soil layers, usually continuing downward to the groundwater table.

Permanent hardness: Hardness that cannot be removed by heating.

Person-rem: A unit of the amount of total radiation received by a population. It is the product of the average radiation dose in rems times the number of people exposed in the population group.

pH: A measure of a solution's hydrogen ion concentration (acidity).

Phreatic zone: The layer the water table down to an impervious layer.

Phreatophytes: Plants that send their roots into or below the capillary fringe to access groundwater.

Phytoplankton: Small drifting plants.

Pier shaft: The part of a pier structure that is supported by the pier foundation.

Piezometer: A nonpumping well, generally of small diameter, that is used to measure the elevation of the water table or potentiometric surface. A piezometer generally has a short well screen through which water can enter.

Piezometer nest: A set of two or more piezometers set close to each other but screened to different depths.

Piezometric level: The level to which water will rise in a pipe due to its own pressure.

Pile bent: A supporting substructure of a bridge consisting of a beam or girder transverse to the roadway and that is supported, in turn, by a group of piles.

Pitot tube traverse: A volume or velocity measurement device that measures the impact energy of an air or liquid flow simultaneously at various locations in the flow area.

Planimeter: A device used to measure the area of a drawn shape.

Plant mix: A paving mixture that is not prepared at the paving site.

Plat: (a) A plan showing a section of land. (b) A small plot of land.

pOH: A measure of a solution's hydroxyl radical concentration (alkalinity).

Point source: A source of pollution that discharges into receiving waters from easily identifiable locations (e.g., a pipe or feedlot). Common point sources are factories and municipal sewage treatment plants. Point sources typically require NPDES permits.

Pollutant: Any solute or cause of change in physical properties that renders water, soil, or air unfit for a given use.

Polymictic: Experiencing numerous or continual overturns. Polymictic lakes, typically in the high mountains of equatorial regions, experience little seasonable temperature change.

Porosity: The ratio of pore volume to total rock, sediment, or formation volume.

Post-chlorination: Addition of chlorine after all other processes have been completed.

Potable: Suitable for human consumption.

Prechlorination: Addition of chlorine prior to sedimentation to help control odors and to aid in grease removal.

Presumptive test: A first-stage test in coliform fermentation. If positive, it is inconclusive without follow-up testing. If negative, it is conclusive.

Prime coat: The initial application of a low-viscosity liquid asphalt to an absorbent surface, preparatory to any subsequent treatment, for the purpose of hardening or toughening the surface and promoting adhesion between it and the superimposed constructed layer.

Probable maximum rainfall: The rainfall corresponding to some given probability (e.g., 1 in 100 years).

Protium: The stable ^{1}H isotope of hydrogen.

Protozoa: Single-celled aquatic animals that reproduce by binary fission. Several classes are known pathogens.

Putrefaction: Anaerobic decomposition of organic matter with accompanying foul odors.

Pycnometer: A closed flask with graduations.

Q

q-curve: A plot of depth of flow versus quantity flowing for a channel with a constant specific energy.

R

Rad: Abbreviation for "radiation absorbed dose." A unit of the amount of energy deposited in or absorbed by a material. A rad is equal to 62.5×10^6 MeV per gram of material.

Radical: A charged group of atoms that act together as a unit in chemical reactions.

Ranger: *See* Wale.

Rapid flow: Flow at less than critical depth, typically occurring on steep slopes.

Rating curve: A plot of quantity flowing versus depth for a natural watercourse.

Reach: A straight section of a channel, or a section that is uniform in shape, depth, slope, and flow quantity.

Redox reaction: A chemical reaction in which oxidation and reduction occur.

Reduction: The loss of oxygen or the gain of electrons in a chemical reaction. Opposite of *oxidation*.

Refractory: Dissolved organic materials that are biologically resistant and difficult to remove.

Regulator: A weir or device that diverts large volume flows into a special high-capacity sewer.

Rem: Abbreviation for "Roentgen equivalent mammal." A unit of the amount of energy absorbed by human tissue. It is the product of the absorbed dose (rad) times the quality factor.

Residual: A chemical that is left over after some of it has been combined or inactivated.

Resilient modulus: The modulus of elasticity of the soil.

Respiration: Any biochemical process in which energy is released. Respiration may be aerobic (in the presence of oxygen) or anaerobic (in the absence of oxygen).

Restraint: Any limitation (e.g., scarcity of resources, government regulation, or nonnegativity requirement) placed on a variable or combination of variables.

Resurfacing: A supplemental surface or replacement placed on an existing pavement to restore its riding qualities or increase its strength.

Retarded flow: A form of varied flow in which the velocity is decreasing and the depth is increasing.

Retrograde solubility: Solubility that decreases with increasing temperature. Typical of calcite (calcium carbonate, $CaCO_3$) and radon.

Right of access: The right of an abutting land owner for entrance to or exit from a public road.

Rigid pavement: A pavement structure having portland cement concrete as one course.

Rip rap: Pieces of broken stone used as lining to protect the sides of waterways from erosion.

Road mix: A low-quality asphalt surfacing produced from liquid asphalts and used when plant mixes are not available or economically feasible and where volume is low.

Roadbed: That portion of a roadway extending from curb line to curb line or from shoulder line to shoulder line. Divided highways are considered to have two roadbeds.

Roentgen: The amount of energy absorbed in air by the passage of gamma or X-rays. A roentgen is equal to 5.4 $\times 10^7$ MeV per gram of air or 0.87 rad per gram of air and 0.96 rad per gram of tissue.

S

Safe yield: The maximum rate of water withdrawal that is economically, hydrologically, and ecologically feasible.

Sag pipe: See Inverted siphon.

Saline: Dominated by anionic carbonate, chloride, and sulfate ions.

Salt: An ionic compound formed by direct union of elements, reactions between acids and bases, reaction of acids and salts, and reactions between different salts.

Sand: Granular material passing through a no. 4 sieve but predominantly retained on a no. 200 sieve.

Sand trap: A section of channel constructed deeper than the rest of the channel to allow sediment to settle out.

Scour: Erosion typically occurring at the exit of an open channel or toe of a spillway.

Scrim: An open-weave, woven or nonwoven, textile product that is encapsulated in a polymer (e.g., polyester) material to provide strength and reinforcement to a watertight membrane.

Seal coat: An asphalt coating, with or without aggregate, applied to the surface of a pavement for the purpose of waterproofing and preserving the surface, altering the surface texture of the pavement, providing delineation, or providing resistance to traffic abrasion.

Second-stage demand: See Nitrogenous demand.

Sedimentary rock: Rocks formed from sediment, broken rocks, or organic matter. Sedimentary rocks are formed when wind or water deposits sediment into layers, which are pressed together by more layers of sediment above.

Seed: The activated sludge initially taken from a secondary settling tank and returned to an aeration tank to start the activated sludge process.

Seep: See Spring.

Seiche, external: An oscillation of the surface of a landlocked body of water.

Seiche, internal: An alternating pattern in the directions of layers of lake water movement.

Sensitivity: The ratio of a soil's undisturbed strength to its disturbed strength.

Separate system: A system with separate sewers for domestic and storm wastewater.

Septic: Produced by putrefaction.

Settling basin: A large, shallow basin through which water passes at low velocity, where most of the suspended sediment settles out.

Sheeted pit: *See* Cased hole.

Shooting flow: *See* Rapid flow.

Sight distance: The length of roadway that a driver can see.

Sinkhole: A natural dip or hole in the ground formed when underground salt or other rocks are dissolved by water and the ground above collapses into the empty space.

Sinuosity: The stream length divided by the valley length.

Slickenside: A surface (plane) in stiff clay that is a potential slip plane.

Slope: The tangent of the angle made by the channel bottom. *See also* Gradient.

Sludge: The precipitated solid matter produced by water and sewage treatment.

Sludge bulking: Failure of suspended solids to completely settle out.

Sludge volume index (SVI): The volume of sludge that settles in 30 min out of an original volume of 1000 mL. May be used as a measure of sludge bulking potential.

Sol: A homogenous suspension or dispersion of colloidal matter in a fluid.

Soldier pile: An upright pile used to hold lagging.

Solution: A homogeneous mixture of solute and solvent.

Sorption: A generic term covering the processes of absorption and adsorption.

Space mean speed: One of the measures of average speed of a number of vehicles over a common (fixed) distance. Determined as the inverse of the average time per unit distance. Usually less than time mean speed.

Spacing: Distance between successive vehicles, measured front bumper to front bumper.

Specific activity: The activity per gram of a radioisotope.

Specific storage: *See* Specific yield.

Specific yield: (a) The ratio of water volume that will drain freely (under gravity) from a sample to the total volume. Specific yield is always less than porosity. (b) The amount of water released from or taken into storage per unit volume of a porous medium per unit change in head.

Split chlorination: Addition of chlorine prior to sedimentation and after final processing.

Spring: A place where water flows or ponds on the surface due to the intersection of an aquifer with the earth surface.

Stadia method: Obtaining horizontal distances and differences in elevation by indirect geometric methods.

Stage: Elevation of flow surface above a fixed datum.

Standing wave: A stationary wave caused by an obstruction in a water course. The wave cannot move (propagate) because the water is flowing at its critical speed.

Steady flow: Flow in which the flow quantity does not vary with time at any location along the channel.

Stilling basin: An excavated pool downstream from a spillway used to decrease tailwater depth and to produce an energy-dissipating hydraulic jump.

Stoichiometry: The study of how elements combine in fixed proportions to form compounds.

Stopping sight distance: The distance that allows a driver traveling at the maximum speed to stop before hitting an observed object.

Stratum: Layer.

Stream gaging: A method of determining the velocity in an open channel.

Stream order: An artificial categorization of stream genealogy. Small streams are first order. Second-order streams are fed by first-order streams, third-order streams are fed by second-order streams, and so on.

Stringer: *See* Wale.

Structural section: The planned layers of specific materials, normally consisting of subbase, base, and pavement, placed over the subbase soil.

Subbase: A layer of aggregate placed on the existing soil as a foundation for the base.

Subcritical flow: Flow with depth greater than the critical depth and velocity less than the critical velocity.

Subgrade: The portion of a roadbed surface that has been prepared as specified, upon which a subbase, base, base course, or pavement is to be constructed.

Submain: *See* Branch sewer.

Substrate: A substance acted upon by an organism, chemical, or enzyme. Sometimes used to mean organic material.

Subsurface runoff: *See* Interflow.

Superchlorination: Chlorination past the breakpoint.

Supercritical flow: Flow with depth less than the critical depth and velocity greater than the critical velocity.

Superelevation: Roadway banking on a horizontal curve for the purpose of allowing vehicles to maintain the traveled speed.

Supernatant: The clarified liquid rising to the top of a sludge layer.

Surcharge: An additional loading. (a) In geotechnical work, any force loading added to the in situ soil load. (b) In water resources, any additional pressurization of a fluid in a pipe.

Surcharged sewer: (a) A sewer that is flowing under pressure (e.g., as a force main). (b) A sewer that is supporting an additional loading (e.g., a truck parked above it).

Surface detention: *See* Surface retention.

Surface retention: The part of a storm that does not contribute to runoff. Retention is made up of depression storage, interception, and evaporation.

Surface runoff: Water flow over the surface that reaches a stream after a storm.

Surficial: Pertaining to the surface.

Swale: (a) A low-lying portion of land, below the general elevation of the surroundings. (b) A natural ditch or long, shallow depression through which accumulated water from adjacent watersheds drains to lower areas.

T

Tack coat: The initial application of asphalt material to an existing asphalt or concrete surface to provide a bond between the existing surface and the new material.

Tail race: An open waterway leading water out of a dam spillway and back to a natural channel.

Tailwater: The water into which a spillway or outfall discharges.

Taxonomy: The description, classification, and naming of organisms.

Temporary hardness: Hardness that can be removed by heating.

Theodolite: A survey instrument used to measure or lay off both horizontal and vertical angles.

Thermocline: The temperature gradient in the metalimnion.

Thermophilic bacteria: Bacteria that thrive in the 45°C to 75°C range. The optimum temperature is near 55°C.

Thixotropy: A property of a soil that regains its strength over time after being disturbed and weakened.

Threshold odor number: A measure of odor strength, typically the number of successive dilutions required to reduce an odorous liquid to undetectable (by humans) level.

Till: *See* Glacial till.

Time mean speed: One of the measures of average speed of a number of vehicles over a common (fixed) distance. Determined as the average vehicular speed over a distance. Usually greater than space mean speed.

Time of concentration: The time required for water to flow from the most distant point on a runoff area to the measurement or collection point.

TON: *See* Threshold odor number.

Topography: Physical features such as hills, valleys, and plains that shape the surface of the earth.

Total float: The amount of time that an activity in the critical path (e.g., project start) can be delayed without delaying the project completion date.

Township: A square parcel of land 6 mi on each side.

Toxin: A toxic or poisonous substance.

Tranquil flow: Flow at greater than the critical depth.

Transmissivity: The rate at which water moves through a unit width of an aquifer or confining bed under a unit hydraulic gradient. It is a function of properties of the liquid, the porous media, and the thickness of the porous media.

Transpiration: The process by which water vapor escapes from living plants (principally from the leaves) and enters the atmosphere.

Traveled way: The portion of the roadway for the movement of vehicles, exclusive of shoulders and auxiliary lanes.

Turbidity: (a) Cloudiness in water caused by suspended colloidal material. (b) A measure of the light-transmitting properties of water.

Turbidity unit: *See* Nephelometric turbidity unit.

Turnout: (a) A location alongside a traveled way where vehicles may stop off of the main road surface without impeding following vehicles. (b) A pipe placed through a canal embankment to carry water from the canal for other uses.

U

Uniform flow: Flow that has constant velocity along a streamline. For an open channel, uniform flow implies constant depth, cross-sectional area, and shape along its course.

Unit process: A process used to change the physical, chemical, or biological characteristics of water or wastewater.

Unit stream power: The product of velocity and slope, representing the rate of energy expenditure per unit mass of water.

Uplift: Elevation or raising of part of the earth's surface through forces within the earth.

UTC: Coordinated Universal Time, the international time standard; previously referred to as Greenwich Meridian Time (GMT).

V

Vadose water: All underground water above the water table, including soil water, gravitational water, and capillary water.

Vadose zone: A zone above the water table containing both saturated and empty soil pores.

Valence: The relative combining capacity of an atom or group of atoms compared to that of the standard hydrogen atom. Essentially equivalent to the oxidation number.

Varied flow: Flow with different depths along the water course.

Varve: A layer of different material in the soil; fine layers of alluvium sediment deposited in glacial lakes.

Vitrification: Encapsulation in or conversion to an extremely stable, insoluble, glasslike solid by melting (usually electrically) and cooling. Used to destroy or immobilize hazardous compounds in soils.

Volatile organic compounds (VOCs): A class of toxic chemicals that easily evaporate or mix with the atmosphere and environment.

Volatile solid: Solid material in a water sample or in sludge that can be burned away or vaporized at high temperature.

Volatilization: The driving off or evaporation of a liquid in a solid or one or more phases in a liquid mixture.

W

Wah gas: *See* Biogas.

Wale: A horizontal brace used to hold timbers in place against the sides of an excavation or to transmit the braced loads to the lagging.

Wasteway: A canal or pipe that returns excess irrigation water to the main channel.

Water table: The piezometric surface of an aquifer, defined as the locus of points where the water pressure is equal to the atmospheric pressure.

Waving the rod: A survey technique used when reading a rod for elevation data. By waving the rod (the rod is actually inclined toward the instrument and then brought more vertical), the lowest rod reading will indicate the point at which the rod is most vertical. The reading at that point is then used to determine the difference in elevation.

Wet well: A short-term storage tank from which liquid is pumped.

Wetted perimeter: The length of the channel cross section that has water contact. The air-water interface is not included in the wetted perimeter.

X

Xeriscape: Creative landscaping for water and energy efficiency and lower maintenance.

Xerophytes: Drought-resistant plants, typically with root systems well above the water table.

Z

Zone of aeration: *See* Vadose zone.

Zone of saturation: *See* Phreatic zone.

Zoogloea: The gelatinous film (i.e., "slime") of aerobic organisms that covers the exposed surfaces of a biological filter.

Zooplankton: Small, drifting animals capable of independent movement.

Index

Index to *PE Civil Reference Manual*

-red clearance period, 73-22
-volatile treatment, 22-26
Allergen, 32-4
Alligator crack, 76-4
Allochthonous material, 25-8
Allowable
    bearing capacity, 36-2
    bearing strength, 59-19, 61-10
    bending stress, 59-15
    compressive strength, steel column,
        61-4
    floor area, 82-7
    height, building, 82-7
    load, fastener, 65-3
    strength design, 58-5, 59-8
    stress, 45-2, 50-3
    stress, bending, 59-15
    stress design, 50-3, 58-5, 59-8
    stress design method, 50-3
    stress rating method, 74-3
    stress, sheet piling (tbl), 39-4
Allowance
    depletion, 87-25
    trade-in, 87-19
Alluvial deposit, G-1
Alluvium, G-1
Alpha, 11-15
    risk, 11-15
Alphabet, Greek, 3-1
Alpha-value, 85-6
Alternate
    depth, 19-17, G-1
    mode, 73-20
    mode operation, 73-20
Alternating current, 84-2, 84-5
Alternating sign series, 3-14
Alternative
    comparison, 87-4
    disinfectants, 25-11
    disinfection, 29-14
    hypothesis, 11-15
    sewer, 28-4
    to chlorination, 25-11, 26-22
Alternator (ftn), 84-12
Altitude valve, 26-26
Alum, 26-8
Aluminum, A-129
    structural, A-143
    structural, properties, A-143
Amber period, 73-22
American
    Insurance Association equation, 26-25
    Insurance Association formula, 26-25
    National Standards Institute, 82-3
    Society of Civil Engineers, 89-1
    wire gauge, 84-9
Americans with Disabilities Act, 82-9
Amictic, G-2
Amide (tbl), 23-2
Amine
    filming, 22-25
    neutralizing, 22-25
    polar, 22-25
Amine (tbl), 23-1, 23-2
Amino acid (tbl), 23-1, 23-2
Ammonia
    in wastewater, 28-15
    in water, 25-9
    nitrogen, in wastewater, 28-15
    removal, in wastewater, 29-3, 29-14
    slip (ftn), 34-20
    stripping, 28-15, 29-14
    toxic effect on fish, 25-9
    un-ionized, in water, 25-9
Amoeba, 27-8
Amoeboid, 27-8
Amortization, 87-25
Amount, factor compound, 87-5
Ampacity, 84-9
Ampere, 84-2
Amperometric sensor, 85-4

Amphoteric behavior, G-2
Amplification
    factor, moment, 53-5
    force, 41-11
Amplified end moment, 53-5
AMU, G-2
Anabolic, 27-9
Anabolism, G-2
Anabranch, G-2
Anaerobe, 27-6, G-2
    digester, 30-17
    obligate, 27-6
Anaerobic, G-2
    decomposition, 27-10
    digester, characteristics, 30-19
    digestion, 30-17
    pond, 29-4
Analysis, (see also type) 47-2
    accident data, 75-11
    ACI coefficient, 51-3
    beam-column, 62-3
    break-even, 87-39
    chemical, 22-7
    column, 61-5
    combustible, 24-2
    comparative, 87-17
    cost-benefit, highway safety, 75-20
    descriptive, safety, 75-2
    dimensional, 1-9
    economic, engineering, 87-2
    economic life, 87-4
    elastic second-order, 53-2
    error, 78-2, 78-3
    first-order, 53-1
    Fourier, 9-7
    frequency (ftn), 9-8
    gravimetric, 22-6, 24-2
    horizon, 87-18
    hydrograph, 20-7
    incremental, 87-18
    inelastic first-order, 47-2
    inelastic second-order, 47-3
    life-cycle cost, 86-7
    nonlinear second-order, 53-2
    numerical, 12-1
    period, 76-20
    plastic, 47-2
    proximate, 24-2
    quantitative predictive, 75-2
    replacement/retirement, 87-4
    risk, 87-46
    sensitivity, 87-45
    signature (ftn), 9-8
    strength, doubly reinforced section,
        50-25
    structural, 47-2
    tension member, 60-4
    tension member (fig), 60-5
    time-series (ftn), 9-8
    ultimate, 22-5, 22-6, 24-2
    uncertainty, 87-46
    value, 87-46
    volumetric, 24-2
Analytic function, 8-1
Analyzer
    FFT, 9-8
    signal, 9-8
    spectrum, 9-8
Anchor pull, 39-5
Anchor trench, 31-4
Anchorage
    masonry, 67-7
    requirement, shear reinforcement (fig),
        50-24
    shear reinforcement, 50-24
Anchored bulkhead, 39-5
Andesite porphyry, 35-33
Andrade's equation, 43-17
Anemometer, 17-29
    hot-wire, 17-29

Angle
    acute, 6-1
    adjacent, 6-2
    adjacent side, 6-2
    banking, 79-8
    bearing (fig), 78-13
    between figures, 7-8
    between lines, 6-1
    complementary, 6-2
    contact, 14-13
    deflection, 78-13, 78-14, 79-4
    direction, 5-2, 7-5
    explement, 78-14
    external friction, 37-6
    face, 6-6
    friction, external (tbl), 37-4
    friction, soil (tbl), 37-4
    function of, 6-2
    helix thread, 45-15
    hypotenuse, 6-2
    interfacial friction, 31-4
    interfacial friction (tbl), 31-5
    interior, 78-14, 79-2
    internal friction, 35-17, 35-26, 40-8,
        43-8, 72-7
    internal friction, pile (fig), 72-7
    intersection, 79-2
    lead, bolt, 45-15
    lift check valve, 16-13
    measurement (fig), 78-13
    measurement equipment, 78-12
    measurement method, 78-12
    miscellaneous formula, 6-4
    obtuse, 6-1
    of depression, 6-2
    of elevation, 6-2
    of intersection, 7-8
    of repose, 72-6, 80-2
    of static friction, 72-6
    opposite side, 6-2
    parking, 73-23
    plane, 6-1
    power, 84-8
    rebound, 72-18
    reflex, 6-1
    related, 6-1
    repose, 40-8
    right, 6-1
    rupture, 43-8
    solid, 6-7
    spiral, 79-20
    stall, 17-42
    station, 78-14
    straight, 6-1
    supplementary, 6-2
    to the right, 78-13
    total deflection, 79-20
    traverse (fig), 78-14
    trihedral, 6-6
    twist, 43-9, 45-16
    type (see also by name), 6-2
    valve, 16-13
    vertex, 6-1
    vertical, 6-2
    wall friction, 37-6
    yaw, 17-30
Angstrom, A-1
Angular
    acceleration, 71-7
    frequency, 84-5
    impulse, 72-15
    momentum, 17-35, 72-3
    momentum (fig), 72-4
    motion, 71-7
    orientation, 5-1
    perspective, 2-3
    position, 71-7
    velocity, 71-7
Angularity, aggregate, 76-16
Anhydride (tbl), 23-3
Anion, 25-1, G-2

INDEX-A

INDEX-C

INDEX-C

of contraction, 17-18, 17-32
of creep, 56-3
of curvature, soil, 35-3
of discharge, 17-19, 17-32
of drag, 17-44
of earth pressure at rest, 37-5
of expansion, thermal (ftn), 14-14
of flow, 17-32
of friction, 72-6, 72-7, 75-6
of friction, static, 15-12
of friction, tire, 75-6
of friction, typical (tbl), 72-7
of gradation, 35-3
of heat transfer, overall, 30-19
of lateral earth pressure, 38-3
of lift, 17-42
of liquid mass transfer, 34-22
of passive earth pressure, 37-5
of permeability, 21-2
of restitution, 72-18
of road adhesion, 75-6
of rolling friction, 72-8
of rolling resistance, 72-8
of secondary consolidation, 40-6
of skidding friction (tbl), 75-6
of the instrument, 17-30
of thermal expansion, piping (tbl),
    34-17
of thermal resistance, 85-7
of transformation, 5-3
of transmissivity, 21-3
of uniformity, 21-6
of variation, 11-14
of velocity, 17-18, 17-31, 18-24
of viscosity, 14-7
of viscous damping, 72-20
orifice (tbl), 17-19
oxygen saturation, 30-9
partition, 27-2, G-12
pipe, 16-9
roughness, 17-9
saturation, 67-5
shear, ACI, 47-20
skin friction, 17-47, 38-3
skin friction (ftn), 17-5
slip, 65-3
smoothing, 87-44
solubility, 22-10
spillway, 19-14
stability, 53-3
steel rainfall, 20-6
storage, 21-3
strength, 43-6, 76-22
thermal expansion, 44-4
torque, 45-15
unit weight, 49-9
valve flow, 17-15
viscous, 72-20
volumetric expansion, 44-4
web plate buckling, 63-2
yield, 27-12, 30-8, 30-14
Cofactor
    expansion by, 4-3
    matrix, 4-1
    of entry, 4-2
Cofferdam, 39-6
    cellular, 39-6
    double-wall, 39-6
    liner-plate, 39-9
    single-wall, 39-7
    vertical-lagging, 39-9
Cogener, 22-2
Coherent unit system (ftn), 1-2
Cohesiometer, Hveem, 76-15
Cohesion, 35-26, 43-8
    clay, 35-27
    gravel, 35-27
    intercept, 35-26
    sand, 35-27
    soil, 35-26
Cohesion (ftn), 14-13

Cohesionless soil, 37-2
Cohesive soil, 37-2
Coke, 24-6
    -oven gas (ftn), 24-8
Cold
    cracking pavement, 76-4
    flow, 43-17
    in-place recycling, 76-29
    planing, 76-29
    train, 76-29
    weather, 49-6
    -weather concrete, 49-6
Colding equation, 31-7
Colebrook equation, 17-6
Coli, Escherichieae, G-7
Coliform, 27-9, G-5
    fecal, 27-9, G-7
Collapsing pressure (ftn), 45-5
Collection
    efficiency, 26-6, 26-8
    efficiency, baghouse, 34-6
    efficiency, ESP, 34-9
    efficiency, generic collector, 34-19
    efficiency, venturi scrubber, 34-19
Collector, 28-3
    data, 78-5
    road, 73-3
    sewer, 28-3
Collinear, 7-2
    force system, 41-6
Collision, 72-18, 75-16
    diagram, 75-12, 75-19
    diagram (fig), 75-19
Colloid, 34-24, G-5
Colloidal material, 14-8
Color in water, 25-9
Column, 45-3
    alignment chart, 53-3
    alignment chart (fig), 61-4
    analysis, 61-5
    base plate, 61-10
    base plate (fig), 61-11
    braced, 45-3, 53-2, 53-5
    concrete, 52-2
    concrete pressure on, 49-8
    design, 61-5
    design strength, 52-4
    eccentricity, 52-3, 52-7
    eccentricity, minimum, 69-3
    effective length, 53-3
    end coefficient, 45-3
    footing, 36-2, 55-3
    ideal, 61-2
    interaction diagrams, A-151, A-152,
        A-153, A-154, A-155, A-156, A-157,
        A-158, A-159, A-160, A-161, A-162,
        A-163, A-164, A-165, A-175, A-176,
        A-177, A-178, A-179, A-180
    intermediate, 45-4, 61-4
    laced (ftn), 61-2
    latticed (ftn), 61-2
    long, 52-1, 53-1, 61-4
    masonry, 69-1
    matrix, 4-1
    rank (ftn), 4-5
    short, 52-1
    slender, 45-3
    spiral, 52-2
    spiral wire, 52-3
    steel, 61-1
    strip, slab, 51-6
    strip moment distribution (tbl), 51-7
    tied, 52-2
    unbraced, 53-2, 53-5
Combination, 11-2
    air valve, 16-15
    direct, 22-2
    section, compression, 61-7
Combined
    ash, 32-5
    footing, 36-2

residual, 25-10, G-5
residual, available, 28-14
residual, unavailable, 28-14
sewer overflow, 29-14
sewer system, 28-2
shear ant tension connection (fig), 65-7
stress, 44-5
stress (fig), 44-6
system, G-5
Combining weight, 22-6
Combustible
    analysis, 24-2
    loss, 24-17
Combustion
    atmospheric fluidized-bed, 24-6
    chamber, primary, 34-13
    chamber, secondary, 34-13
    complete, 24-13
    data (tbl), 24-11
    efficiency, 24-17
    fluidized-bed, 24-5
    heat equivalents, A-105
    heat of, 24-15, A-105
    incomplete, 24-12
    loss, 24-17
    pressurized fluidized-bed, 24-6
    product, incomplete, 34-14
    reaction, 24-9, 24-10
    reaction, ideal (tbl), 24-9
    staged, 34-21
    stoichiometric, 24-10
    temperature, 24-16
Combustor
    bubbling-bed, 34-10
    fluidized-bed, 34-10
    fluidized-bed (fig), 34-10
Comfort, curve length, 79-18
Comity, 90-2
Comminutor, 29-7, G-5
Common
    bolt, 65-1
    borrow, 80-2
    curvature, point of, 79-7
    excavation, 80-2
    ion effect, 22-14
    logarithm, 3-5
    ratio, 3-12
    size financial statement (ftn), 87-35
Communication factor, 26-25
Commutative
    law, addition, 3-2, 4-4
    law, multiplication, 3-2
    set, law, 11-2
Compact
    beam, 59-3
    section, 59-3, 61-7
Compacted
    cubic yard, 80-1
    -measure, 80-1
Compaction, G-2
    equipment (fig), 35-19
    factor, 31-2
    relative, 35-18
    relative, suggested, 35-20
Compactness, 59-3
Compactor
    gyratory, Superpave, 76-16
    kneading, 76-14
Company health, 87-36
Comparative
    analysis, 87-17
    negligence, 88-6
Comparison
    alternative, 87-4
    test, 3-13
Compartment, water quality, 32-15
Compass rule, 78-15
Compatibility method, 46-2
Compensation
    footing, 36-10
    level, G-5

INDEX-E

INDEX-E

INDEX-F

INDEX-H

INDEX-H

Industrial
exemption, 90-1
wastewater, 28-2, 28-9, 29-2
wastewater, standards, 29-2
wastewater pollution (tbl), 34-23
Industry classification system (ftn), 83-2
Inelastic
analysis, 47-2
buckling, 61-4
design, 59-8
first-order analysis, 47-2
impact, 72-18
second-order analysis, 47-3
strain, 43-3
Inequality, 4-4
Inertia, 70-2
area moment, 42-4
centroidal mass moment, 70-2
centroidal moment of, 42-4
cracked, transformed, 50-15
mass moment of, 70-2, A-181
moment, A-127
moment of, 42-4, A-181
polar moment of, 42-6
principal moment of, 42-8
product of, 42-7, 70-2
rectangular moment of, 42-4
resistance, vehicle, 75-3
vector, 70-2
vector (ftn), 72-6
Inertial
force, 70-2, 72-6
frame of reference, 71-10
frame of reference, Newtonian, 71-10
resisting force, vehicle, 75-3
separator, 32-8, 34-7
survey system, 78-7
Infant mortality, 11-9
Infective dose, 27-11
Infiltration, 21-10, G-9
capacity, 21-10
sewer, 28-2
test, double-ring, 21-10
Infiltration (ftn), 20-7
Infinite
series, 3-13, 87-7
series (ftn), 3-13
Inflation, 87-41
effect on cost, 86-7
Inflection
point, 7-2, 8-2
point, assumed, 47-19
point, deflection, 44-18
point, flexible bulkhead, 39-5
Inflow, 28-2
Influence
Influence, radius of, 21-6, 31-6
chart, 40-2, 40-3
cone, 40-2
diagram, 41-10, 46-8, 46-9, 46-11, 46-12, 46-13
graph, 41-10
line, 41-10
value, 40-3
zone of, 40-2
Influent, G-9
stream, G-9
Information
cluster, 81-1
header, 81-1
Infrared
incineration, 34-14
testing, 43-13
Inhalable fraction, 32-7
Initial
abstraction, 20-2, 20-21
loss, G-9
modulus, 48-5
period, 73-21
rate of absorption, 67-5

value, 9-4, 10-1
value problem, 10-1
Injection
sorbent, 34-20
well, 34-11
well installation (fig), 34-11
Injured party, 88-6
Injury
incidence rate, 83-2
log (ftn), 83-2
occupational, 83-2
Inlet
control, 19-29
curb, 28-5
grate, 28-5
gutter, 28-5
pressure, net positive (ftn), 18-3
In-line filtration, 26-2
Inline sampling, 22-25
Inner transition element, 22-3
Innovative method, 32-16
Inorganic
chemical, 28-15
compound, volatile, 32-17
salt, removal, in wastewater, 29-3
Insecticide (see also Pesticide), 32-13
Insensitivity, 85-2
experiment, 11-12
Inspection
polynomial, 3-4
radiographic, 82-4
Instability, 85-2
Instant center, 71-13
Instantaneous
center, 71-13
center, constrained cylinder (fig), 72-12
center of acceleration, 71-14
center of rotation, 66-8
center of rotation method, 65-6
deflection, 50-16
growth, 3-13
reorder inventory (fig), 87-45
values, 71-2
Instrument
coefficient of, 17-30
factor, 78-8
height, 78-11, 81-4
person, 81-4
telescopic, 78-5
Insurance, 88-8
Services Office, 26-24
Intangible property (ftn), 87-21
Integral, 9-1, 9-2
convolution, 10-7
cosine function, 9-9
definite, 9-4
definite (ftn), 9-1
double, 9-3
elliptic function (ftn), 9-9
exponential function, 9-9
Fresnel function (ftn), 9-9
function, 9-9
gamma function (ftn), 9-9
indefinite, 9-1, 9-4, A-11
of combination of functions, 9-2
of hyperbolic function, 9-2
of transcendental function, 9-1
sine function, 9-9
triple, 9-3
Integrand, 9-1
Integrated gasification/combined cycle, 24-6
Integrating factor, 10-2
Integration, 9-1
by parts, 9-2
by separation of terms, 9-3
constant of, 9-1, 10-1
method, 42-1, 42-4
Intensity
-duration-frequency curve, 20-6
rainfall, 20-5

Interaction
AISC equation (steel), 59-15
diagram, 52-5, 52-6, 52-7, A-154, A-155, A-156, A-157, A-158, A-159, A-160, A-162, A-163, A-164, A-165, A-175, A-176, A-177, A-180
diagram, masonry wall, 68-11
diagram, reinforced concrete, A-151, A-152, A-153, A-161, A-178, A-179
equation, 62-2
equation, flexural/force, 62-2
Intercept, 7-5
cohesion, 35-26
form, 7-5
Interception, G-9
Interceptor, 28-3
Interchange (see also type)
adaptability (fig), 73-27
all-directional, 73-25
cloverleaf, 73-25
diamond, 73-25
directional, 73-25
highway, 73-25
highway (see also type), parclos, 73-25
partial cloverleaf, 73-25
single-point diamond, 73-25
single-point urban, 73-25
trumpet, 73-25
type (fig), 73-26
urban, 73-25
Interest
compound, 87-12
rate, effective, 87-5, 87-28
rate, effective, per period, 87-28
rate, effective annual, 87-5
rate, nominal, 87-29
simple, 87-12
Interfacial
area, 34-22
friction angle, 31-4
friction angle (tbl), 31-5
Interference, 45-7
fit, 45-7
Interflow, G-9
Intergranular, 22-21
attack, 22-21
corrosion, 22-20
Interior
angle, 78-14, 79-2
angle, curve, 79-2
intermediate stiffener, 63-3
Intermediate
clarifier, 29-13
column, 45-4, 61-4
metals, law of, 85-8
stiffener, 44-20, 59-18, 63-4
stiffener (fig), 63-4
stiffener design, 63-4
temperatures, law of, 85-8
Intermittent
duty, 84-15
sand filter, 29-13
Intern engineer (ftn), 90-1
Internal
energy, 13-4
force, 41-2, 41-14, 72-2
friction, angle of, 35-17, 35-26, 40-8, 43-8, 72-7
work, 13-2, 43-7
Internal rate of return, 87-13
International
Building Code, 58-5, 82-1
Building Code (ftn), 58-5
Code Council, 82-1
Fire Code, 82-2
Mechanical Code, 82-2
Plumbing Code, 82-2
Residency Code, 82-2
standard atmosphere, A-66
standard metric condition, 24-2

INDEX-L

INDEX-L

Maximum
achievable control technology, 34-2
capacity, 73-7
capacity, traffic, 73-7
contaminant level, 25-5, 28-16
contaminant level goal, 25-5
dry density, 35-18
flame temperature (ftn), 24-16
flood, probable, 20-7
flow, 73-7
fluid velocity, 17-3
freeway service flow rate (tbl), 73-9
moment, 44-9
moment condition (fig), 50-9
period, 73-22
plastic moment (tbl), 59-12
point, 8-2
precipitation, probable, 20-7
prestress, 56-7
rainfall, probable, G-13
service flow rate, 73-5
shear envelope, 47-21
shear stress theory, 43-9
slenderness ratio, steel tension member, 60-4
specific gravity, 76-9
specific growth rate, 30-7
specific growth rate coefficient, 28-7
stress, with and without shoring, 64-2
theoretical combustion temperature, 24-16
total uniform load table, 59-8
value, sinusoid, 84-5
velocity, open channel, 19-28
velocity in pipe, 17-3
water-cement ratio (tbl), 77-3
yield coefficient, 28-7
MCL, 25-5, 28-16
MCLG, 25-5
Mean, 11-12, 34-9
annual air temperature, 76-17
arithmetic, 11-12
cell residence time, 30-4, 30-8
depth, hydraulic (ftn), 19-3
effective pressure, 85-15
fourth moment of, 11-14
free path, air, 34-9
free path length, gas, 34-9
geometric, 11-13
harmonic, 11-13
number, M/M/s system (fig), 73-31
residence time, 29-8, 32-10
sea level, 78-8
slip coefficient, 65-3
speed, 73-4, 73-5
speed, space, G-15
standard error of, 11-15
stress, 43-10
third moment about, 11-14
time before failure (ftn), 11-9
time to failure, 11-9
velocity, open channel, 19-2
Meander corner, G-10
Meandering stream, G-10
Measure
bank-, 80-1
compacted-, 80-1
loose-, 80-1
Measurement
angle (fig), 78-13
areal, 1-9
board foot, 1-8
crow's foot, 81-3
distance, 78-8, 78-9
elevation, 78-9, 78-10, 78-12
flow, 17-28, 26-4
ground stake, 81-1
guard stake, 81-3
horizontal stadia (fig), 78-9
hub stake, 81-1
inclined stadia (fig), 78-9

offset stake, 81-3
reference point stake, 81-1
reliable, 85-2
ton, 1-7
witness stake, 81-3
Mechanic's lien, 88-6
Mechanical
advantage, 15-15, 41-11
advantage, pulley block, 41-12
advantage, rope-operated machine (tbl), 41-12
property, high-performance steel plate (tbl), 58-7
property, steel, 48-9
seal, 34-16
similarity, 17-48
Mechanics
engineering, 41-2
material, 44-2
space, 72-22
Mechanism, 41-21, 47-2
method, 59-8
two-dimensional, 41-21
Mechanistic
-empirical design, 77-11
-empirical method, 76-17
method, 76-17
Media
factor, 29-12
packing, 34-21
Median, 11-12, G-11
barrier, heavy vehicle, 75-14
barrier, heavy vehicle (fig), 75-14
lane, G-11
speed, 73-5
Medium
cure, asphalt, 76-2
screen, wastewater, 29-5
Megagram, 1-6
Meinzer unit, 21-3
Member
axial, 41-12
axial, force, 41-12
circular, 50-22
force, 41-18
pin-connected, axial, 41-13
redundant, 41-8, 41-14
set, 11-1
tension, staggered holes (fig), 60-2
tension, uniform thickness, unstaggered holes (fig), 60-2
tension, unstaggered row of holes (fig), 60-2
tension (fig), 60-1
three-force, 41-7
two-force, 41-7, 41-13
zero-force, 41-14
zero-force (fig), 41-14
Membership interest, 88-3
Membrane
cell, 27-1
pumping, 27-3
reinforced, 31-4
semipermeable, 14-11
support, 40-11
supported, 31-4
synthetic, 31-4
unreinforced, 31-4
unsupported, 31-4
Meniscus, 14-13
Mensuration, 7-1, A-8
of area, A-6
of volume, A-10
pile, 80-2
three-dimensional, A-10
two-dimensional, A-8
MEP, 85-15
Mer, 26-9
Mercaptan compound, 27-10
Mercury, density (tbl), 14-3
Merging taper, 73-33

Meridian, 78-12, 78-21 (fig), G-11
assumed, 78-12
grid, 78-12
magnetic, 78-12
true, 78-12
Meromictic lake, G-11
Mesa, G-11
Mesh, steel, A-150
Mesokurtic distribution, 11-14
Mesophile (tbl), 27-6
Mesophyllic bacteria, G-11
Metabolism, 27-9, G-11
Metacenter, 15-18
Metacentric height, 15-18
Metal, 22-2, 22-3
active gas welding, 66-2
alkali, 22-3
alkaline earth, 22-3
base, 66-1
concentration, 28-16
deck, filled, 57-4
deck, orthotropic, 57-5
deck, partially filled, 57-4
deck, unfilled composite, 57-4
deck system, 57-4
filler, 66-1
heavy, 22-3, 28-16
heavy, in wastewater, 28-16
inert gas welding, 66-2
light, 22-3
mechanical properties, A-129
properties, A-129
total, in wastewater, 28-16
transition, 22-3
transition (ftn), 22-3
weld, 66-1
Metalimnion, G-11
Metallic property, 22-2
Metalloid, 22-3
Metamorphic rock, 35-33, G-11
Metathesis (ftn), 22-7
Meteoric water, G-11
Meter
constant, 17-31
current, 17-29
density-on-the-run, 76-9
displacement, 17-28
normal cubic, 24-2
obstruction, 17-28
orifice, 17-32
orifice (fig), 17-32
pressure (ftn), 14-3
torque, 85-14
turbine, 17-29
variable-area, 17-28
venturi, 17-31
venturi (fig), 17-31
Metering pump, 18-4
Methane
gas, 30-18
heating value, 30-18
landfill, 31-5
properties (tbl), 31-6
series, 24-1
sludge, 30-18
Methanol, 24-8
Methemoglobinemia, 25-9
Method (*see also type*)
60-degree, 40-2
α-, 38-3
absolute volume, concrete, 77-2
accelerated depreciation, 87-23
allowable stress design, 45-2, 50-3
angle measurement, 78-12
annual cost, 87-16
annual return, 87-16
area transformation, 44-21
average cost, 87-39
average end area, 80-4
base exchange, 26-18

INDEX-N

INDEX-P

-relief valve, 17-41
safe bearing, 36-2
saturation, 14-10
soil, at-rest, 37-5
soil, backward, 37-2
soil, compressed, 37-2
soil, depth, 40-2
soil, forward, 37-2
soil, tensioned, 37-2
standard temperature and, 24-2
surface, 15-6, 15-7, 15-9, 15-10
total, 16-3, 37-10
transverse, 17-30
unit (fig), 14-2
uplift, 21-9
vapor, 14-10
water service, 26-26
Pressurized
fluidized-bed combustion, 24-6
liquid, 15-14
tank, 17-19
Prestress
loss, 56-3
maximum, 56-7
tendon, 56-3
Prestressed
beam, analysis, 56-11
beam deflection, 56-5
concrete, 56-2
concrete cylinder pipe, 16-11
concrete pavement, 77-2
concrete pipe, 16-11
section, shear, 56-11
Prestressing
effect on simple beam (fig), 56-2
fully, 56-3
midspan deflections from (fig), 56-5
partial, 56-3
steel ratio, 56-10
tendon, ASTM (tbl), 56-6
tendon, stress, 56-7
Presumptive test, G-13
Pretension
bolt, 65-4
loss, creep, 56-4
loss, shrinkage, 56-3
Pretensioned
concrete, 56-1, 56-2
connection, 65-2
construction, 56-2
Pretensioning, DTI washer, 58-4
Pretreatment, 26-3
Prevention, pollution, 32-1
Prewash, air, 26-14
Price-earnings ratio, 87-37
Pricing, congestion, 73-29
Primacy, 75-11
Primary
combustion chamber, 34-13
consolidation, 40-4, 40-5
consolidation rate, 40-5
creep, 43-17
dimension, 1-8
structure, 47-7
treatment, 29-3
unit, 1-8
Prime
coat, G-13
contractor, 88-5
cost, 86-9, 87-34
mover (ftn), 18-10
Primer, red, 16-12
Principal
axis, 42-8, 59-15, 70-2
contract, 88-3
moment of inertia, 42-8
organic hazardous constituent, 34-13
point, 78-20
point, photograph, 78-19
stress, 44-5
view, 2-2

Principle
Archimedes', 15-15
D'Alembert, 72-6
impulse-momentum, 17-35, 72-16
Le Châtelier's, 22-13
of proportionality, 47-2
perpendicular line, 2-1
virtual work, 47-3
work-energy, 13-3, 13-4, 45-22
Priority, ethical, 89-2
Prism, 80-4
level, 78-10
test method, 67-4
Prismoid, 80-4
Prismoidal formula, 80-4
Privity
of contract, 88-7
of contract, of flood, 20-6
of contract, probability, 11-3
Probabilistic problem, 87-31
Probability
complementary, 11-4
density function, 11-4
of failure, conditional, 11-9
of flood, 20-7
success, 11-3
theory, 11-3
Probable
error, 78-2
error of the mean, 78-2
maximum flood, 20-7
maximum precipitation, 20-7
maximum rainfall, G-13
ratio of precision, 78-2
value, 78-2
Probe
direction-sensing, 17-29
static pressure, 17-28
Problem
initial value, 10-1
loan repayment, 87-4
mixing, 10-9
rate of return, 87-4
three-reservoir, 17-24
Procedure
diagnosis, transportation network, 75-19
Johnson, 45-4
Marshall mix test, 76-12
Process
absorption, 34-2
anabolic, 27-9
base exchange, 22-23
bioactivation, G-3
biofilm, 26-15
biosorption, G-3
catabolic, 27-9
control, statistical, 11-17
cost accounting, 87-37
curing, 49-6
economic analysis, 75-20
efficiency, BOD, 30-6
group, 86-1
ion exchange, 22-23, 26-18
Krause, G-10
lime-soda ash, 26-16
nitrification/denitrification, 29-13
reacted asphalt, 76-30
rubberized asphalt, 76-30
SAW, 66-2
SMAW, 66-1
unit, G-17
waste, 32-2
zeolite, 22-23, 26-18
Processing
attention and information, 75-9
joint, 28-2
Prochloraz, 32-13
Proctor test, 35-18, 35-19

Producer
gas (ftn), 24-8
risk, 11-15
Product, 22-7
cross, 41-3
dot, 5-3
incomplete combustion, 34-14
ion, 22-16
mixed triple, 5-5
of inertia, 42-7, 70-2
scalar, 5-5
solubility, 22-18
triple cross, 5-5
triple, scalar, 5-5
triple, vector, 5-5
vector, cross, 5-4
vector, cross (fig), 5-4
vector, dot, 5-3
vector, dot (fig), 5-3
waste decomposition, 27-10
Professional
corporation, 88-2
engineer, 90-1
engineering exam, 90-1
limited liability company, 88-3
services, cost, 86-5
Profile
auxiliary view, 2-2
depth, 19-23
diagram, 80-7
drag, 17-43
velocity, 16-9
Profiling, 76-29
Profit, 87-39
and loss statement, 87-35
and loss statement, simplified (fig), 87-36
flow through, 88-2
gross, 87-39
margin, 87-37
margin ratio, 87-37
net, 87-39
project, 86-7
Program evaluation and review technique, 86-14
Programming, 86-8
Progressive method, landfill, 31-2
Project
coordination, 86-15
documentation, 86-16
management, 86-1
monitoring, 86-15
overhead, 86-7
scheduling, 86-8
transportation, 75-16
Projected bearing area, 60-6
Projectile, 71-4
motion, 71-4
motion equations (tbl), 71-5
Projection, 2-3
cabinet, 2-3
cavalier, 2-3
clinographic, 2-3
factor, growth, 76-18
fee, 86-2
method, 87-10
Projector, 2-1
Prokaryote, 27-1
Prompt-NOx, 32-10
Prony brake, 85-14
Proof
alcohol, 24-8
load, bolt, 45-10, 45-14
strength, 45-10, 45-14
stress (ftn), 43-4
test, 11-9
test interval, 11-9
Proper
polynomial fraction (ftn), 3-6
subset, 11-1

INDEX-R

INDEX-R

INDEX-R

INDEX-R

INDEX-S

INDEX-S

INDEX-S

INDEX-T

INDEX-T

INDEX-W

INDEX-Z

FUNDAMENTAL AND PHYSICAL CONSTANTS

quantity	symbol	U.S.	SI
Density			
air [STP] [32°F (0°C)]		0.0805 lbm/ft^3	1.29 kg/m^3
air [70°F (20°C), 1 atm]		0.0749 lbm/ft^3	1.20 kg/m^3
earth [mean]		345 lbm/ft^3	5520 kg/m^3
mercury		849 lbm/ft^3	1.360×10^4 kg/m^3
seawater		64.0 lbm/ft^3	1025 kg/m^3
water [mean]		62.4 lbm/ft^3	1000 kg/m^3
Gravitational Acceleration			
earth [mean]	g	32.174 (32.2) ft/sec^2	9.8067 (9.81) m/s^2
moon [mean]		5.47 ft/sec^2	1.67 m/s^2
Pressure, atmospheric		14.696 (14.7) lbf/in^2	1.0133×10^5 Pa
Temperature, standard		32°F (492°R)	0°C (273K)
Fundamental Constants			
Avogadro's number	N_A		6.022×10^{23} mol^{-1}
gravitational constant	g_c	32.174 lbm-ft/lbf-sec^2	n.a.
gravitational constant	G	3.440×10^{-8} ft^4/lbf-sec^4	6.674×10^{-11} N·m^2/kg^2
specific gas constant, air	R	53.35 ft-lbf/lbm-°R	287.03 J/kg·K
specific gas constant, methane	R_{CH_4}	96.32 ft-lbf/lbm-°R	518.3 J/kg·K
triple point, water		32.02°F, 0.0888 psia	0.01109°C, 0.6123 kPa
universal gas constant	R^*	1545.35 ft-lbf/lbmol-°R	8314.47 J/kmol·K
	R^*	1.986 Btu/lbmol-°R	0.08206 atm·L/mol·K

THE GREEK ALPHABET

A	α	alpha	N	ν	nu
B	β	beta	Ξ	ξ	xi
Γ	γ	gamma	O	o	omicron
Δ	δ	delta	Π	π	pi
E	ϵ	epsilon	P	ρ	rho
Z	ζ	zeta	Σ	σ	sigma
H	η	eta	T	τ	tau
Θ	θ	theta	Υ	υ	upsilon
I	ι	iota	Φ	ϕ	phi
K	κ	kappa	X	χ	chi
Λ	λ	lambda	Ψ	ψ	psi
M	μ	mu	Ω	ω	omega

SI PREFIXES

symbol	prefix	value
a	atto	10^{-18}
f	femto	10^{-15}
p	pico	10^{-12}
n	nano	10^{-9}
μ	micro	10^{-6}
m	milli	10^{-3}
c	centi	10^{-2}
d	deci	10^{-1}
da	deka	10
h	hecto	10^2
k	kilo	10^3
M	mega	10^6
G	giga	10^9
T	tera	10^{12}
P	peta	10^{15}
E	exa	10^{18}

EQUIVALENT UNITS OF DERIVED AND COMMON SI UNITS

symbol	equivalent units					
A	C/S	W/V	V/Ω	J/s·V	N/T·m	Wb/H
C	A·s	J/V	N·m/V	V·F		
F	C/V	C^2/J	C^2/N·m	A^2·s^4/kg·m^2	kg·m^2/V·s^2	
F/m	C/V·m	C^2/J·m	C^2/N·m^2	A^2·s^4/kg·m^3	kg·m^2/V·s^3	
H	Wb/A	V·s/A	T·m^2/A	kg·m^2/A^2·s^2	Ω·s	
Hz	1/s					
J	N·m	V·C	W·s	C^2/F	kg·m^2/s^2	
m^2/s^2	V·C/kg					
N	J/m	V·C/m	kg·m/s^2			
N/A^2	Wb/N·m^2	kg·m/A^2·s^2				
Pa	N/m^2	kg/m·s^2				
Ω	V/A	kg·m^2/A^2·s^3				
S	A/V	A^2·s^3/kg·m^2				
T	Wb/m^2	N/A·m	N·s/C·m	kg/A·s^2		
V	W/A	C/F	J/C	kg·m^2/A·s^3		
V/m	N/C	J/m·C	kg·m/s^2·C	kg·m/A·s^3		
W	J/s	V·A	kg·m^2/s^3	N·m/s		
Wb	V·s	H·A	T·m^2	kg·m^2/A·s^2		